Tailoring of Engineering Material Properties through Laser Cladding

In this comprehensive guide to laser cladding of engineering materials, expert contributors provide a detailed yet easy-to-follow explanation of the process, its use for surface modification and the benefits and applications of this technique in different environments and for different purposes.

The optimization of an engineering material's properties and behavior is vital for economic, safety and quality assurance reasons when these materials are applied in various industrial settings. This book therefore explains how laser cladding can be used to tailor and control a material's surface layer properties, such as its elastic and plastic deformation and permanent detachment in the form of debris during the contact of mating parts. Through individual chapter contributions from experts in various elements related to laser cladding, this book guides readers through the evaluation of different structures in laser-cladded engineering materials. The book addresses three key questions:

- What is the influence of micro- and nano-level structures on mechanical properties?
- How does laser cladding improve a material's wear resistance at the micron and nano level and determine the material's suitability for particular applications?
- What challenges are related to the laser cladding of different engineering materials?

In addressing these questions, the book enables readers to determine the metallurgical, mechanical, tribological and corrosion behavior of any type of engineering material. Readers will also be able to make informed decisions based on their knowledge of the properties and industrial applications of different laser-cladded materials.

This book is essential for all manufacturing industry personnel who work with materials in various industrial settings and need to control their surface properties and enhance their behavior at the micron and nano levels. Students who are building experience prior to entering industry will also benefit from the detailed descriptions, step-by-step approach and focus on practical application.

Tailoring of Engineering Material Properties through Laser Cladding

Metallurgical, Mechanical and Tribological Properties

Edited by
Natarajan Jeyaprakash, Govindarajan Prabu
and Che-Hua Yang

CRC Press
Taylor & Francis Group
Boca Raton London New York

CRC Press is an imprint of the
Taylor & Francis Group, an **informa** business

Designed cover image: Getty Images

First edition published 2025
by CRC Press
2385 NW Executive Center Drive, Suite 320, Boca Raton FL 33431

and by CRC Press
4 Park Square, Milton Park, Abingdon, Oxon, OX14 4RN

CRC Press is an imprint of Taylor & Francis Group, LLC

First edition published by Routledge 2025

ISBN: 978-1-032-69354-5 (hbk)
ISBN: 978-1-032-69356-9 (pbk)
ISBN: 978-1-032-69357-6 (ebk)

DOI: 10.1201/9781032693576

Typeset in Palatino
by codeMantra

Contents

Editors

Natarajan Jeyaprakash is an Associate Professor at the Mechanical Engineering Department, Global College of Engineering and Technology, Muscat, Oman. He earned bachelor's and master's degrees at Anna University, Chennai, India, in 2006 and 2009, respectively, and a PhD in production engineering in 2018 at the National Institute of Technology, Tiruchirappalli, India. He teaches various postgraduate courses in automotive, mechanical, manufacturing, and laser ultrasound and conducts academic research and consultancy.

Govindarajan Prabu is a Postdoctoral Research Associate at the Graduate Institute of Manufacturing Technology, National Taipei University of Technology. He earned MTech and doctoral degrees at the National Institute of Technology, Tiruchirappalli, India. His areas of research include tribology, wear, friction, laser surface treatment, and mechanical characterization.

Che-Hua Yang is a Professor at the Graduate Institute of Manufacturing Technology, National Taipei University of Technology. He earned a bachelor's degree at the National Taiwan University in 1984 and master's and PhD degrees in 1989 and 1991, respectively, at Johns Hopkins University. He has previously been Vice President of National Taipei University of Technology and Director of its Additive Manufacturing Center for Mass Customization Production.

Contributors

Ibrahim A. Alnaser serves as the Director of the Center of Excellence for Research in Engineering Materials and is an Associate Professor of Mechanical Engineering at King Saud University in Riyadh, Saudi Arabia. His research interests focus on the characterization, analysis, manufacturing, testing, and applications of various advanced materials, including metallic and nonmetallic composite materials. He has led and supervised various research initiatives and consultancy projects. Over the course of his career, Dr. Alnaser has published over 67 peer-reviewed research articles and three book chapters.

K. Ananthakumar is a Deputy Manager in the Metallurgist Process Engineering Department at Motherson Health and Medical Systems, Kanchipuram, India. Before this, he worked as an Associate Professor in the Department of Mechanical Engineering at Karpagam College of Engineering, Coimbatore. He earned a PhD in metallurgical and materials engineering at the National Institute of Technology, Tiruchirappalli. His research interests are solid-state welding and powder metallurgy. He has published a number of research articles in international journals, conferences, and book chapters.

P. Dinesh Babu is an Associate Professor at the School of Mechanical Engineering, SASTRA Deemed University, Thanjavur, India. He earned a PhD in mechanical engineering in 2013 at the National Institute of Technology, Tiruchirappalli, India. He has published more than 60 peer-reviewed research articles with a cumulative impact factor of around 90. He also published one book chapter titled "Laser Surface Hardening" in 2015. He also completed two funded projects from DRDO, India, and he produced three PhD candidates. His research interests primarily focus on laser surface engineering, material characterization, laser-based manufacturing, and metal additive manufacturing.

Raj Kumar Durairaj is a Professor at the School of Mechanical Engineering, Sri Ramakrishna College of Engineering, Perambalur, India. He earned a bachelor's degree at Jayaram College of Engineering and Technology, Trichy, India, in 2002 and a master's degree at MIET Engineering College, Trichy, India, in 2009 and a PhD in mechanical engineering in 2018 at Anna University, Chennai, India. Dr. Durairaj has published more than 30 peer-reviewed research articles and 5 book chapters. He teaches various postgraduate courses in mechanical and manufacturing, and conducts academic research and consultancy.

N. Jayanth is an Assistant Professor in the Centre of Excellence in Product Design and Smart Manufacturing at Maulana Azad National Institute of Technology, Bhopal. He earned a BE in mechanical engineering at Anna University, Chennai, and an ME in computer-integrated manufacturing at PSG College of Technology, Coimbatore, and a PhD in additive manufacturing at the National Institute of Technology, Tiruchirappalli. He has organized an international conference and a research internship for undergraduate students. Dr. Jayanth has been an invited expert speaker in the field of additive manufacturing. His areas of research are additive manufacturing of multifunctional composites and their applications. He also has published research articles in reputed journals and international conferences.

Natarajan Jeyaprakash is an Associate Professor at the Mechanical Engineering Department, Global College of Engineering and Technology, Muscat, Oman. He earned bachelor's and master's degrees at Anna University, Chennai, India, in 2006 and 2009, respectively, and a PhD in production engineering in 2018 at the National Institute of Technology, Tiruchirappalli, India. Dr. Jeyaprakash has published more than 120 peer-reviewed research articles, 8 book chapters, and 2 books, titled *Laser Surface Treatments for Tribological Application* and *Advances in Additive Manufacturing Processes* in 2021. He has edited 2 books for CRC Press, Taylor & Francis. He teaches various postgraduate courses in automotive, mechanical, manufacturing, and laser ultrasound, and conducts academic research and consultancy.

Sundara Subramanian Karuppasamy is a PhD student at the Institute of Manufacturing Technology at the National Taipei University of Technology, Taiwan. He earned a BE in mechanical engineering at Velammal College of Engineering and Technology (affiliated with Anna University, Chennai, India) in 2017 and an MS in manufacturing technology at the National Taipei University of Technology, Taiwan, in 2020. He has published more than 10 articles and 5 book chapters in Scopus-indexed journals.

Aditya Kumar is an Assistant Professor in the Department of Energy and Environment, NIT Trichy. He did one and a half years of postdoc research in the Mechanical Engineering Department at IIT Bombay. He earned PhD and MTech degrees from the oldest institute in Asia at the Sustainable Power and Energy System Lab, Department of Mechanical and Industrial Engineering, IIT Roorkee. He is an experienced researcher in the area of heat transfer application of nanofluids. Dr. Kumar has published and contributed to several leading international peer-reviewed journals. He attended and presented papers at several national and international conferences on various topics related to nanofluids. His areas of interest include natural convection heat transfer, magnetic nanofluids, thermal energy storage, fuel cell, and phase change materials.

K. Nirmal Kumar is a Research Assistant at the School of Mechanical Engineering, SASTRA Deemed to be University. He earned an ME in CAD/CAM at Kumaraguru College of Technology, Coimbatore, which is affiliated with Anna University, Chennai, in 2013. He is enrolled in a PhD program at the School of Mechanical Engineering at SASTRA Deemed University in Thanjavur, Tamil Nadu, India. His research interests focus on natural fiber-reinforced polymer composites, mechanical behaviors of materials, material characterization, and machining characteristics of composite materials.

S. Pradeep Kumar is a Researcher in the field of mechanical engineering with expertise in metal additive manufacturing, material characterization, and laser-based manufacturing. He earned a PhD at the School of Mechanical Engineering, SASTRA Deemed University, Thanjavur, in 2024. He has published 7 peer-reviewed research articles with a cumulative impact factor of 23 and has served as a reviewer for renowned journals, including *3D Printing and Additive Manufacturing*. Dr. Kumar is passionate about advancing research in metal additive manufacturing and continues to explore innovative approaches in material science and laser-based techniques.

T. Satish Kumar is an Assistant Professor (SlGr) in the Department of Mechanical Engineering, Amrita School of Engineering, Coimbatore, Amrita Vishwa Vidyapeetham, India. He earned a BE in mechanical engineering at SV University—Tirupati in 2009, a master's and a PhD degree in mechanical engineering at Anna University, Chennai, India, in 2011 and 2016, respectively. He has published more than 54 peer-reviewed research articles and guided many research scholars. He teaches various postgraduate and undergraduate courses in composites, materials science and metallurgy, and additive manufacturing and also conducts academic research and consultancy.

V. Lakshmanan is an Assistant Professor in the Department of Mechanical Engineering at SRM TRP Engineering College, and a multi-disciplinary researcher. He earned a PhD in production engineering at NIT Trichy (MHRD fellowship, full-time scholar). He is also a graduate engineer in mechanical engineering at Thigarajar College of Engineering, Madurai, a postgraduate at the Government College of Technology, Coimbatore. He holds memberships in various professional bodies such as IEI and SAE. He worked on the green project scheme of CIPET Chennai funding project. He has published several articles in reputed journals, enhancing dissimilar welding and corrosion science.

B Mallikarjuna is an Assistant Professor in the Production Engineering Department, National Institute of Technology Tiruchirappalli. He earned a PhD in mechanical engineering at the National Institute of Technology, Surathkal, India, in 2020, and a master of technology in computer-integrated manufacturing in 2013. He has organized workshop and research internships on additive manufacturing funded by the Government of India. He has been an invited guest speaker, technical committee member, and session chair and judge at international conferences. He is an active researcher in the areas of additive manufacturing and finite element analysis, and he has published articles in several referred journals and conference publications.

Durairaj Manikandan is an Assistant Professor in the Department of Mechanical Engineering, SRM TRP Engineering College. He earned bachelor's and master's degrees at Anna University in Chennai, Tamil Nadu, India, in 2007 and 2010, respectively. He has authored more than 12 peer-reviewed research articles and contributed to five book chapters. He has conducted a thorough examination of additional book chapters and esteemed journals. He teaches a range of postgraduate courses in automotive, mechanical, and manufacturing disciplines.

P. Marimuthu is an Associate Professor in the School of Mechanical Engineering, SASTRA Deemed University, Thanjavur. He earned a PhD in mechanical engineering at the Indian Institute of Technology, Madras, Chennai, in 2016. His research interests include the asymmetric spur gear design to improve the load-carrying capacity of power transmission systems. He has published 30 papers in reputed international journals and earned four external funded projects from DST-SERB and DRDO (GTRE). He teaches undergraduate and postgraduate courses in automotive and mechanical engineering.

Govindarajan Prabu is a Postdoctoral Research Associate at the Graduate Institute of Manufacturing Technology, National Taipei University of Technology. He earned an MTech and a PhD degree at the National Institute of Technology, Tiruchirappalli, India. He has published more than 15 SCI papers in reputed journals in the areas of tribology, wear, friction, laser surface treatment, and mechanical characterization.

S. Hari Prasadh is an Assistant Professor in mechatronics engineering at K.S. Rangasamy College of Technology. His expertise includes drone technology and electric vehicles. He has published research papers on LUO converters, KY BOOST converters, and smart agricultural drones and has filed multiple patents. He has secured over ₹57.88 lakhs in grants for projects, including autonomous UAV development. He also mentors student projects and actively participates in professional development programs.

T. Sathies is an Assistant Professor in the Department of Mechanical Engineering, Amrita School of Engineering, Amrita Vishwa Vidhyapeetham Amaravati, Andhra Pradesh, India. Dr. Sathies served as a postdoctoral fellow at the Central Research Facility, National Institute of Technology Karnataka

(NITK) for two years and worked on the development of a wearable piezoelectric energy harvester. He earned a bachelor's degree in mechanical engineering at Anna University, Chennai, India, in 2013, and a master's degree in manufacturing engineering in 2017 at Anna University, Chennai, India. He earned a PhD in additive manufacturing in 2021 at the National Institute of Technology Tiruchirappalli (NIT Trichy), India. He has 10 publications in Web of Science-indexed journals and 4 book chapters. His research interests include additive manufacturing of multifunctional components, post-processing of additively manufactured components, and development and characterization of functional polymer composites.

Titus Thankachan is an Associate Professor in Karpagam College of Engineering, Coimbatore. He earned a PhD in mechanical engineering at Anna University, Chennai, India, in 2017. He has published more than 34 articles in international journals and 6 book chapters in the areas of advanced materials and manufacturing.

S. Martin Vinoth is an Assistant Professor in the Department of Mechanical Engineering at SASTRA University. He earned a bachelor's degree in mechatronics at SASTRA University in 2008 and a master's degree in 2013 from the Government College of Technology. He is enrolled in a PhD program at SASTRA University, specializing in laser cladding, while teaching undergraduate courses in mechanical engineering.

Che-Hua Yang is a Professor at the Graduate Institute of Manufacturing Technology, National Taipei University of Technology. He earned a bachelor's degree at National Taiwan University in 1984 and master's and PhD degrees in 1989 and 1991, respectively, at Johns Hopkins University. He has been Vice President of the National Taipei University of Technology and Director of its Additive Manufacturing Center for Mass Customization Production. Dr. Yang has published more than 120 peer-reviewed research articles, book chapters, and two books, titled *Laser Surface Treatments for Tribological Application* and *Advances in Additive Manufacturing Processes* in 2021.

1

Introduction

Raj Kumar Durairaj, Natarajan Jeyaprakash, and Che-Hua Yang

1.1 Introduction

Every developed material has its own advantages and disadvantages from the time of invention of the material to this time. Increasing advantages and applications while minimizing disadvantages and limitations is a critical endeavor. Therefore, all the professional scientists and researchers know the importance of the material and are continuously working to increase its functionality. Moreover, the material can cause wear, friction, and corrosion, which is all an important problem. This problem can be solved in two ways: (i) conventional cladding and (ii) laser cladding (LC). Technically speaking, the poor wear resistance and corrosion resistance materials are enhanced by LC methods [1]. In conventional cladding, thermal spraying and arc welding are used, but LC has more advantages than conventional cladding. The LC has a fusion bond, low porosity, low dilution, low heat-affected zone, low thermal distortion, and reduced post-cladding processing costs in comparison with conventional cladding. Mainly, the service life of the components is increased by LC [2]. Tungsten carbide, alumina, zirconia, aluminium, cobalt-chromium, titanium, stainless steel, and nickel-based alloys are commonly utilized as cladding and substrate materials. LC is commonly processed with CO_2, Nd:YAG, and fiber lasers [3]. Any metal powders, metal strips, or wires can be melted and deposited on substrate using these three types of lasers. LC is a powder injection technique that is used to deposit metals or alloys on substrates and is used in industrial applications such as part repair, rapid manufacturing, innovative alloy development, and surface coating. In this process, the selected material is fed on the substrate under the laser by powder blow, wire feed, and preplaced powder coating. The desired properties of LC can be obtained by properly selecting the clad material and controlling processing parameters such as laser beam power density, laser beam diameter, laser beam, and travel speed. The LC process can produce coatings ranging in thickness from 50 μm to 2 mm. The book *Tailoring of Engineering Material Properties through Laser Cladding* covers these topics in various aspects over 12 chapters.

DOI: 10.1201/9781032693576-1

1.2 Laser-Cladding Process

LC integrates three technologies, namely, laser technology, manufacturing, and its control. Three technological benefits are shown in the cladding process. Additionally, a complex plays a significant role in bonding, metallurgical properties, and chemical processes.

1.3 Process Principle

LC is to increase the wear resistance properties on a low-wear-resistance substrate, which is a clad layer formed on substrate using a high-power laser [4]. Powder feeding plays a main role in the LC process. This powder-feeding method is divided into four types: (i) wire feed system, (ii) off-axis powder system, (iii) preplace powder system, and (iv) coaxial powder system. Among these four feeding systems, coaxial powder method and preplaced powder method are mostly used. The schematic diagram of coaxial powder method, preplaced powder method, and a cladded sample of IN78 with Ni nanoparticles is shown in Figure 1.1.

When a high-power laser beam strikes the substrate, it creates a molten pool. Clad powder, delivered through a nozzle by gas, is then deposited into the molten pool, forming a clad layer on the substrate. The nozzle acts linear to laser beam for clad layer formation. The benefit of coaxial system is a rate of utilization, and it needs quality of cladding equipment. The preplaced powder method is different one to form the clad layer in which the clad material is replaced on the substrate. Due to the laser scanning, the preplaced powder is melted and cooled rapidly to form a clad layer on substrate. The benefit of preplaced powder method is better clad quality and simple process [5]. The drawback of this system is high dilution and complex control on penetration depth. The cladded substrates have four zones, namely, the substrate (SUB), Heat-Affected Zone (HAZ), Interfacial Zone (IZ), and Cladding Zone (CZ).

1.4 Simulation Analysis

The LC simulation is carried out using the finite element analysis (FEA). The MATLAB software is used to perform the FEA. The model shows the interactive effect of laser beam, clad material, and substrate [6]. The simulation is clearly displaying the temperature distribution, stress, and flow analysis of

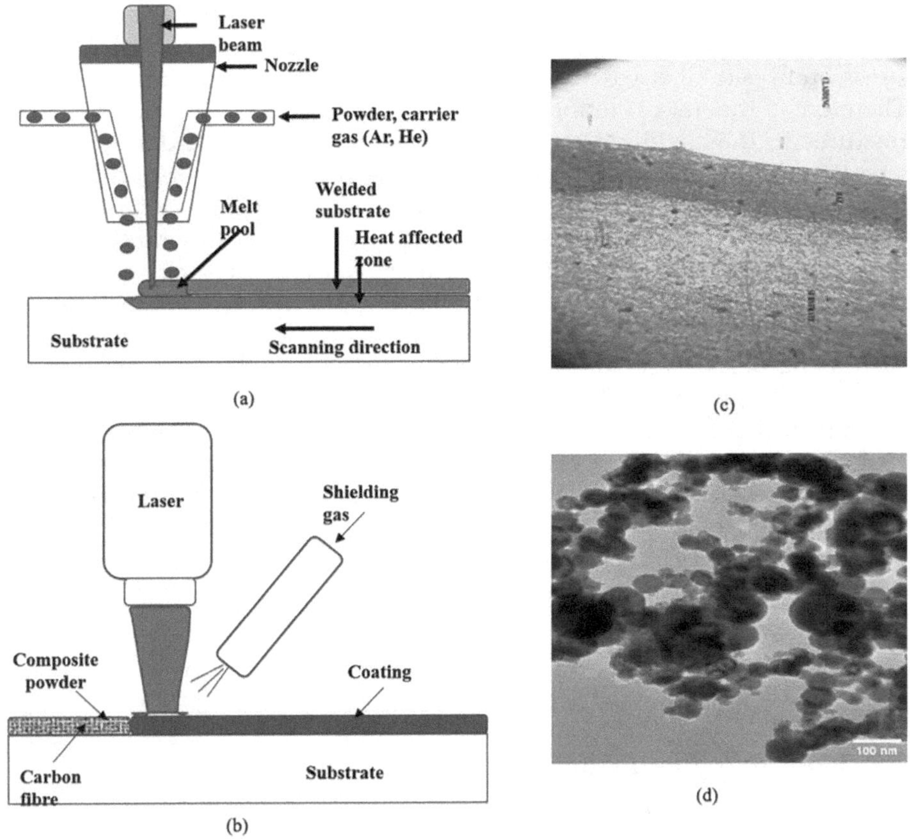

FIGURE 1.1
Schematic diagram of: (a) coaxial powder method, (b) preplaced powder, (c) macroscopy image of sample, and (d) ZrO_2 particles.

material under various conditions. The major benefits of using simulation in the LC on substrate are to show the morphology of clad surface, microstructure of clad zone, and performance of clad layer. Hence, the FEA is used in the LC. Based on the hydrodynamic and phase field processes, the microstructure of clad layer, stress field, temperature field, and powder deposition process can be simulated. The efficiency of clad layer is highly depending upon the factors such as powder size, laser power, substrate material, nozzle velocity, and air pressure. Lower particle heating is associated with good powder jet focusing. The mass flow on substrate's plane is influenced by where it is in relation to the nozzle. Additionally, the gas and the associated powder are significantly influenced by the flow of protective gas [7]. In the interaction between powder and laser, the powder absorbs, reflects, and scatters the laser's energy, raising the temperature of the flowing powder.

The link between the laser power and the separation between the laser focus and the powder's temperature distribution is quite strong. Thermal stress and residual stress would both be produced during the LC process. Therefore, a theoretical foundation for successfully reducing flaws such as fractures in the cladding layer is provided by the simulated study of the stress fields. In addition to introducing the plasticity theory with its LC Inconel 718 model, Ghorashi et al. [8] developed a thermo-mechanical finite element model to precisely predict residual stresses in the multiple beads laser cladding process. The analysis integrates cyclic plasticity by employing nonlinear kinematic hardening behavior. The results demonstrate that nonlinear cyclic plasticity modeling effectively captures critical phenomena such as hardening and residual stress, offering valuable insights into the laser cladding process. Zhang et al. [9] developed a three-dimensional (3D) uncoupled thermo-mechanical finite element (FE) model to simulate stress evolution during the laser cladding of cobalt-based coatings on mild steel A36. The stress distributions were thoroughly analyzed for three scenarios: a single track on a flat substrate, a double track on a flat substrate, and a double track on a cylindrical substrate. The experimental results exhibited strong agreement with the model's predictions.

1.5 Monitoring and Control

Although LC has become more and more popular as a surface strengthening and mending technique, it occasionally has issues with the consistency and quality of clad employed. However, advancements in computers and sensors have enhanced the monitoring of melt pool temperature, accounting for the effects of powder and laser parameters. The microstructure, flaw, and geometrical accuracy are related with above techniques. Temperature signals, image signals, and spectrum data are used to understand LC in the view of challenging metallurgical process. The flow and distribution of powder may be observed using an image signal or spectral data, and by optimizing the nozzle settings, the efficiency of powder utilization is increased. The CMOS digital camera, spectrometer, high-speed camera, and infrared pyrometers are highly used to monitor and control the flow velocity of the molten pool, melt pool life, cooling rate, microstructure, wettability in clad layer, melt pool life, and the cooling rate in good wettability. The advanced sensor and controlling equipment need to monitor the dilution rate and thermal stress. A high-speed camera capable of recording up to 67,000 frames per second was employed by Wirth et al. [10] to examine the melt pool. Tracking the un-melted WC particles will yield the flow velocity field on the melt pool's surface, which is approximately consistent with the melt pool's flow velocity. Concurrently, an analysis was conducted on the melt pool's flow velocity across various process parameters. However, because the brightness of the interest region varies greatly, it might be challenging to capture a video of an acceptable quality under certain procedure conditions. Consequently, it may be thought of using band-pass-filtered illumination lasers in the future. Infrared pyrometers were employed by Muvvala et al. [11] to track the melt pool's thermal evolution as they investigated the cladding layer's wettability, microstructure, cooling rate, and melt pool life. The location of the solidification rack may be found online by analyzing the temperature history graph of composite coating, which was acquired using an infrared pyrometer. An efficient indication for online state evaluation of TiC particle conditions under various process settings is the shift in the solidification frame slope of the particles in the melt pool.

1.6 Optimization

In the LC process, the laser power, scanning-rate, powder-feeding rate, scanning technique, defocus amount, and other process parameters are directly connected to the dilution, aspect ratio, microstructure, and mechanical characteristics of the cladding layer [12]. Many academics have dedicated their time to studying the process parameters from many angles to produce a better microstructure, better composition, and better mechanical properties. Insufficient laser power will result in poor-quality and ineffective cladding layers while reducing fractures and voids. In the view of the high laser power, the cladding layer cracked and distorted, and when the laser power is too low, local pill and voiding are caused because the powder is not totally melted [13]. The impact of laser power on the coating's micromorphology and microstructure is shown in Figure 1.2.

It has been discovered that when laser power increases, the cladding layer's height, breadth, and penetration grow. The majority of cracks begin in the HAZ and run perpendicular to the joint surface. As the laser power increases, columnar dendrites, uniform dendrites, and a small quantity of equiaxed crystals are formed, and their grain development occurs at the bottom of the cladding layer. This is due to a negative correlation between grain size and power, which causes the cooling rate to decrease as power increases. The creation of the cladding layer is significantly influenced by the laser speed. An extremely high scanning speed results in insufficient melting of the substrate and powder, whereas a sluggish scanning speed overheats the layer. A graphite microstructure is formed in the bonding zone between the cladding layer and the substrate after an analysis of the effects of various speeds and powers on the microstructure of the layer. According to the study [14], when the scan rate lowers, the graphite structure gets finer under a constant laser intensity. This is because carbon atoms may diffuse more easily into the

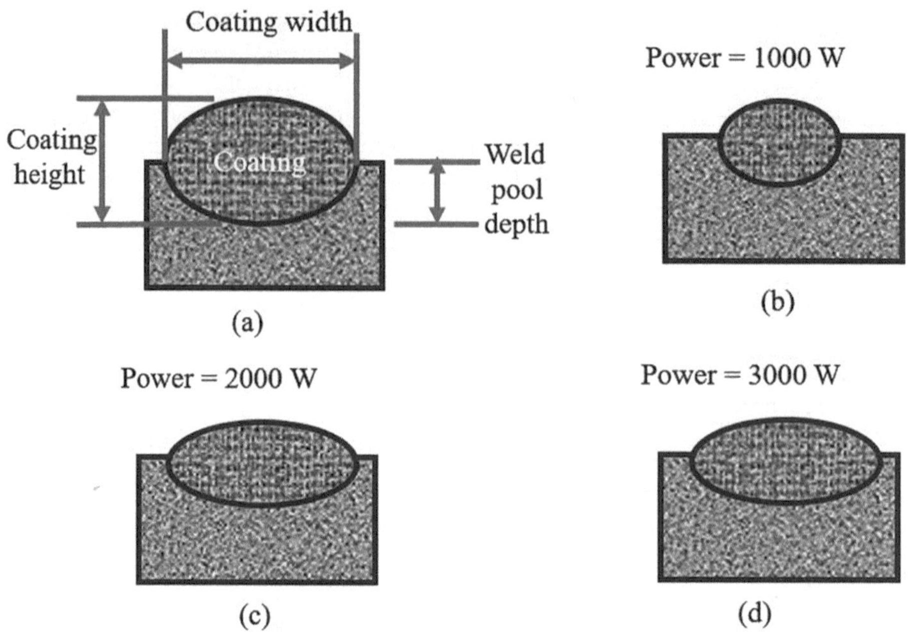

FIGURE 1.2
The macro-morphology of the cross section at different laser powers.

Gr in bonding zone. Yu et al. [15] used the Taguchi-Gray correlation analysis to optimize the LC process parameters. The chosen process parameters for optimization are power, speed, and feed, with the optimal response goals being clad height and rate of dilution. The optimal set of process parameters is identified to get the desired dilution rate, lowest melting height, and maximum melting breadth. Experiments are conducted to verify the optimal combination of parameters, and a 0.1533282 increase in the gray correlation value is observed. A mathematical assist in parameter graph based on IN78 cladding layer was formed by Bax et al. [16]. The derived semi-empirical connection also provides the graph, and it relates laser process parameters. It is relevant to one track, though more work has to be done on studying numerous tracks. Reddy et al. [17] used an optimization test on LC amorphous Fe-Cr-B alloy and found the optimum values.

1.7 External-Supported Laser Cladding

To provide a covering with good characteristics and considerably lessen the influence of these imperfections on the microstructure, the development of induction heating, laser cladding, ultrasonic-assisted, electromagnetic-assisted, and other tools in recent years has been facilitated by the collaboration of numerous academics working on LC. An induction preheating can lower the temperature gradient. The study looks the cracks in the heat zone to high-temperature induction preheating. The absence of cracks in heat zone in the thick substrate is explained by the induction preheating temperature's effect on the HAZ's microstructure, which alters the crack-signaling characteristics. The maximal deposition rate with laser efficiency are also significantly impacted by the induction preheat temperature. The maximal deposition rate and laser energy efficiency rise with an increase in induction preheating at constant laser power and scanning speed. Ultrasonic vibration has a significant impact on the formation with solidification of microstructures in the pool, as well as the distribution of element, acting as an external physical field. The magnetic field primarily cooperates with material's electron and has an impact on chemical reactions, which then have an impact on the microstructure and element distribution. Zhai et al. [18] investigated the dilution in cladding under various magnetic fields and discovered that a steady magnetic field greatly reduces the dilution rate with coating, although a magnetic field plays a smaller impact. The examination of the coating phases revealed that the phases varied very little between layers. This is because the magnetic field has no influence with heating conditions. A steady magnetic field hindered the dispersion of the elements, according to a cross-sectional analysis of coating. As an outcome, a bigger CrB hard forms in coating z2, resulting in the greatest average microhardness (876 $HV_{0.2}$) and the best corrosion resistance. Zhang et al. [19] investigated how magnetic fields affect the porosity of the laser-clad layer. The equivalent gravitational acceleration increases when the ampere-force direction is used. As a result, the consequent buoyancy reacts in the pores. The porosity is eventually decreased, and there are more optimization tools used for LC [20,21].

1.8 Laser-Cladding Material

The physical and chemical characteristics of the chosen clad metal have a significant impact on the microstructure and surface quality of the clad layer during the LC process, in addition to the process parameters. In general, the clad metal should be compatible and wettable to the substrate in addition to evaluating how well it performs [22]. The list of clad material is shown in Table 1.1.

LC materials have now evolved from being made of a single metal or ceramic to being made of many alloys or ceramics. Additionally, metal matrix composites, which have excellent toughness and hardness, are frequently employed. Research is now concentrating on materials including

TABLE 1.1

List of Clad Materials

Metal powders	Chrome base, nickel base, cobalt base, Fe base
Ceramic powders	WC, SiC, TiC, Al_2O_3, ZrO_2

high-entropy alloys (HEAs), amorphous alloys, and crystal alloys. Single metal powder, which is the fundamental component in LC in this case, plays a significant role on LC.

Self-fluxing powders based on iron, nickel, and cobalt are being extensively used in LC. Particularly Ni-related powders, which provide the better wear resistance, corrosion with affordable pricing. Intermetallic compounds and other multicomponent alloys have also been employed. As shown in Figure 1.3, the microstructure of coatings made of iron, nickel, and cobalt is

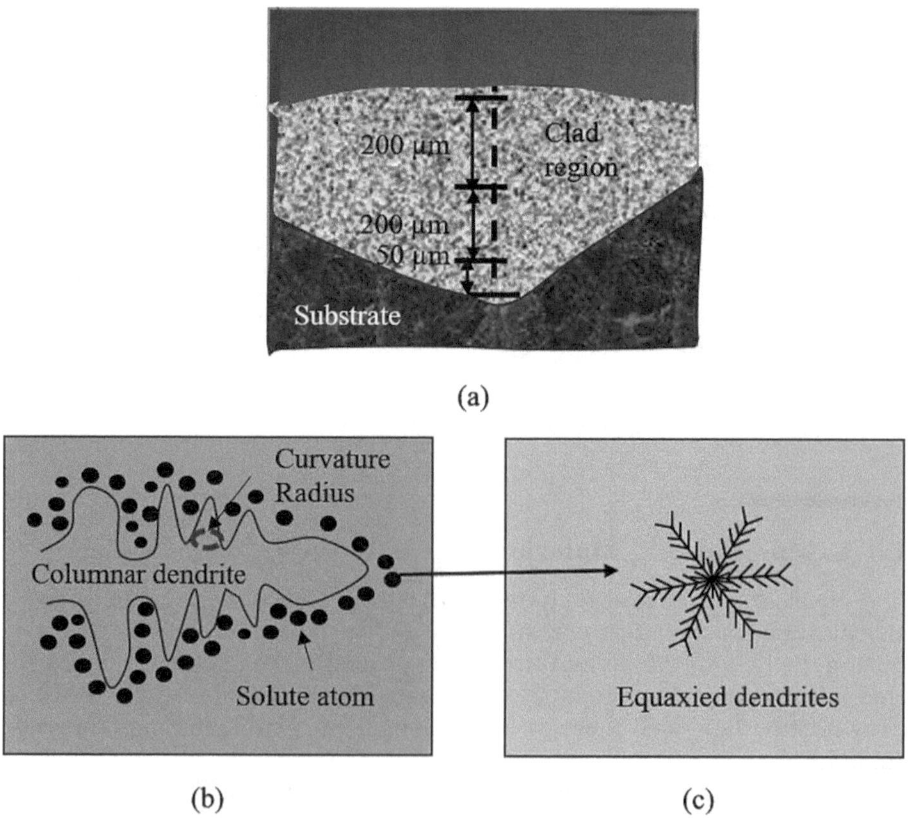

FIGURE 1.3
(a) Melt pool at laser-clad region, (b) columnar dendrites, and (c) equiaxed dendrites.

significantly influenced by many process variables, which in turn alters the coatings' wear resistance and corrosion resistance.

Intermetallic compounds provide benefits including high specific strength, strong oxidation, and resistance to heat [23]. The components of the coating frequently interact with those of the base material, creating a reinforced phase that increases the coating's microhardness. Alloys are amorphous when the atoms arrive during solidification so late to form an ordered crystal, resulting in the creation of amorphous alloys. The resultant solid alloy lacks both the crystal grains and grain boundaries of the crystalline alloy and has a disordered structure. A superior hardness, strong corrosion, and a high strain limit are all features of amorphous alloys. In the meanwhile, it is as stable as ceramics and as durable as metals. It is feasible to form an amorphous coating because of the LC's incredibly rapid heating and cooling speeds. Nickel-based alloys exhibit greater creep strength, low-cycle fatigue resistance, and thermal fatigue resistance compared to polycrystalline alloys. Consequently, these alloys are highly favored for aerospace applications. HEAs, also known as multi-principle alloys, deviate from the usual alloy design concept of using only one or two constituents [24]. The benefits of HEA include strength and hardness, wear and corrosion, and high-temperature softening resistance, according to studies. It has extensively employed in LC recently in the domains of industry, aircraft, and other things. Different elements have various characteristics. For instance, aluminum (Al), titanium (Ti), molybdenum (Mo), and other similar elements favor the creation of the body-centered cubic (BCC) solid solution, while copper (Cu), cobalt (Co), nickel (Ni), and other elements promote the formation of the solid solution. Out of them, the inclusion of Co element was increased the fluidity and wear, while the addition of Cu element results in a nanophase. Ni, in the meantime, strengthens the alloy's resistance to oxidation at high temperatures and corrosion. Ceramic coatings are widely used in aircraft, chemical machinery, and other applications due to their excellent abrasion resistance, corrosion resistance, and high-temperature oxidation resistance [25]. Advancements in laser cladding (LC) technology have led to the growing popularity of ceramic-reinforced metal matrix composites and oxide ceramics. Due to their high strength and excellent wear resistance, metal matrix composites (MMCs) have also emerged as one of the material systems used for LC. Particle-reinforced phase and fiber-reinforced phase now make up the majority of MMCs' reinforced phases in LC. The primary ceramics used in the particle-reinforced phases are WC, NbC, TiN, and others. Additionally, the choice of metal powders and particle-reinforced materials is crucial in LC. In general, it is important to take into account not only the physical properties of the materials but also the solubility and lack of violent chemical reactions of the metal matrix and particle-reinforced phase. LC particle-reinforced MMCs are now the subject of much research. The metal matrix for LC MMCs is increasingly being replaced by recently developed

HEAs with MMC materials including nickel-, iron-, and cobalt-based ones. According to the study [26], a small quantity of rare earth elements may considerably improve the microstructure, wear, and corrosion of LC coatings. Additionally, it can be accomplished by adding certain alloying elements. Due to their unique electrical structure and chemical activity, rare earth elements and their oxides can be added to coatings to increase their microstructure and laser absorptivity. By incorporating a little quantity of rare earth elements into the LC material, the performance of the coating is considerably enhanced and flaws are decreased. Nazari et al. [27] generated a Ti-Fe cladding layer with Ti6Al4V, and the coating contained TiC, reinforcing phase, and Fe_2O_3 and Fe_2C iron phases. Coating hardness is roughly double that of the substrate. However, more testing is needed to determine the wear resistance of this coating made from recycled titanium powder. Nickel-based layers were created utilizing a high-power ytterbium-doped yttrium aluminum garnet (Yb:YAG) disk laser on 316L stainless steel substrate. As shown in Figure 1.4, the mixing of a Co peak with other metastable phases, such as Cr_7C_3 and $Cr_{23}C_6$, was seen on cobalt coating, whereas the most dominating peak in nickel-based coating was a Ni-rich matrix.

The microhardness of the cladding layers is significantly higher than that of the substrate. Nickel cladding outperformed cobalt cladding and substrate sample in terms of wear resistance, indicating that nickel-based LC plays a role in wear resistance [28]. Gonz'alez et al. [29] coated an AISI4340 substrate with LC phosphor bronze. The results demonstrate that the coating is made of eutectoid and dendritic, and its hardness is much higher than that of the substrate. The stability of phosphor bronze coatings without dilution requires more research. Through simulation and experimentation, Liu et al. [30] investigated the effects of changing the angle and scanning direction on the melt zone and found the microstructure changes in the cladding layer. The results suggest that tilting the coaxial nozzle minimizes melt pool height. Wang et al. [31] used theoretical and experimental approaches to investigate the structure of the cladding layer in superalloys. The size in this melt pool, as well as the development direction and temperature gradient of dendrites, was modeled and forecasted using the vectorization approach, and the experimental results validated these. The dendrite's epitaxial growth direction is bent by more than 20°, and new grain boundaries form, which may be described properly by the flow field of the melt pool. Zhou et al. [32] investigated the crack process of LC-based superalloys and discovered that the sensitivity of cracks elevated as the grain boundary misalignment angle increased. Solidification in cracks is caused by dendritic carbides with a high melting point, whereas cracks in liquefaction are caused by a low melting point. Hence, further research into crack formation mechanisms is required before LC can be used for surface strengthening and repair. The influence of process factors on the microstructure and characteristics of LC CrMnFeCoNi HEA

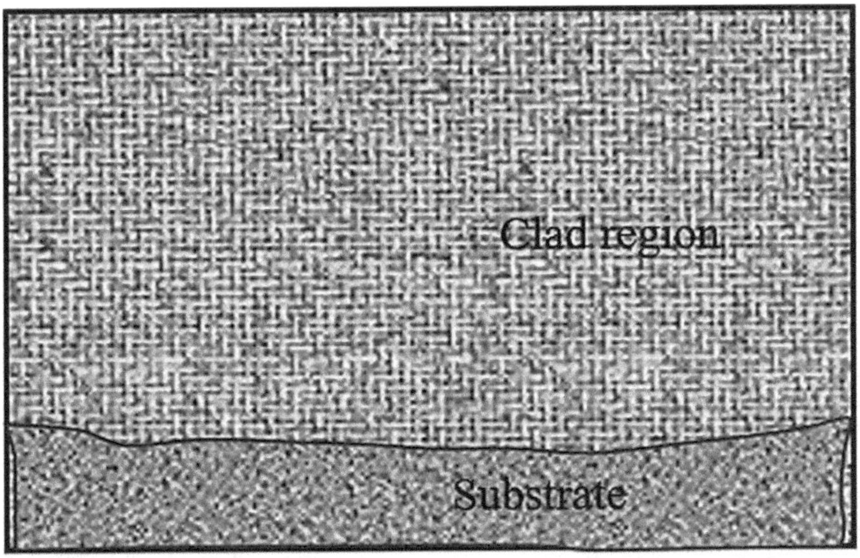

(a)

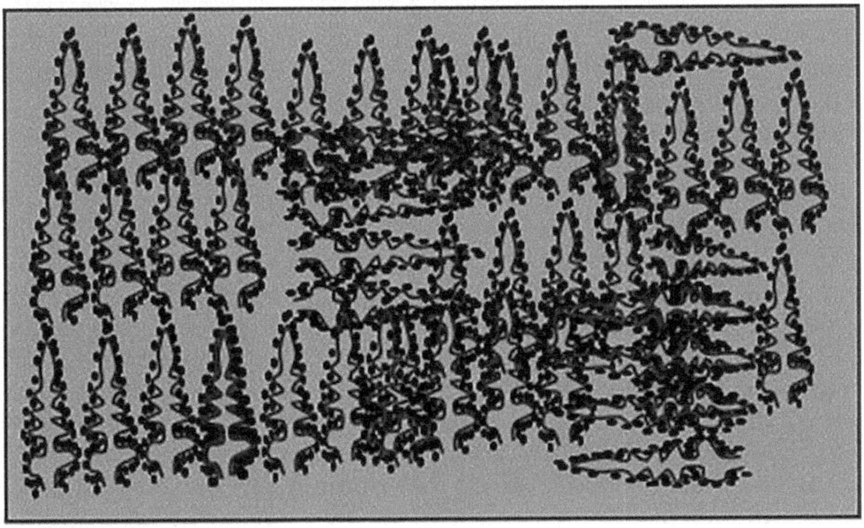

(b)

FIGURE 1.4
(a) Schematic micrograph of nickel-alloy coated 316L substrate showing various structures and (b) micrograph of cladding center region with fishbone structure.

was researched by Xiang et al. [33], and the findings reveal that the structure is primarily equiaxed with columnar. The laser power influences the temperature and heat direction, altering the microstructure of the coating. Because of the LC's extremely high cooling rate, strengthening occurs, and element segregation is lessened, resulting in a denser LC structure than that of cast structure, and the tensile characteristics are enhanced. Qiu et al. [34] investigated the corrosion of LC $Al_2CrFeCoxCuNiTi$ HEA coatings with Q235 steel, and revealed that the cladding layer had high corrosion rates in 3.5 wt.% NaCl and 1 mol/L NaOH solutions. Corrosion current density is lowered by three orders of magnitude in 1 mol/L NaOH solution compared to Q235 steel, while HEAs coating ($x=0.5$) shows no pitting corrosion in 3.5 wt. % NaCl solution. The corrosion resistance of the cladding layer is affected by microstructure. Li et al. [35] investigated the effects of Al_2O_3 on the geometric shape, microstructure, and hardness of Al_2O_3-TiB_2-TiC ceramic on carbon surfaces. According to the results, the cladding layer is made up of white and block Al_2O_3 and Fe phases. The cladding layer's hardness is higher when the Al_2O_3 percentage is low (10%), and the wear surface of the coating is better when the Al_2O_3 level is between 10% and 30%. Cladding layer has the best wear at 30%. However, when the Al_2O_3 concentration was raised to 50%, spalling pits formed in the wear morphology. Liu et al. [36] investigated the structure and mechanical of Al_2O_3/ZrO_2-directed solidification eutectic ceramics synthesized, as well as the modification of eutectic with growing conditions and solidification shrinkage creation process. By optimizing process settings, shrinkage flaws could be avoided, and the optimized fracture toughness and average hardness were 4.5 MPa $m^{1/2}$ and 16.7 GPa, respectively. Lian et al. [37] developed a mathematical model to predict the hardness and wear of TiC-reinforced nickel-based composite coatings based on LC process parameters. The model errors were 1.87% and 6.33%, respectively, according to the experimental data. However, the efficacy of this approach for rapidly and reliably determining the ideal parameter combination could be enhanced. Fan et al. [38] evaluated the optimal process parameters of LC WC (40 wt%) particle-reinforced cobalt-based composite coatings using single-factor and orthogonal tests, using coating dilution rate and hardness as indicators. The optimal process parameters are 2.4 kW laser power, 0.5 g/s powder-feeding speed, and 7 mm/s scanning speed. Lin et al. [39] investigated the influence of Mo on the structure and characteristics of LC NiAl coatings on medium-carbon steel surfaces. According to the findings, when the Mo concentration increases, the coating grains become more polished, and wear improves. Two different alloy powders, NiCrFeMoNb and FeCrMoVC, were deposited on nickel-based superalloy in this work. The generated layers were also evaluated for hardness, phase formation, and microstructure. The wear resistance of both layers was tested using a ball-on-disk wear tester. In addition, worn surfaces of deposition and roughness were studied. The results reveal that both laser depositions produced a homogeneous structure with enhanced toughness. When compared to NiCrFeMoNb cladding and base material, the wear resistance of FeCrMoVC cladding has improved, indicating that FeCrMoVC coating plays an important role in tribological qualities. A main wear process was discovered as abrasion, adhesion, and tribo-oxidation [40]. The different superalloy materials have also performed in the LC to reduce the wear [41]. On using plasma transferred arc (PTA) technique, the LC is carried out on Ti with FeCoCrNiAl particles. As shown in Figure 1.5, it is found that the different phases are formed, leading to enhanced mechanical and wear behavior properties [42]. Similarly, the FeCoCrNiAlCuMn particles alloyed on Inconel-718 substrate using PTA, as shown in Figure 1.6. The mechanical properties improved, and EDS of the surface is shown in Table 1.2.

1.9 Laser Cladding Applications

Due to its low dilution rate and strong metallurgical bonding with the substrate [43], LC is widely employed in the space, chemical equipment, and rest of areas [44]. It is helpful for us to repair machinery components since it can be used to create a surface coating with superior performance, such as greater wear, better corrosion, and temperature oxidation. The coatings may be broadly classified as wear, lubricating, corrosion, high-temperature, and bioaffinity coatings based on their respective functions [45].

1.10 Wear Resistance and Self-Lubricating Coatings

Recent advancements in nuclear and aerospace engineering have raised standards for component wear resistance under extremely demanding environments such as high temperatures and tremendous loads. Consequently, it is an economical way for LC to manufacture a self-lubricating and wear-resistant coating on the part's surface. Wear-resistant coatings research currently focuses on adding ceramic-reinforced particles to metals or alloys that have enough hardness and wear resistance [46]. These ceramics fall into two categories: oxide ceramics, such as Al_2O_3, ZrO_2, and TiO_2, and non-oxide ceramics, such as WC, TiC, ZrN, SiN, TiB, and TiSi. Liquid lubricants would be far less effective at high temperatures; hence, lubricating coatings often add solid lubricants with alloys. Generally speaking, there are three types of commonly used solid lubricants: layered solids (such as graphite and h-BN), sulfides (such as MoS_2 and WS_2), fluorides (such as CaF_2 and BaF_2), and soft metals (such as Ag and Au). In many applications, wear resistance

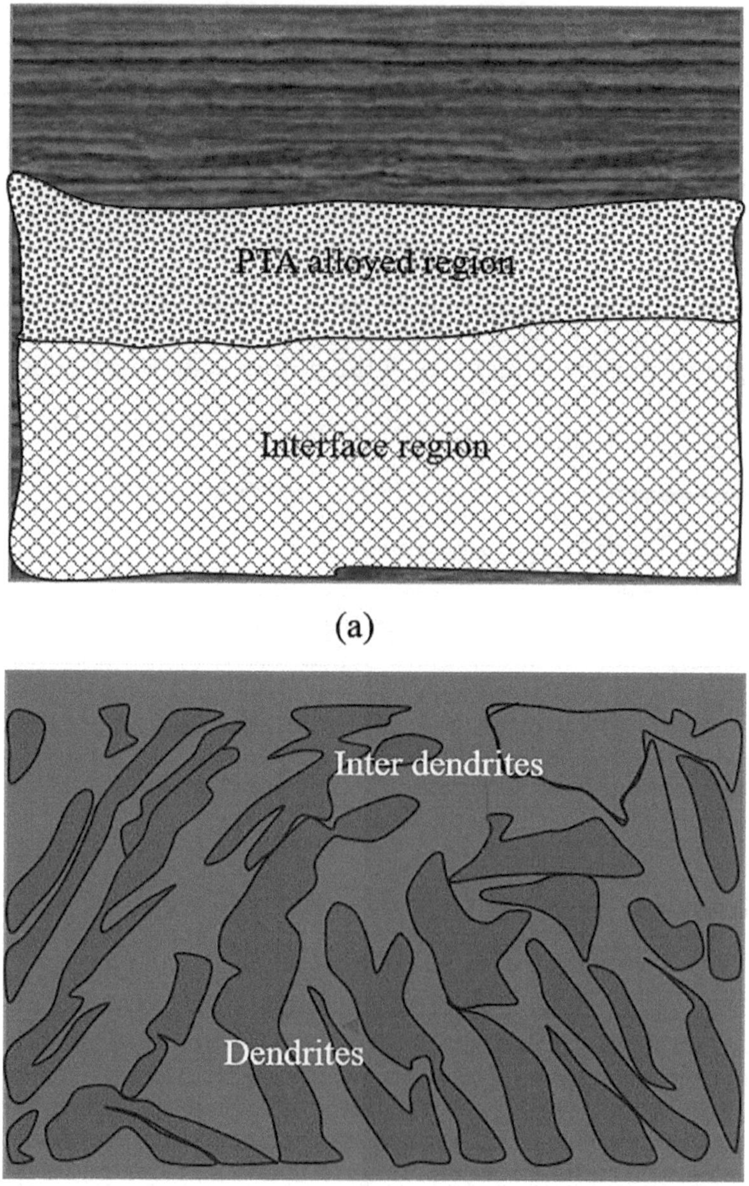

FIGURE 1.5
(a) Schematic of various regions of Ti substrate and PTA and (b) inter-dendrites and dendrites.

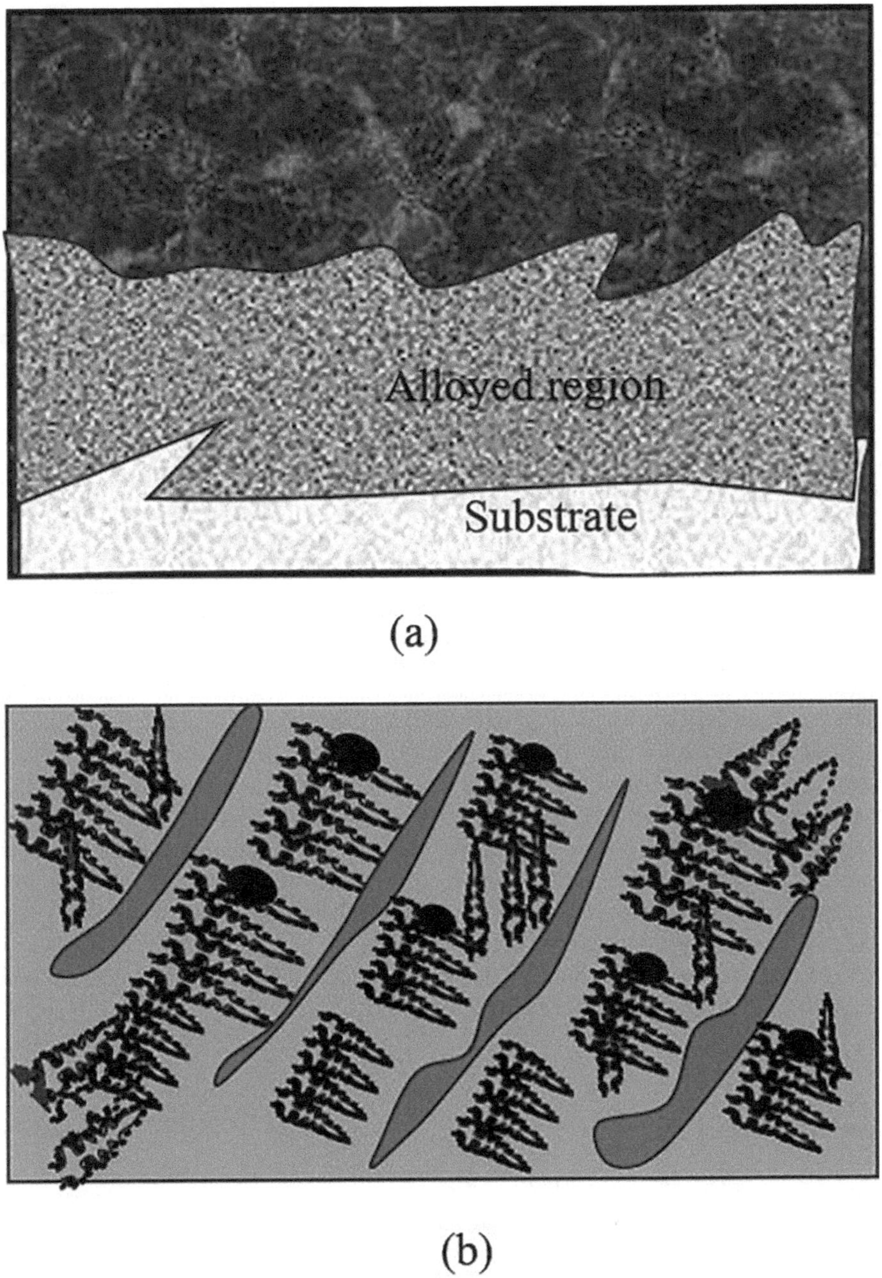

FIGURE 1.6
Schematic of various regions of IN718 substrate and PTA and (b) dendrites.

TABLE 1.2
EDS Results of PTA FeCoCrNiAlCuMn Alloyed Region

Presence of elements in the powder	Al	Co	Cr	Fe	Ni	Cu	Mn
EDS analysis (wt. %)	22.79	13.41	15.28	14.36	12.62	12.94	8.60

and lubrication are regarded as essential qualities. Because layered solids have atoms arranged in layers, adding a lubricant can enhance the coating's lubricating properties. The interatomic forces are weak in separate layers and high in the same layer. Because of the simple relative sliding between layers, the coating offers high self-lubricating qualities. Because of the brittle-tough transition at 500°C, the coating applied. Because the soft metal softens thermally throughout the friction process, it lowers the coating's friction coefficient.

1.11 Oxidation-Resistant and Corrosion-Resistant Coatings

Parts are frequently exposed to extreme conditions, such high temperatures and high pressure, in the petrochemical, aircraft, and other industries. To prolong the service life of components and lower their production costs, it is crucial to provide high-temperature oxidation and corrosion-resistant coatings for their surfaces. Numerous academics have created a variety of material coatings that are resistant to corrosion and oxidation at high temperatures. Because of their strong resistance to high-temperature oxidation, oxide ceramics and intermetallic compounds, such as Al_2O_3, ZrO_2, TiSi, TiAl, and NiCr, are frequently used in the creation of coatings that are resistant to oxidation at high temperatures [47].

1.12 Bio-Ceramic Coatings

Since biomedical metals have high mechanical qualities and resistance to corrosion, they are frequently used in the replacement or repair of human tissues. However, they are often bio-inert materials. Bioactive ceramics are biocompatible and have high bioactivity, but their application in bone-bearing locations is limited by their brittleness, low strength, and better toughness [48]. Therefore, the advantages of bio-ceramic materials can be utilized through laser cladding (LC) to create bio-ceramic coatings with biological functionality on the surface of biomedical metals, offering significant application value. Today, hydroxyapatite, fluoroapatite, 13-tricalcium phosphate, and other similar minerals are often used in bio-ceramics.

1.13 Repair of Parts

Aero turbine blades, machine element surface, shafts, and rest of mechanical assemblies that have been subjected to high temperatures, high pressures, or severe loads over an extended period of time are now the primary applications for LC. Restoring the workpiece to its intended shape and functionality is the primary aim of restoration [49]. With LC's innovative surface forming and repair technology, broken structures may now function better and have lower structural costs because damaged components do not need to be replaced as often. Turbine and aeroengine blades are frequently subjected to extreme temperatures, pressures, and external impacts during operating. It is weak to fatigue harm creep failure, fractures, etc., even if it can withstand mechanical and thermal stress. LC and laser remelting are used to create the turbine blade tips for the nickel-based superalloys like CMSX-4 and PWA 1426. The solidification structures of notable dimensions are created on the plane and tops of turbine blades with appropriate LC process parameters, offering a foundation for single-crystal blade repair. However, further research is yet required on the process parameters of the blade's trailing edge and other components. The study focused on LC and remelting single-crystal blade damage restoration, including fractures and other types of damage [50]. The temperature field simulation is used to calculate the process parameters, which are subsequently confirmed by experiments to provide the desired directional temperature gradient. By varying the laser power, the optimal unit energy input for every individual cladding layer is found. Its efficacy is then assessed on five distinct CMSX-4 cladding track shapes. More investigation is necessary, though, on the use of LC in conjunction with remelting to fix the blade's leading and trailing edges. For the purpose of repairing blade tips, an adaptive LC process parameter optimization approach is used, which considered the blade's real geometry. This study determined that laser power is the most useful process parameter. The investigation discovered that the equation might be used to calculate the laser power needed to fix the variable geometry blade tip. The suggested method's correctness and reasonableness have been confirmed by the LC tests conducted to repair the Ti6Al4V compressor blades [51]. The agreement of the calculated milling forces with surface topography with the observations is demonstrated by process simulations, according to a study on LC and milling repairs on polycrystalline aerospace material Rene 80. Further research is necessary to see whether this model is universal for use with other materials. Railway steel wheel disks are among the heavy machinery parts that frequently carry significant alternating loads during operation. As a result, their surfaces are vulnerable to wear and tear, plastic deformation, and cracking. A study examined the rolling contact fatigue and wear characteristics of the three cladding layers on the rail wheel disk surfaces – Hastelloy-C, Inconel 625, and Stellite 21. There has been an 83%, 42%, and 29% decrease in the cladding layer's wear, correspondingly. The best resistance to wear is found in Stellite 21; however, Hastelloy-C cladding performs best when both the rolling contact fatique and the cladding boundary's wear are taken as into account. Nonetheless, the cladding layer's performance is also significantly influenced by the parameters of the LC process. As a result, it is crucial to examine how well railway wheel's function following various material repairs under various process conditions [52]. The wear characteristics of stainless steel grades 316L, 410, and 420 are studied for use as cladding materials for rail wheel disks. The study discovered that wear lowers with increasing cladding material hardness. On the surface devoid of LC, small surface cracks started to emerge, and deep fissures between the substrate and the cladding started to spread along the border from the point at where the highest shear stress subsided. Future studies must, however, also assess the mechanical characteristics of railways wheel repair components, such as their bending and shear strengths. At the same time, big slippage rates or pure slip tests should be carried out to improve the practical application of the restored railway wheel hub. In the exiting problems, despite the fact that the LC is now used in the surface hardening and restoration of mechanical parts, several issues remain unresolved. Microstructure-wise, pore flaws in the material are easily produced during the LC process because of things like damp powder or impure material, too much protective gas, and how quickly the materials solidify and cool down. Large-area LC is also susceptible to cracking because of residual stresses like thermal stress. Simultaneously, there is the issue of element segregation within the microstructure. The macroscopic sizes and surface polish of LC coatings require more research and improvement in terms of their macroscopic morphology. Large, tiny, and complicated curved elements are frequently strengthened and repaired in the actual use of LC. Thus, the question of how to automate and produce high-quality LC surface hardening or repair remains critical.

1.14 Development of a New Material

Future research should concentrate on developing novel material systems, such as multi-principle HEA, multi-performance composite materials, and amorphous and nanocrystalline materials, in accordance with the material selection principle. As of now, functionally graded coatings that are primarily made possible by material composition and structural gradation can lessen internal stresses, minimize fractures, and enhance the metallurgy combination of coatings and substrates [53]. As a result, the creation of gradient coatings is crucial for strengthening and repairing mechanical parts' surfaces. Future study should concentrate on optimizing the LC process parameters to provide coatings with a precise gradient distribution of the material content and high performance. Simultaneously, it must be done to enhance the advancement of a mechanism that accurately regulates the powder flow rate of different component powders. The benefits of amorphous materials include a wide elastic strain limit, excellent corrosion resistance, and high hardness [54]. In addition, the LC procedure promotes the development of amorphous forms. As a result, in recent years, it has emerged as a hub for research. Nonetheless, further research on the following areas needs to be done in the future: (i) Amorphous coating composition design and control: The actual composition in LC differs from the planned composition because of the transfer of mass of the melting pool and the impact of the composition of the growth of the epitaxial layer on the substrate. The alloying elements will oxidize and burn at the same time, so it's important to design a suitable material composition based on the features of the LC process. (ii) Future research should concentrate on a thorough examination of the microstructure growth mechanism as well as parameter optimization for LC amorphous/nanocrystalline coatings. (iii) Another problem is the LC's ability to prepare huge volumes of effectively performing amorphous materials. Excellent qualities of HEA include high resistance in wear, corrosion, and high temperatures [55]. The theory underlying the crystallization and solidifying of HEA coatings, as well as how it relates to their mechanical and physical characteristics, still needs investigation. An amorphous phase can arise as a result of the properties of the LC process, and the HEA's properties also encourage the amorphization of the structure as a whole. As a result, another hotspot for current research is the preparation of the HEA amorphous coating using LC.

1.15 Computer Simulation, Monitoring, and Control

A complicated metallurgical process is the LC process. A study of the outside temperature field, stress, and flow of the LC process is now very important due to advancements in computer technology [56]. Computer simulation in the LC model is formed based on the melt pool, convection and process parameters. It is capable of doing a quantitative analysis of the effects of temperature variations and fluid movement in the pool of melt on the cladding layer's microstructure. Consequently, the integrated simulation studies of the microstructure, stress field, and temperature field are highly significant.

To create a cladding layer with a suitable microstructure and performance, surface thickening or repair requires close observation and management of the LC process. Currently, the most detectable information is related to the ambient temperature of the melt pool and the powder-cladding process. Future studies should focus on dilution rates, informational content, and the identification and management of voids. Mega data, artificial intelligence, and LC should be integrated at the same time. The LC is evolved toward perception, adaptation, and decision intelligence by creating an empirical database, which may greatly enhance the cladding layer's quality and performance. This is also where development is headed in the future [57].

1.16 Optimization of Laser Cladding Process

The cladding layer's macro-morphology and microstructure characteristics are directly determined by the LC process. The existing LC technology, however, has low power that is challenging to suit actual demands in big area LC of mechanical components. Therefore, to increase the efficiency in the future, further research into high-power LC devices has to be done. To reinforce and fix the surfaces of small precise parts, it is necessary to investigate compact, portable, and in-situ laser compasses. Key elements, such as LC nozzles, should be improved concurrently to enhance cladding quality and powder utilization. It is recommended that LC technology be integrated with robotics and control technologies to enhance automation and monitoring for the surface fortification and repair of intricately curved components [58]. Although LC is gradually implementing technologies, such as thermal induction aid, electromagnetic help, and ultrasonic vibration, there are still a lot of unresolved issues. As an illustration, consider the impact of electromagnetic and ultrasonic fields on the microstructure of LC, the use of ultrasonic frequency tracking equipment throughout the LC process, and the corresponding arrangement of auxiliary and LC devices for intricate components.

1.17 Subsequent Processing of Laser Cladding

A clad layer with an acceptable dilution and a good metallurgical connection with the substrate may be formed using the LC process. Subsequent processing is necessary since it still has issues with residual stress, and surface quality that don't fulfill real demands. The benefits of the present ultrasonic-assisted cutting method include reduced surface roughness and subsurface damage, as well as strong processing capabilities for brittle and hard materials, MMCs, and fiber-reinforced material [59]. It is anticipated that the combination of LC with ultrasonic-assisted cutting would produce refined grains and improved surface hardness because these materials are also primarily used in LC. The benefits of laser-assisted cutting over traditional cutting technologies include lower cutting force, less tool wear, and better surface machining quality. Furthermore, the use of ultrasonic impact is crucial in mitigating the cladding layer's surface roughness, residual stress, and enhancing fatigue characteristics, among other benefits. To accomplish the systematic and integrated growth of LC technology in the future, the coupling of LC with ultrasonic, laser, ultrasonic impact, and other hybrid processing techniques enhanced [60].

1.18 Conclusions

The LC technique gives potential improvement in metals and composites based on the tribological properties and performance measures. The powder-feeding system provides a great effort on coating metal properties. The various optimization techniques are used to improve both the performance and tribological properties. The ceramic powder gives better wear resistance compared with metal powder coatings. The repaired parts performance is improved in both metal powder and ceramic powder coatings. The HEA coatings are currently developed on metals/superalloys/composites. The modeling and simulation are also used to improve the performance measures and wear resistance properties.

Acknowledgment

The editor N. Jeyaprakash provided valuable suggestions for improving this book chapter.

References

1. Zhang, X., Yin, X., Liu, C., & Liu, C. (2023). Laser melting of prefabrication AlOOH-activated film on the surface of nodular cast iron and its associated properties. *Materials*, 16(15), 5486.

2. Zhao, C., Ma, C., Yang, J., Li, M., Zhao, Q., Ma, H., & Jia, X. (2023). Numerical simulation study of multi-field coupling for laser cladding of shaft parts. *Micromachines*, 14(2), 493.
3. Wood, D. L., Quass, J. D., Li, J., Ahmed, S., Ventola, D., & Daniel, C. (2018). Technical and economic analysis of solvent-based lithium-ion electrode drying with water and NMP. *Drying Technology*, 36(2), 234–244.
4. Kannan, V. S., Lenin, K., Srinivasan, D., & Raj Kumar, D. (2023). Analysis of microstructural, mechanical and surface properties of aluminium hybrid composites obtained through stir casting. *Journal of the Institution of Engineers (India): Series* D, 105, 733–744.
5. Jeyaprakash, N., Yang, C. H., & Kumar, D. R. (2020). Laser surface modification of materials. In Practical Applications of Laser Ablation.
6. Lusthaus, N., Raichlin, S., Nadiv, R. & Regev, O. (2025). Polymer composite cladding for selective frequency filtration: An experimental and modeling study. *Composites Science and Technology*, 259, 110929.
7. Wang, C., Zhou, J., Zhang, T., Meng, X., Li, P., & Huang, S. (2022). Numerical simulation and solidification characteristics for laser cladding of Inconel 718. *Optics & Laser Technology*, 149, 107843.
8. Ghorashi, M. S., Farrahi, G. H., & Movahhedy, M. R. (2019). Considering cyclic plasticity to predict residual stresses in laser cladding of Inconel 718 multi bead samples. *Journal of Manufacturing Processes*, 42, 149–158.
9. Zhang, Z., & Kovacevic, R. (2019). A thermo-mechanical model for simulating the temperature and stress distribution during laser cladding process. *The International Journal of Advanced Manufacturing Technology*, 102, 457–472.
10. Wirth, F., Arpagaus, S., & Wegener, K. (2018). Analysis of melt pool dynamics in laser cladding and direct metal deposition by automated high-speed camera image evaluation. *Additive Manufacturing*, 21, 369–382.
11. Muvvala, G., Karmakar, D. P., & Nath, A. K. (2017). Online monitoring of thermo-cycles and its correlation with microstructure in laser cladding of nickel based super alloy. *Optics and Lasers in Engineering*, 88, 139–152.
12. Kannan, V. S., Lenin, K., Srinivasan, D., & Kumar, D. R. (2021). Investigation on laser square hole drilling of AA7475/SiC/$ZrSiO_4$ composites. *Silicon*, 14, 4557–4574.
13. Ramanujam, N., Dhanabalan, S., Raj Kumar, D., & Jeyaprakash, N. (2021). Investigation of micro-hole quality in drilled CFRP laminates through CO_2 laser. *Arabian Journal for Science and Engineering*, 46(8), 7557–7575.
14. Jeyaprakash, N., Yang, C. H., Karuppasamy, S. S., & Radhika, N. (2023). Microstructural evolution and anti-corrosion characteristics of as-cast and LPBF printed Inconel plates in seawater environment. *Surface Review and Letters*, 31, 2450032.
15. Yu, T., Yang, L., Zhao, Y., Sun, J., & Li, B. (2018). Experimental research and multi-response multi-parameter optimization of laser cladding Fe313. *Optics & Laser Technology*, 108, 321–332.
16. Bax, B., Rajput, R., Kellet, R., & Reisacher, M. (2018). Systematic evaluation of process parameter maps for laser cladding and directed energy deposition. *Additive Manufacturing*, 21, 487–494.
17. Reddy, L., Preston, S. P., Shipway, P. H., Davis, C., & Hussain, T. (2018). Process parameter optimization of laser clad iron-based alloy: Predictive models of deposition efficiency, porosity and dilution. *Surface and Coatings Technology*, 349, 198–207.

18. Zhai, L. L., Ban, C. Y., & Zhang, J. W. (2019). Microstructure, microhardness and corrosion resistance of NiCrBSi coatings under electromagnetic field auxiliary laser cladding. *Surface and Coatings Technology*, 358, 531–538.
19. Zhang, N., Liu, W., Deng, D., Tang, Z., Liu, X., Yan, Z., & Zhang, H. (2018). Effect of electric-magnetic compound field on the pore distribution in laser cladding process. *Optics & Laser Technology*, 108, 247–254.
20. Raj Kumar, D., Jeyaprakash, N., Yang, C. H., & Ramkumar, K. R. (2020). Investigation on drilling behavior of CFRP composites using optimization technique. *Arabian Journal for Science and Engineering*, 45, 8999–9014.
21. Kumar, D. R., Jeyaprakash, N., Yang, C. H., & Sivasankaran, S. (2021). Optimization of drilling process on carbon-fiber reinforced plastics using genetic algorithm. *Surface Review and Letters*, 28(03), 2050056.
22. Ragunath, S., Radhika, N., Krishna, S. A., & Jeyaprakash, N. (2023). Enhancing microstructural, mechanical and tribological behaviour of AlSiBeTiV high entropy alloy reinforced SS410 through friction stir processing. *Tribology International*, 188, 108840.
23. Ma, K., Cheng, Y., Jeyaprakash, N., Zhou, J., Wan, Y., & Yang, W. (2023). Temperature gradient and solidification rate simulation model of the microstructure of laser-cladded 27SiMn. *Metals*, 13(10), 1682.
24. Jeyaprakash, N., Prabu, G., & Yang, C. H. (2022). The influence of different phases on the microstructure and wear of inconel-718 surface alloyed with AlCuNiFeCr hard particles using plasma transferred arc. *Journal of Materials Engineering and Performance*, 31(12), 9921–9934.
25. Yeh, C. H., Jeyaprakash, N., & Yang, C. H. (2020). Nondestructive characterization of elastic modulus of APS Ni–5Al/10hBN coating on stainless steel 304 under high temperature. *Archives of Civil and Mechanical Engineering*, 20, 1–15.
26. Wang, C., Gao, Y., Zeng, Z., & Fu, Y. (2017). Effect of rare-earth on friction and wear properties of laser cladding Ni-based coatings on 6063Al. *Journal of Alloys and Compounds*, 727, 278–285.
27. Nazari, K. A., Rashid, R. R., Palanisamy, S., Xia, K., & Dargusch, M. S. (2018). A novel Ti-Fe composite coating deposited using laser cladding of low cost recycled nano-crystalline titanium powder. *Materials Letters*, 229, 301–304.
28. Jeyaprakash, N., Yang, C. H., & Sivasankaran, S. (2020). Laser cladding process of Cobalt and Nickel based hard-micron-layers on 316L-stainless-steel-substrate. *Materials and Manufacturing Processes*, 35(2), 142–151.
29. Arias-González, F., Del Val, J., Comesaña, R., Penide, J., Lusquiños, F., Quintero, F., Riveiro, A., Boutinguiza, M., & Pou, J. (2017). Laser cladding of phosphor bronze. *Surface and Coatings Technology*, 313, 248–254.
30. Liu, Z., Jiang, L., Wang, Z., & Song, L. (2018). Mathematical modeling of transport phenomena in multi-track and multi-layer laser powder deposition of single-crystal superalloy. *Metallurgical and Materials Transactions A*, 49, 6533–6543.
31. Wang, G., Liang, J., Yang, Y., Shi, Y., Zhou, Y., Jin, T., & Sun, X. (2018). Effects of scanning speed on microstructure in laser surface-melted single crystal superalloy and theoretical analysis. *Journal of Materials Science & Technology*, 34(8), 1315–1324.
32. Zhou, Z., Huang, L., Shang, Y., Li, Y., Jiang, L., & Lei, Q. (2018). Causes analysis on cracks in nickel-based single crystal superalloy fabricated by laser powder deposition additive manufacturing. *Materials & Design*, 160, 1238–1249.

33. Xiang, S., Li, J., Luan, H., Amar, A., Lu, S., Li, K., Zhang, L., Liu, X., Le, G., Wang, X., & Qu, F. (2019). Effects of process parameters on microstructures and tensile properties of laser melting deposited CrMnFeCoNi high entropy alloys. *Materials Science and Engineering: A*, 743, 412–417.
34. Qiu, X. W. (2019). Corrosion behavior of $Al_2CrFeCo_xCuNiTi$ high-entropy alloy coating in alkaline solution and salt solution. *Results in Physics*, 12, 1737–1741.
35. Li, Z., Wei, M., Xiao, K., Bai, Z., Xue, W., Dong, C., Wei, D., & Li, X. (2019). Microhardness and wear resistance of Al_2O_3-TiB_2-TiC ceramic coatings on carbon steel fabricated by laser cladding. *Ceramics International*, 45(1), 115–121.
36. Liu, Z., Song, K., Gao, B., Tian, T., Yang, H., Lin, X., & Huang, W. (2016). Microstructure and mechanical properties of Al_2O_3/ZrO_2 directionally solidified eutectic ceramic prepared by laser 3D printing. *Journal of Materials Science & Technology*, 32(4), 320–325.
37. Lian, G., Zhang, H., Zhang, Y., Yao, M., Huang, X., & Chen, C. (2019). Computational and experimental investigation of micro-hardness and wear resistance of Ni-based alloy and TiC composite coating obtained by laser cladding. *Materials*, 12(5), 793.
38. Fan, P., & Zhang, G. (2020). Study on process optimization of WC-Co_{50} cermet composite coating by laser cladding. *International Journal of Refractory Metals and Hard Materials*, 87, 105133.
39. Lin, C. M., Kai, W. Y., Su, C. Y., & Key, K. H. (2017). Empirical alloys-by-design theory calculations to the microstructure evolution mechanical properties of Mo-doped laser cladding NiAl composite coatings on medium carbon steel substrates. *Journal of Alloys and Compounds*, 702, 679–686.
40. Jeyaprakash, N., & Yang, C. H. (2020). Comparative study of NiCrFeMoNb/FeCrMoVC laser cladding process on nickel-based superalloy. *Materials and Manufacturing Processes*, 35(12), 1383–1391.
41. Manikandan, K., & Palanikumar, K. (2020). Machinability evaluation and comparison of Incoloy 825, Inconel 603 XL, Monel K400 and Inconel 600 super alloys in wire electrical discharge machining. *Journal of Materials Research and Technology*, 9(6), 12260–12272.
42. Jeyaprakash, N., Prabu, G., & Yang, C. H. (2022). Formation of different phases and their influences on the mechanical and tribological properties of surface alloyed FeCoCrNiAl particles using PTA technique. *Intermetallics*, 142, 107457.
43. Jeyaprakash, N., Prabu, G., & Yang, C. H. (2021). Microstructure and tribological behaviour of FeCoCrNiAlCuMn particles alloyed on Inconel-718 substrate using plasma transferred arc technique. *Intermetallics*, 139, 107360.
44. Balasubramaniam, V., Rajkumar, D., Ranjithkumar, P., & Narayanan, C. S. (2020). Comparative study of mechanical technologies over laser technology for drilling carbon fiber reinforced polymer materials. *Indian Journal of Engineering and Material Sciences*, 27, 19–32.
45. Jeyaprakash, N., Yang, C. H., Duraiselvam, M., Prabu, G., Tseng, S. P., & Kumar, D. R. (2019). Investigation of high temperature wear performance on laser processed nodular iron using optimization technique. *Results in Physics*, 15, 102585.
46. Jeyaprakash, N., Yang, C. H., & Kumar, D. R. (2020). Minimum cutting thickness and surface roughness achieving during micromachining of aluminium 19000 using CNC machine. *Materials Today: Proceedings*, 21, 755–761.

47. Jeyaprakash, N., Yang, C. H., & Ramkumar, K. R. (2020). Microstructure and wear resistance of laser cladded Inconel 625 and Colmonoy 6 depositions on Inconel 625 substrate. *Applied Physics A*, 126, 1–11.
48. Al-Amin, M., Abdul Rani, A. M., Abdu Aliyu, A. A., Bryant, M. G., Danish, M., & Ahmad, A. (2020). Bio-ceramic coatings adhesion and roughness of biomaterials through PM-EDM: A comprehensive review. *Materials and Manufacturing Processes*, 35(11), 1157–1180.
49. Liu, H., Hu, Z., Qin, X., Wang, Y., Zhang, J., & Huang, S. (2017). Parameter optimization and experimental study of the sprocket repairing using laser cladding. *The International Journal of Advanced Manufacturing Technology*, 91, 3967–3975.
50. Liu, Z., Jiang, Q., Li, T., Dong, S., Yan, S., Zhang, H., & Xu, B. (2016). Environmental benefits of remanufacturing: A case study of cylinder heads remanufactured through laser cladding. *Journal of Cleaner Production*, 133, 1027–1033.
51. Jeyaprakash, N., & Yang, C. H. (2020). Microstructure and wear behaviour of SS420 micron layers on Ti–6Al–4V substrate using laser cladding process. *Transactions of the Indian Institute of Metals*, 73, 1527–1533.
52. Seo, J. W., Kim, J. C., Kwon, S. J., & Jun, H. K. (2019). Effects of laser cladding for repairing and improving wear of rails. *International Journal of Precision Engineering and Manufacturing*, 20, 1207–1217.
53. Lai, Q., Abrahams, R., Yan, W., Qiu, C., Mutton, P., Paradowska, A., & Soodi, M. (2017). Investigation of a novel functionally graded material for the repair of premium hypereutectoid rails using laser cladding technology. *Composites Part B: Engineering*, 130, 174–191.
54. Rajkumar, D., Ranjithkumar, P., & Jenarthanan, M. P. (2017). Experimental investigation and analysis of factors influencing delamination and thrust force during drilling of carbon-fibre reinforced polymer composites. *Pigment & Resin Technology*, 46(6), 507–524.
55. Yeh, C. H., Jeyaprakash, N., & Yang, C. H. (2020). Temperature dependent elastic modulus of HVOF sprayed Ni-5% Al on 304 stainless steel using nondestructive laser ultrasound technique. *Surface and Coatings Technology*, 385, 125404.
56. Li, C., Yu, Z., Gao, J., Zhao, J., & Han, X. (2019). Numerical simulation and experimental study of cladding Fe60 on an ASTM 1045 substrate by laser cladding. *Surface and Coatings Technology*, 357, 965–977.
57. Gouge, M. F., Heigel, J. C., Michaleris, P., & Palmer, T. A. (2015). Modeling forced convection in the thermal simulation of laser cladding processes. *The International Journal of Advanced Manufacturing Technology*, 79, 307–320.
58. Jeyaprakash, N., Yang, C. H., & Tseng, S. P. (2020). Optimization of tribological parameters over WC-12% Co laser alloyed pearlitic ductile iron using Taguchi based Grey relational analysis. *Journal of Central South University*, 27(3), 736–751.
59. Rajkumar, D., Ranjithkumar, P., & Narayanan, C. S. (2017). Optimization of machining parameters on microdrilling of CFRP composites by Taguchi based desirability function analysis. *Indian Journal of Engineering and Material Sciences*, 24, 331–338.
60. Jeyaprakash, N., & Yang, C. H. (2020). Improvement of tribo-mechanical properties of directionally solidified CM-247 LC nickel-based super alloy through laser material processing. *The International Journal of Advanced Manufacturing Technology*, 106, 4805–4814.

2

Overall View of the Laser Cladding Process

S. Pradeep Kumar, K. Nirmal Kumar, S. Martin Vinoth, P. Marimuthu, and P. Dinesh Babu

2.1 Introduction

In recent times, the industry has sought crucial components that exhibit exceptional resistance to high temperatures, wear, and corrosion. Laser cladding treatment is a type of surface modification technique superior to others due to its distinctive qualities, including high energy density and good metallurgical bonding between the coatings and substrates, which can be used to meet the current industrial requirements. High-power lasers are used for the laser cladding process. Depending on how the powder is fed, cladding methods can be classified into four types: coaxial, pre-placed, off-axis powder systems, and wire feeding systems. Coaxial and pre-placed powder systems are the most widely used cladding techniques [1].

The process parameters such as laser power, powder feed rate, and scanning rate are important and decide the quality of the clad layer formed. Optimisation of the process parameters will help to achieve the required height, width, and clad penetration and to minimise the crack and void formation in the clad layer [2]. The cladding materials can be prepared in powder, paste, wire, and rod forms. Among them, powder form is used most widely. Ni-, Co-, and Fe-based powders give good toughness, oxidation, and corrosion resistance [3]. Composite powders have excellent properties for wide application, high-entropy alloy (HEA) powders will give a simple phase structure with good mechanical properties, and rare-earth oxide powders have excellent fine-grain strengthening effects. Ceramic powders are used for their excellent high-temperature wear, corrosion, and oxidation resistance [4]. In addition, copper, titanium, aluminium, magnesium, zirconium, chromium base and intermetallic compound base powders are also used to achieve good wear, corrosion resistance, friction reduction, electrical conductivity, and other functions. This chapter will summarise the different laser cladding systems, materials used, and characterisation required for cladding. In addition, the applications, challenges, and future trends of laser cladding are also discussed in this chapter.

DOI: 10.1201/9781032693576-2

2.2 Different Laser Cladding Methods and Equipment Used for Cladding

2.2.1 Two-Step Method

The two-step laser cladding method, often called the pre-placed process, involves applying powder to the substrate surface using adhesives or thermal spraying. When the laser beam scans over this surface, the pre-applied powder and a portion of the matrix melt, forming the cladding layer. Notably, this technique doesn't require a powder-feeding system, making it cost-effective [5]. However, the pre-deposited powder must strongly adhere to both the substrate and itself. To prevent powder particles from escaping during melting due to gas flow, chemical adhesives are typically added to the powders. These additives maintain powder cohesion during the process [6].

The laser beam melts the pre-applied powder in this two-step process, forming a molten pool. Heat conduction extends the pool to the substrate's edge, where it fuses with the substrate material. Precise heat control prevents excessive dilution, where too much substrate material is mixed into the clad region. This limitation often confines the two-step laser cladding method to single-track applications [7].

2.2.2 One-Step Method

One-step laser cladding is a single-stage process that involves simultaneous deposition and fusion. The laser beam dissolves the surface of the substrate while cladding material (often in powder or wire form) is introduced directly into the laser-produced molten pool. This approach offers precise control over cladding layer thickness, high material efficiency, and suitability for rapid prototyping of complex components. It includes the powder injection, wire feeding, and paste laser cladding processes [8].

2.2.2.1 Laser Cladding with Powder Feeding

The fine cladding powder material is introduced into the melt pool using a carrier gas in powder injection laser cladding. The powder is directed as a targeted jet onto the melt pool's surface at a specific angle. Upon impact, the laser beam heats and melts the incoming powder and a portion of the substrate, forming a clad track as the substrate is scanned. Melted powder particles contribute to the clad layer, while those hitting the solid material may bounce off or adhere, leading to surface unevenness [9]. The powder-feeding method can be classified into off-axis and coaxial systems. Scanning electron microscopy (SEM) images of powder particles and their particle distribution are shown in Figure 2.1.

Off-axis powder feeding strategically positions the nozzle in the front for efficient powder collection. Adjusting the feeding angle enhances this

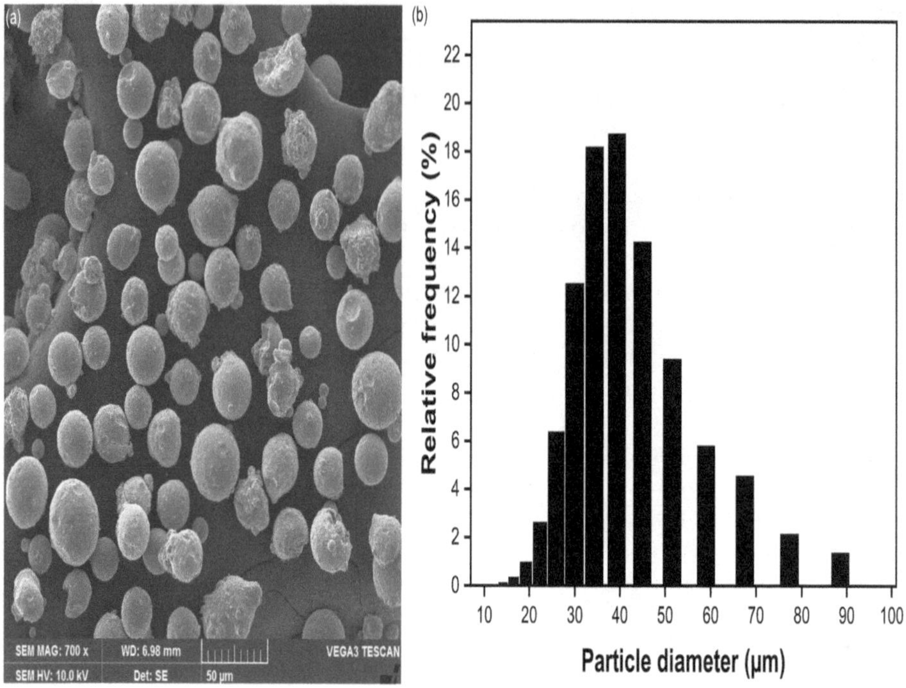

FIGURE 2.1
(a) Spherical morphology and (b) particle size distribution of clad powder particles.

efficiency. While it offers precise control over powder flow, improving application accuracy, it's sensitive to powder placement variations, leading to inconsistent bead properties. This method is precious for incorporating hard materials like carbides. However, it relies on cladding direction, and altering it can affect local cladding conditions unless the nozzle is adjusted [10].

Using a cone-shaped powder nozzle, coaxial laser cladding offers advantages over off-axis feeding (Figure 2.2). It's omnidirectional, not dependent on cladding direction, and provides longer powder–beam interaction times, improving pre-heating efficiency. Laser power reduction depends on particle velocity, powder feed rate, density, size, and angle. This method offers higher energy efficiency due to extended interaction times and multiple reflections within the powder cloud [11].

2.2.2.2 Laser Cladding with Wire Feeding

Laser cladding by wire feeding is an advanced manufacturing technique that employs a high-energy laser beam to melt a continuously supplied wire onto a substrate. It offers advantages such as one-step processing and high material efficiency. Typically held within a ceramic cylinder, the wire must remain straightened to ensure stable transport without vibrations [12].

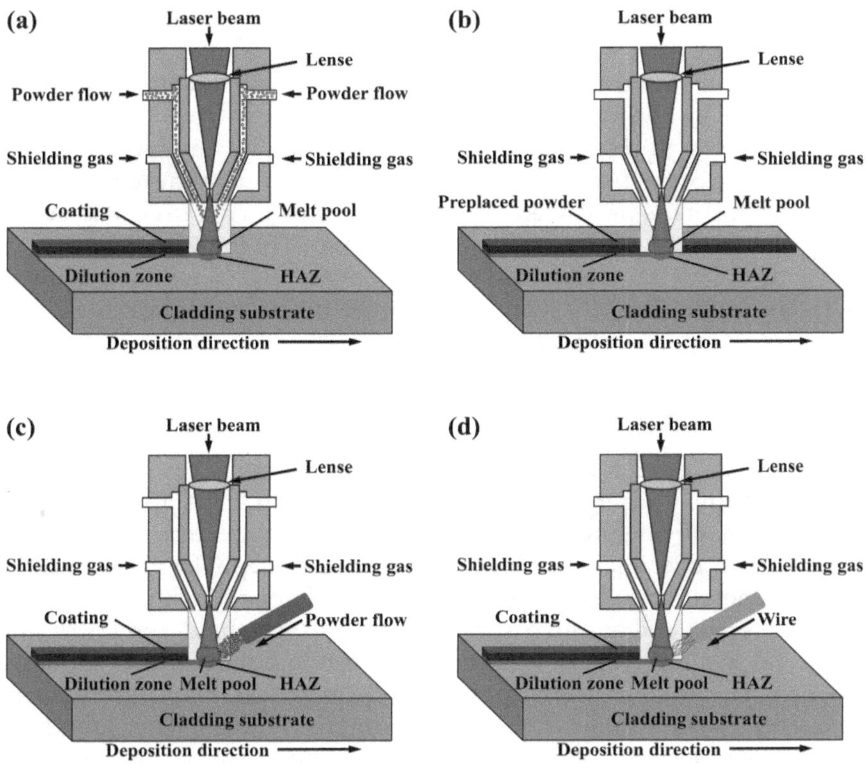

FIGURE 2.2
Schematic diagrams of the laser cladding processing mode: (a) coaxial powder-feeding method, (b) pre-placed powder method, (c) off-axis powder-feeding method and (d) wire feeding method. (Reproduced with permission from Elsevier.)

Wire feeding has distinct benefits compared to powder injection cladding, including adaptability to various cladding positions and reduced material costs and waste. However, it can lead to surface quality issues, weakened bonding, porosity, fractures, and occasional material loss. Pre-heating the wire to 1,000°C with partial laser power significantly reduces overall power requirements, improving material utilisation but potentially increasing dilution and process complexity [13].

Careful wire feeding direction, positioning, and inclined angles for the wire and laser beam can help reduce splattering during the process. While wire feeding is more energy-efficient than powder injection, it has challenges in regulating material feed rates and may suffer from inefficient energy transmission and shadowing effects on the surface [14].

2.2.2.3 Laser Cladding with Paste-Bound Powders

Laser cladding with paste-bound particles is a specialised manufacturing process for surface modification, restoration, and complex component production. It involves applying a paste of cladding material mixed with a binding agent onto the substrate's surface. A high-energy laser beam is directed to melt both the pre-placed paste and the substrate simultaneously, forming a clad layer [15].

This process requires a paste delivery system and a cooling mechanism to protect the paste from thermal effects in the operating environment. Precise control of the paste feed rate and substrate speed is crucial to achieve high-quality cladding. Insufficient paste or excessive processing speeds can increase dilution and reduce track height. Conversely, using too much paste may increase pore formation due to hindered binder evaporation, resulting in more significant material loss [16].

2.2.2.4 Direct Metal Deposition

Direct metal deposition (DMD) is an advanced manufacturing process that uses a high-power laser to create a molten pool on a metal surface while introducing metal filler material, often in powder or wire form, into this pool. Precise control over material flow and laser power enables the addition of functional metal layers to surfaces [17]. DMD is applied in surface cladding, metal part repair, and additive manufacturing, especially in aerospace, automotive, and tooling industries. The laser focuses just above the substrate surface where deposition is needed, and a coaxial powder stream enters the melt pool, melting upon contact. A computer numerical control (CNC) multi-axis system moves the substrate to deposit two-dimensional layers, building a three-dimensional object. Recent DMD advancements allow the monitoring and controlling of critical parameters like geometry, temperature, composition, and microstructure based on data collected from the laser–material interaction zone [18].

2.2.3 Laser Cladding Equipment

There are numerous specialised instruments and devices utilised for laser cladding. This apparatus is designed to precisely deposit and dissolve cladding material onto a substrate to produce a clad layer with enhanced properties. Here are the essential components typically found in laser cladding equipment: laser sources, cladding nozzle, particle feeder, and positioning devices. The configuration and efficacy of laser cladding apparatus may vary based on the specific application and employed materials [19].

2.2.3.1 Laser Source

The primary apparatus component, the laser source, emits a high-intensity laser beam. It is renowned for being monochromatic, coherent, directional, and focused. In laser technology, CO_2, Nd:YAG, fibre lasers, excimer, and diode lasers are common. The selection of a laser system for laser cladding is contingent on factors such as the extent of the area to be covered and the thickness of the clad material. Laser cladding requires considerable laser power, typically 10^2 to 10^4W/cm^2, with contact periods ranging from 10^{-3} to 0.5 seconds. Nd: YAG lasers and high-power diode lasers (HPDLs) are preferred for smaller areas and precision energy control, particularly in complex 3D components. For precision control, Nd: YAG lasers are favoured over CO_2 lasers. CO_2 lasers are commonly used for cladding applications [20]. Continuous wave beams are commonly used in laser cladding because they maximise coverage rate and suitability for various applications. However, pulsed lasers can be used in some situations where reduced energy inputs are required to limit distortion.

2.2.3.2 Cladding Nozzle

This component directs the flow of material, typically small particles, into the path of the laser beam. It can predominantly be designed in two ways:

Lateral nozzle – A lateral laser cladding nozzle is a specialised nozzle used in laser cladding procedures. In contrast to conventional nozzles, which are typically coaxial with the laser beam's path, a lateral nozzle is positioned obliquely or at an angle to the laser beam's path. This enables the transfer of additional materials from a different angle or direction, such as particulate or vapours. In laser cladding applications that demand precise control over material delivery and laser beam interaction, lateral nozzles are utilised [7].

Coaxial nozzle – The path independence of this orifice is a significant advantage. In addition, it permits enhanced powder efficiency. It is essential to analyse particle flow structure to create a precise and efficient clad in coaxial laser cladding. The coaxial nozzle contains three streams that interact with the solid surface. These streams include shield gas, particle stream, and forming gas [10].

2.2.3.3 Powder Feeder

A powder feeder is necessary when powder is used as a cladding material. It regulates the rate at which powder is added to the process, ensuring a uniform layer. In industrial processes, numerous types of particulate feeders are utilised, and they can be classified according to their operational principles:

1. Gravity-Based Feeders
2. Mechanical Wheel Feeders
3. Fluidised-Bed Feeders
4. Vibrating Feeders

Certain powder feeders use components from these categories to increase the powder flow's stability. Regardless of the type of feeder, it is necessary to provide a carrier gas that transports pulverised substances from their source to their destination [19].

2.2.3.4 Positioning Devices

Positioning devices in laser cladding systems must provide ample space and flexibility for required trajectories and assure the proper velocity, a crucial process parameter. These devices can be CNC workstations or autonomous manipulators with multiple degrees of freedom. In both instances, dynamic constraints must be considered to accomplish the intended velocities and accelerations at various points along the trajectory. Managing joint singularities is a crucial constraint when designing the intended trajectory for robotic systems [19].

2.3 Materials Used in the Cladding Process

2.3.1 Mild Steels

These steels have low carbon content (0.05%–0.25%C) and contain ferrite and pearlite, making them suitable for laser cladding without requiring additional pre-heating. During the heating process in laser cladding, a portion of the base material changes austenite and pearlite dissolution, leading to martensite formation in the heat-affected zone (HAZ) during rapid cooling. The low carbon content helps prevent hardening, cold or hydrogen cracks, and decreased ductility [21]. Rapid thermal cycles and minimal heat input can impede impurity segregation and the development of cracks. Elevated carbon levels diminish the weldability of medium- and high-carbon steels, requiring extra pre-heating to decelerate cooling rates and avoid the creation of brittle martensite in the HAZ [22].

2.3.2 Alloy Steel

Alloy steels, which include elements like chromium (Cr), molybdenum (Mo), vanadium (V), and nickel (Ni), in addition to carbon (C) and manganese (Mn), possess enhanced hardening capabilities [23]. However, they also exhibit a higher susceptibility to cold cracks in the HAZ. Quenched and tempered steels, a common type of alloy steel, find application in various industries such as power plants, diesel engines, and paper mills, making them potential candidates for laser cladding [24].

2.3.3 Stainless Steel

Stainless steels primarily consist of iron and chromium, with chromium content typically exceeding 10% to impart corrosion resistance. Austenitic stainless steels are well suited for laser cladding due to their resistance to unwanted phase transformations in the HAZ, eliminating the need for pre-heating [25]. Their high coefficient of thermal expansion counters tensile residual stresses during cooling, reducing the risk of hard coating cracks. Additionally, their lower thermal conductivity results in higher dilution levels than cladding onto mild steel with similar parameters [26].

Martensitic stainless steel, commonly used in laser cladding, offers moderate corrosion resistance and high strength. However, it tends to harden when exposed to air, leading to martensite formation in the HAZ during laser cladding. These considerations are crucial during cladding and subsequent processing steps [27].

2.3.4 Tool Steels

Tool steels are frequently employed in parts subjected to significant loads due to their outstanding strength, toughness, and wear resistance. Tool steels are categorised into hot and cold work varieties depending on the temperatures at which they are intended to operate [28]. Hot work tool steels exhibit enhanced cladability in a tempered state owing to their heightened ductility. Nevertheless, akin to martensitic stainless steels, the majority of tool steel grades undergo hardening upon exposure to air. This mandates pre-heating, prolonged cooling, and subsequent heat treatment to avert cold cracks in the HAZ and guarantee ductility under operational conditions [29].

2.3.5 Cast Irons

Cast irons comprise various types: ductile, grey, white, malleable, and alloy cast irons. Each type has unique microstructures and mechanical properties that impact their suitability for laser cladding. Grey, white, and alloy cast irons pose challenges for laser cladding due to their limited ductility, increasing the risk of crack formation in the HAZ due to tensile stresses during cooling [30].

Ductile and malleable cast irons, on the other hand, exhibit better stress tolerance, making laser cladding feasible without pre-heating. The hardening of the HAZ in cast irons primarily depends on the carbon content, particularly the amount of pearlite present. This is because graphite, a component of cast irons, doesn't readily dissolve during the rapid thermal cycles of laser cladding [25].

2.3.6 Aluminium

Aluminium alloys, encompassing both cast and wrought types, find widespread use in the automotive, marine, and aerospace industries owing to their lightweight properties, impressive strength-to-weight ratio, and resistance to corrosion. Nevertheless, their inherent limitations in hardness and wear resistance have spurred the advancement of laser coatings to augment their overall performance [31]. These coatings encompass various aluminium-based alloys, metal matrix composites (MMCs), and the injection of carbides like SiC and TiC. Laser cladding of aluminium alloys presents challenges, as they tend to form brittle intermetallic compounds with common coating elements, and controlling dilution can be problematic due to differences in melting temperatures [32].

2.3.7 Nickel

Ni-based superalloys are exceptionally corrosion-resistant and mechanically robust, well suited for demanding applications like gas turbines, diesel engines, and offshore equipment. Materials and laser technology advancements have enabled the repair and 3D printing of intricate components made from Ni-based superalloys. These superalloys are essential for the efficient operation of gas turbines. However, their repair demands precise control of the cladding process due to their moderate melting temperatures, moderate thermal expansion, and low thermal conductivity. To prevent liquidation cracks in the HAZ, low heat input or external cooling techniques may be necessary [33].

2.3.8 Titanium

Titanium and its alloys, known for their strength, lightweight nature, biocompatibility, and corrosion resistance, serve in various applications, including aerospace, automotive, marine, medical, and chemical industries. However, they suffer from poor wear resistance due to high friction coefficients. Laser cladding techniques have been explored to address this issue, with effective methods involving Ti-based alloy and direct injection of hard particles. Proper inert gas shielding is crucial due to titanium's reactivity [34].

2.3.9 Magnesium

Magnesium alloys, renowned for their exceptionally low density and impressive strength-to-weight ratios, hold substantial promise for structural applications in industries such as automotive and aerospace, where weight reduction is of utmost importance. Nevertheless, they face challenges related to poor wear resistance and susceptibility to corrosion, especially from chlorides and certain acids [35]. To address these issues, laser cladding has emerged as an effective solution. Typically, coatings employed in this process are Al-based alloys or MMCs due to their compatible melting temperature ranges and some solid solubility with Mg. Alternatively, direct injection of carbides, such as SiC and Cr_3C_2, without a metal matrix has been explored [36].

2.3.10 Copper

Copper and its various alloys, including brasses, bronzes, aluminium bronzes, and cupro-nickels, are valued for their exceptional thermal and electrical conductivity, low friction coefficient, pleasing appearance, resistance to microbiological attacks, and corrosion resilience, particularly in marine environments. However, they often lack hardness, wear resistance, and resistance to cavitation erosion [37]. To address these shortcomings, laser cladding is employed with these materials. Ni-based alloys have proven effective as coatings for copper-based materials due to their ability to form a solid solution with copper without brittle intermetallics [38].

2.3.11 High-Entropy Alloys

Laser cladding is a promising technique for synthesising HEA coatings (LC-HEACs). LC consistently demonstrates superior surface properties compared to conventional methods such as magnetron sputtering, plasma arc cladding, electro-spark processes, and casting. HEAs are deposited layer by layer onto various substrates through laser cladding, resulting in clads with thicknesses ranging from a few microns to several mm [39]. LC-HEACs fall into three categories based on composition: laser-cladded refractory HEA coatings (LC-RHEACs) with high-melting elements; transition metal-based LC-HEACs containing elements such as Al, Cr, Co, Mn, Cu, Ni, Fe, and Ti; and LC-HEA ceramic-based coatings, blending HEAs with other elements. HEA elements serve as matrices reinforced with lightweight alloys (Al, Mg) or ceramic particles (TiN, NbC, TiC) in LC-HEA-based composite coatings [40].

2.3.12 Coating Materials

2.3.12.1 Cobalt

Cobalt (Co)-based hard-facing alloys are the preferred choice for laser cladding, prized for their performance under sliding wear conditions, high temperatures, and corrosive surroundings. These alloys derive outstanding properties from Co and various alloying elements such as carbon, chromium, molybdenum, tungsten, nickel, and silicon. They are classified into three groups: solid solution-strengthened, carbide, and intermetallic alloys. Low-carbon variants, particularly solid solution-strengthened types with elements like chromium, molybdenum, and tungsten, are popular due to their ductility, enabling the application of crack-free coatings on significant components without needing pre-heating [41]. Maintaining low dilution is crucial when applying these alloys to steel substrates. Compared to carbide-type Co alloys, these alloys offer superior resistance to crevice corrosion in solutions containing chlorides. They possess moderate coefficients of thermal expansion and slightly lower melting temperatures [42].

2.3.12.2 Nickel Based

This significant category of laser coating materials can be further categorised based on their intended use. Several alloys come into play in applications requiring high strength, exceptional corrosion resistance, and the ability to withstand high-temperature oxidation [43]. These include NiCr, Monel, and a range of superalloys like Inconel, Nimonic, Incoloy, Hastelloy, and advanced single-crystal alloys like CMSX-4. When applied to Fe-based base materials, they exhibit excellent cladding characteristics because nickel can form a solid solution with iron. Except for cast irons, these alloys offer high ductility and have lower melting points than Fe-based base materials [44].

2.3.12.3 Iron

Fe-based coating materials can be categorised into several groups, including stainless steels, tool steels, Fe-based hard-facing alloys, and Hadfield steels, each tailored for specific purposes in laser cladding. Austenitic stainless steels are utilised to enhance corrosion resistance in laser cladding applications. Their high ductility and rapid solidification characteristics promote austenite formation while suppressing harmful ferrite content, negatively affecting corrosion resistance and ductility [45].

Martensitic stainless steels offer a cost-effective option for applications requiring moderate corrosion and wear resistance. These steels transform into martensite during cladding, reducing the risk of cracking during cooling. Though less studied, ferritic stainless steels show potential in laser cladding processes. Tool steels are valued for their high hardness and excellent wear properties, primarily due to the presence of carbides [46].

2.3.12.4 Copper

Cu-based alloys are favoured for laser cladding to enhance Al-based alloys' wear and high-temperature properties. Copper's higher solubility in aluminium and the lower melting point of Cu-based alloys make them suitable by narrowing the interface zone and reducing the formation of brittle intermetallics. Cu-based alloys also offer common friction properties in sliding contacts. Cu and CuNi coatings on Fe-based base materials have also shown promise, with low electrical resistivity and minimal interface defects or embrittlement, potentially alleviated using Ni as an intermediate layer [47].

2.3.12.5 Aluminium

Aluminium-based coatings, mainly AlSi alloys, enhance hardness and wear resistance on aluminium-based materials. Another promising option is AlSn-based bearing alloys known for low friction. However, applying Al-based coatings on steel is challenging due to the formation of brittle intermetallics. Gilkes introduced laser plasma deposition as an innovative method to apply Al coatings onto steel without melting the steel [48].

2.3.12.6 Titanium

Ti-based alloys in laser cladding are mainly used to repair and build gas turbine components and provide high-quality titanium coatings for components in various industries. These alloys have challenges like high melting points and intermetallic formation, making direct coating on materials like steel impossible. To address this, Ni intermediate layers are used. Ti alloys also react with atmospheric elements, so inert gas shielding with Ar or He is crucial for quality coatings, typically achieved using trailing shoes or closed chambers [49].

2.3.12.7 Metal Matrix Composites

MMCs are extensively studied coating materials in laser cladding, consisting of a ductile metal matrix holding rigid reinforcements like carbides. They are used in various equipment, offering wear resistance. Production methods include mixing powders and injecting them into the melt pool. In situ synthesis, where carbides form during cooling, is another option, providing cleaner interfaces between carbides and the matrix [50].

Production of MMCs in laser cladding involves various methods. A commonly employed method involves producing a mechanical powder blend comprising hard particles and a metal matrix, which is then fed through a cladding nozzle from a single hopper. Alternatively, hard particles can be directly introduced into the molten base material without the inclusion of a metal matrix. In situ synthesis of hard reinforcement is accomplished through cooling reactions, wherein metal or compound powders that have a tendency to form carbides, along with graphite powder, are concurrently introduced into the metal matrix. In situ-formed MMCs offer cleaner interfaces without brittle phases compared to those produced using external addition methods [51].

2.3.12.8 Functionally Graded Materials

Using multiple powder hoppers and wire feeders simultaneously enables the production of laser-clad MMCs or monolithic beads with graded compositions along their length or depth. To create these functionally graded materials (FGM), the feed rates of each feeding unit are synchronised with the translation movement, allowing for controlled variation [52]. For instance, some studies have demonstrated a gradual increase in the volume fraction of hard particulates along the depth of the coating. This can be achieved within a single bead or through multiple consecutive layers. The graded structures can be tailored to meet specific functional requirements in various applications and address differences in properties between the clad and base materials [53].

2.3.12.9 Intermetallic Compounds

Intermetallic compounds and hard particles added externally or synthesised in situ are essential for reinforcing metal matrices in laser cladding. These intermetallics include CoMoSi Laves phases, nickel silicides, chromium, tungsten, molybdenum, and titanium silicides. Another notable category of intermetallics employed in laser cladding comprises low-density Ni-aluminides, Fe-aluminides, and Ti-aluminides. These materials find applications in high-temperature environments, including but not limited to gas turbines, heat exchangers, turbocharger components, valves, and furnace elements [54].

2.3.12.10 Other Materials

The rapid solidification rates in laser glazing enable the creation of thin layers of amorphous structure characterised by distinctive properties on metal and ceramic surfaces. Thick amorphous metal coatings are also interesting but can crystallise in continuous layers due to overlapping. Oxide ceramics are challenging to laser-clad on metal bases due to limited solid solubility, brittleness, and melting temperature differences. Some success was achieved with Al_2O_3/TiO_2 layers on aluminium through chemical and mechanical bonding. Another approach involved exothermic reactions to produce Al_2O_3 layers on aluminium. Bioceramic calcium phosphate coatings have the capability to be laser-clad onto human implant materials such as Ti6Al4V and 316L. This process is utilised to encourage bone growth on the implant surfaces [55].

2.4 Characterisation of Laser-Clad Materials

2.4.1 Microstructural Analysis

2.4.1.1 Optical Microscopy

Optical microscopes are vital in analysing and characterising materials [56]. They use visible light to magnify and visualise the microstructure of a material. These microscopes are preliminary inspection tools, offering a macroscopic view of laser-cladded materials. Laser-clad materials must be cut, polished, and etched before optical microscopy. They reveal the grain size, shape, distribution, and macroscopic flaws such as pores and cracks. Optical microscopy enables scientists and engineers to comprehend how laser cladding affects material properties by examining the microstructure of laser-clad materials. It optimises process parameters for microstructures of your choosing. A microscope reveals how laser power (P) and powder-feeding rate (F/S) affect the dilation and characteristics of the laser-clad track (Figure 2.3). This information assists in optimising laser cladding for material and application needs [57].

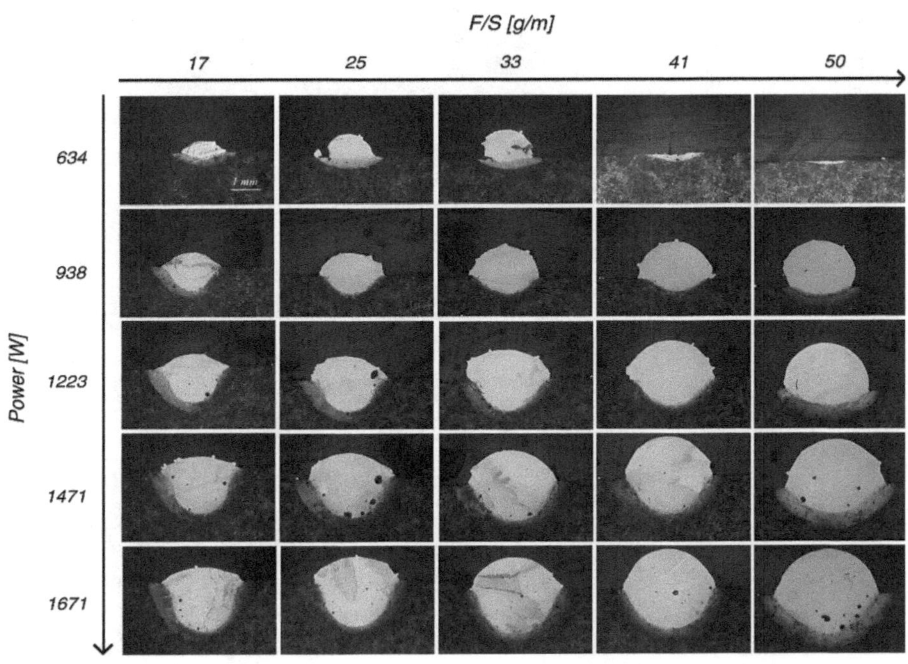

FIGURE 2.3
Transverse region of Co-based alloy cladding on grey cast iron substrate on the horizontal axis, F/S determines the amount of material fed per unit track length. (Reproduced with permission from Elsevier.)

2.4.1.2 Scanning Electron Microscopy Analysis with Energy-Dispersive X-Ray Spectroscopy

By directing a focused electron beam over the specimen's surface, SEM produces high-resolution images. These microscopes analyse the surface topography, grain boundary formation, sub-grain structure, and microstructural defects such as pores and voids of laser-clad materials [58]. Furthermore, when coupled with energy-dispersive X-ray spectroscopy (EDS), the elemental composition of laser-clad regions can be identified using elemental mapping or line scan methods.

The microstructure of laser-clad materials can be viewed using SEM at high resolution (Figure 2.4). The SEM can even reveal the fine dendritic microstructure formed during the solidification process during laser cladding [59]. The voids, surface roughness, cracks, intertrack porosity, bonding quality, and other surface irregularities can be identified using SEM analysis. Surface quality is essential for applications where smoothness or texture play an important role. The thickness and uniformity of the laser cladding coating can be precisely measured using SEM by examining cross-sections.

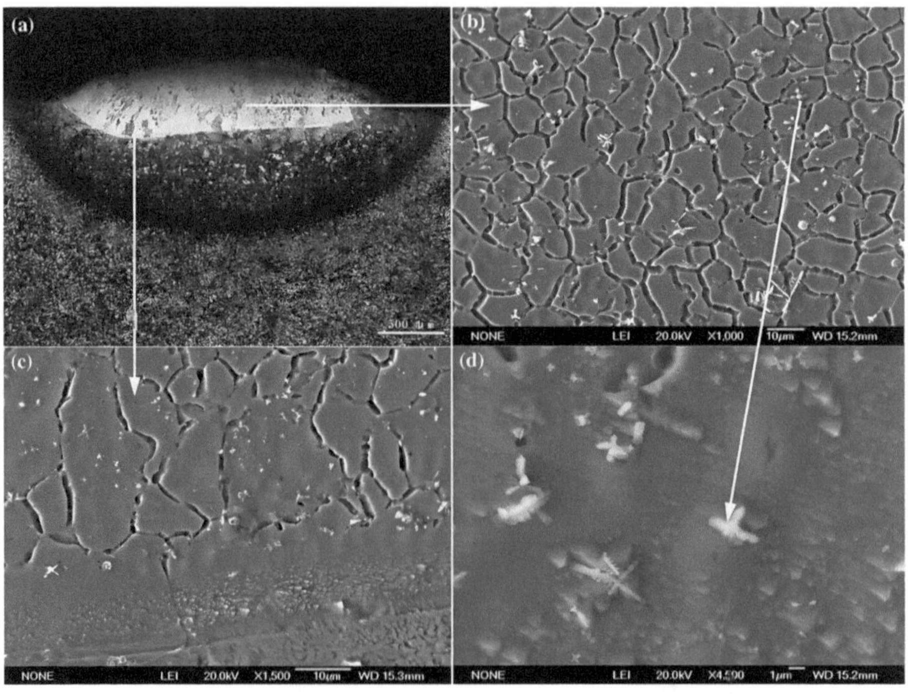

FIGURE 2.4
Microstructure morphologies of a high-entropy alloy: (a) macroscopic feature, (b) cladding zone, (c) bounding zone and (d) high-magnification feature of the cladding zone. (Reproduced with permission from Elsevier.)

EDS analysis, often coupled with SEM, identifies the elemental composition of the laser-cladded material. It can detect the presence of alloying elements, impurities, and inclusions. It determines the distribution and concentration of various elements. This is essential for verifying that the desired composition is achieved. It can help identify different phases within the material, especially in cases of multi-material laser cladding.

It can also be used to determine how variations in laser cladding process parameters affect the material's microstructure, surface quality, and chemical composition. By performing a line scan analysis on the cross section of a laser-clad material, changes in elemental composition are identified to measure the thickness of the coating by determining the distance between the transitions at the coating-substrate interface.

2.4.1.3 Transmission Electron Microscopy

Transmission electron microscopy (TEM) provides high-resolution images and diffraction patterns of ultrathin specimens at the atomic scale by transmitting electrons. It can reveal the internal microstructure at the atomic level and identify the dislocation of lattice planes, crystallographic features, defects, and grain orientation at nanometre scales [60]. TEM can also be used to identify the various phase transformations during the laser cladding process of metallic materials.

2.4.1.4 X-Ray Diffraction

X-ray diffraction is an essential, non-destructive technique for determining the crystal structures of metals, alloys, minerals, and polymers. XRD is indispensable in laser cladding research for revealing materials' crystallographic structure, phase composition, and residual stresses. This method evaluates the diffraction patterns produced when X-rays interact with a material's crystal lattice. It helps identify phases present in a substance and provides insight into its properties [39]. Additionally, XRD reveals crystallographic orientations, particularly in textured materials, and measures residual lattice strains from localised heating and rapid cooling.

2.4.1.5 Electron Backscatter Diffraction

Electron backscatter diffraction (EBSD) is a powerful microstructural characterisation technique incorporated with SEM. It reveals crystallographic details, including grain orientation, grain size distribution, phase identification, crystallographic texture, low- and high-angle grain boundaries, and residual strain.

EBSD facilitates the examination of grain structure and orientation, providing valuable insights into anisotropy and mechanical properties via

FIGURE 2.5
IPF map of grain morphology at various regions: (a) laser cladding layer, (b) heat-affected zone and (c) base material. (Reproduced with permission from Elsevier.)

techniques such as inverse pole figures and kernel average misorientation mapping (Figure 2.5). EBSD shows a variety of material phases, detects flaws such as fractures and irregularities, and analyses structural changes brought about by the laser cladding process [61].

2.4.1.6 Atomic Force Microscopy

Atomic force microscopy (AFM) measures surface forces at the atomic or nanometre scale by scanning the surface with a sharp tip. Surface qualities, including surface morphology, roughness, and nanoscale features, can be measured using these microscopes. They provide accurate 3D topographical data down to the nanoscale level. Engineers and scientists may utilise this information to fine-tune the procedure so that the final product has the necessary surface properties. The thickness of the laser-clad layer may be determined using AFM. Scanning the surface and taking readings at various spots can produce accurate estimations of thickness. The clad layer's quality and integrity may be determined using AFM, which is sensitive enough to identify nanoscale defects such as surface irregularities, fractures, voids, and other problems. High-resolution pictures of the oxide layer's surface in the laser-clad area may be obtained using AFM in tapping mode [62].

2.4.1.7 Confocal Microscope

Confocal microscopes utilise laser illumination and a pinhole detector to generate high-resolution 3D images. These microscopes are perfect for analysing the topography, surface irregularities, micro-level defects and depth of laser cladding and coated surfaces. Most confocal microscopes employ a laser beam for illumination, scanned across the imaging surface. Confocal microscopes can measure the area of the wear track and the depth and morphology of wear features such as wear debris, enabling a thorough evaluation of tribological behaviour. Using confocal microscopes such as the Leica DCM 3D, it is possible to determine the volume of the wear track in laser-cladded materials [63].

2.4.2 Mechanical Testing Methods

2.4.2.1 Hardness Testing

Most laser cladding research focuses on enhancing the surface hardness of components. The hardness of the clad layer determines its use in applications requiring wear resistance. The yield and tensile strength, wear and scratch resistance of a material is significantly influenced by its hardness. Hardness testing is a technique for determining a material's resistance to plastic deformation, typically via indentation. A Micro-Vickers Hardness tester (Figure 2.6), such as Shimadzu HV-100DT, is used to analyse the hardness of laser-clad specimens. This analysis evaluates the micromechanical properties of micron-thick laser-clad layers. It measures a material's resistance to surface deformation and diamond indenter penetration. In accordance with ISO 22826:2005, the hardness test is typically conducted at room temperature with a 1 kg applied load and a 10-second dwell time.

The common hardness testing methods include the Vickers hardness test, Rockwell hardness test, Brinell hardness test, and Knoop hardness test. Rockwell and Vickers hardness testings are commonly used for most laser-clad metals due to their versatility and accuracy. The Vickers hardness (HV) is calculated using the following Vickers hardness formula:

$$\mathrm{HV} = 1.864\ \frac{F}{d^2}\ \text{approximately}$$

where

F=Force (kgf)

d=Mean diagonal length d (mm)

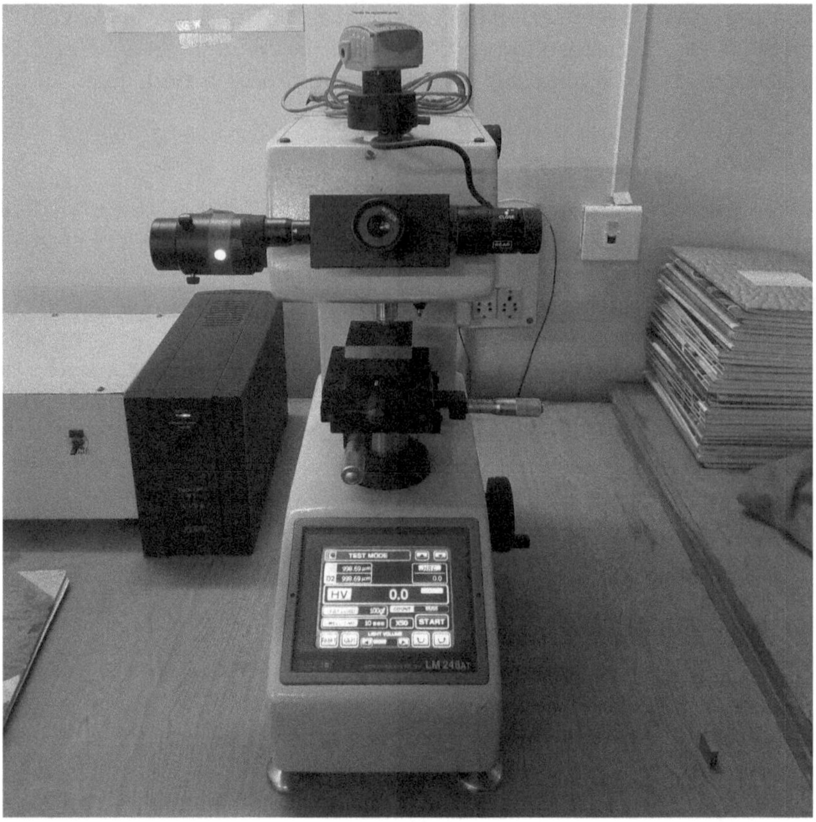

FIGURE 2.6
Micro-Vickers Hardness Tester.

The Knoop hardness test has a rhombic-based diamond indenter with a specific geometry that produces long, narrow impressions well suited for microhardness measurements, making it suitable for evaluating the hardness of laser-clad surfaces.

2.4.2.2 Tensile Testing and Fractography Analysis

Tensile testing evaluates the mechanical properties of a material, such as its tensile strength, ductility, and deformation behaviour. The sample dimensions for tensile testing should comply with standard tensile testing specifications, such as ASTM E08 or ISO 6892 [64]. The rapid cooling rate during laser cladding strengthens the solid solution and mitigates element segregation, resulting in a more uniform and dense structure than traditional casting techniques [1]. The ability to strengthen and homogenise the material through laser cladding is essential for applications where superior mechanical strength is of greatest significance, such as aerospace components, automotive parts, and tools [65] The enhanced tensile strength enables laser-clad materials to withstand greater loads and stresses, enhancing their durability and reliability in critical engineering environments. Understanding how FGM perform mechanically requires tensile testing, which helps optimise their strength and overall functionality. SEM fractography enables a close examination of the laser-clad regions that fracture during the test. The detailed SEM images can reveal microscopic features, such as inclusions and precipitates, acting as stress points and affecting the material's ability to withstand tensile forces. SEM is particularly useful for distinguishing ductile and brittle breaks. Ductile fractures appear as dimples, indicating that the material can be bent slightly before breaking. In contrast, brittle fractures exhibit cleavage planes, indicating a catastrophic and sudden failure without much elongation. SEM's ability to capture these fine details helps identify potential weak points and stress concentrations, providing valuable insights for improving the integrity of laser cladding materials.

Researchers use controlled tensile forces on samples to evaluate strength, ductility, and resilience, optimising laser cladding parameters and enhancing its performance in real-world applications.

2.4.2.3 Wear Testing

Wear testing is essential to validate the performance of laser cladding layers in practical scenarios. It helps assess whether the added layer effectively resists wear and maintains its integrity under varying conditions. The laser cladding layers often find applications in aerospace, automotive, and tooling industries, where wear resistance is critical. Wear testing verifies the capacity of laser cladding layers to endure real-world challenges, specifically addressing friction, abrasion, and sliding wear, and is crucial for informing their effective integration into practical applications. The pin/ball-on-disc sliding wear tests, in particular, analyse the frictional and wear characteristics of laser cladding materials, providing critical details about how well they perform under various operating circumstances [66]. The common wear testing methods are listed below as follows:

1. Ball-on-disc test – This method is commonly used for testing the wear resistance of laser-clad layers. A laser-clad disc is pressed against a rotating pin to simulate sliding wear conditions. Controlling precise parameters such as applied load, sliding speed, and test duration. This ensures that the wear test simulates relevant application-specific operating conditions. The laser-clad spinning disc with a known surface finish comes into contact with a ball. When the ball presses against the rotating disc, friction and wear result. The amount of

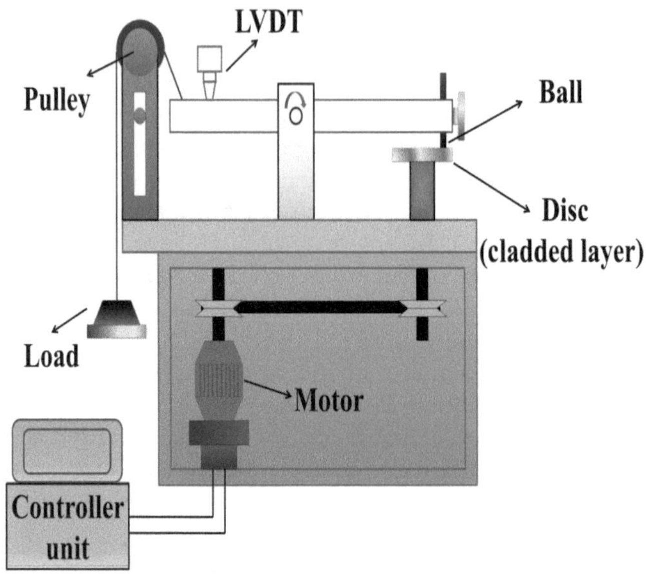

FIGURE 2.7
Disc wear tester used to test the wear resistance of the cladding layer.

material worn off the disc and the wear track on the disc are measured to determine wear rate and behaviour. Understanding wear mechanisms is essential for optimising the material's wear resistance. The worn surface will be examined under a microscope to obtain images and data that may reveal surface characteristics, wear patterns, and damage. Wear mechanisms, such as abrasion, adhesion, or fatigue, can be identified by analysing wear patterns. Correlating the results of wear tests with laser cladding parameters helps optimise parameters for increased wear resistance in laser-clad layers. Based on the results of wear tests, iterative studies should be conducted to adjust the laser cladding process or material composition to continuously enhance wear resistance [67] (Figure 2.7).

2. Abrasion testing involves exposing the laser-cladded surface to abrasive particles or papers under controlled conditions. The clad layer's weight loss or volume loss is measured to assess its resistance to abrasion.
3. Slurry erosion testing – Slurry erosion tests immerse the laser-cladded surface in a slurry of abrasive particles in a controlled flow. The choice of abrasive particles is crucial, and it should ideally mimic the real-life environment and application where the material or component is expected to perform. The selection of abrasive particles

is influenced by particle size, shape, hardness, and concentration, ensuring that the test conditions accurately simulate the wear that the material might experience in its intended operational environment. The rate of material loss and formation of wear tracks are analysed to determine the resistance to slurry erosion.

2.4.2.4 Shear Testing

Shear strength is the maximum force applied to a material until it fails under shear stress. A known shear force is applied to a test specimen, typically a laser-clad layer adhered to a substrate during shear testing. An adhesive is frequently used to create a consistent shear bond. It provides insight into the material's adhesion and cohesion properties [68].

2.4.2.5 Acoustic Emission Testing

Acoustic emission (AE) testing is a non-destructive evaluation method used to analyse laser-clad materials for improved quality control. AE testing enables real-time material monitoring during laser cladding. It detects acoustic signals, such as stress waves and microcrack formation, and provides immediate feedback on the material's behaviour. The recorded signals are then processed with AE detection devices, such as PAC Spartan AT/72 [69]. Data processing makes it possible to determine the time and location of crack initiation and growth, providing valuable insights into the laser-clad material's integrity.

2.4.2.6 Ultrasonic Testing

Ultrasonic testing (UT) is an effective non-destructive testing (NDT) technique that utilises sound waves to analyse clad layer thickness, number of defects, and material stress. UT functions by transmitting and receiving high-frequency sound waves through a material. By analysing both the time it takes for sound waves to traverse the material and the amplitude of the reflected waves, valuable information regarding the material's defects, including cracks, voids, and integrity, can be acquired [70]. Time-of-flight diffraction (TOFD), phased-array ultrasound, through-transmission ultrasound, and pulse-echo ultrasound are crucial in defect detection, thickness measurement, and material characterisation.

2.4.2.7 Fatigue Testing and Fracture Surface Analysis

Fatigue behaviour involves the progressive growth of cracks under repeated loading, resulting in material failure at a lower stress level than its ultimate strength. This is essential for applications in which components undergo dynamic mechanical forces during service life. The fatigue test results provide valuable information regarding the material's durability and mechanical performance after laser cladding. Low-cycle fatigue (LCF) testing involves laser-clad components to a limited number of stress cycles, typically between 10^2 and 10^3, emphasising high-stress amplitudes and plastic deformation. High-cycle fatigue (HCF) testing subjects' laser-clad materials to many stress cycles, typically 10^3 to 10^6, with lower stress amplitudes and elastic deformation. ASTM E466 defines the standard procedure for performing force-controlled axial fatigue tests with constant amplitude on metallic materials [71].

Fracture behaviour, the way a material breaks under high stress, is crucially analysed through post-fatigue testing, including laser-cladded materials. This analysis provides insights into failure modes, crack propagation, and material behaviour. It aids in identifying failure mechanisms, offering critical information for prevention and design improvements. As a quality assessment tool, fracture surface analysis reveals manufacturing defects or anomalies in laser cladding components. It also contributes to material characterisation, detailing properties like ductility and toughness. Common techniques include SEM for high-magnification images, optical microscopy (OM) for broader views, and X-ray microtomography (X-ray CT) for three-dimensional imaging. Overall, fracture surface analysis optimises the material selection process and ensures laser cladding components' reliability in demanding applications.

2.4.2.8 Corrosion Testing

Corrosion testing helps in material selection by evaluating resistance to specific environments and estimating laser-clad components' service life under simulated real-world conditions. Various testing methods, including salt spray, electrochemical, immersion, and cyclic corrosion testing, help simulate different environmental conditions and assess corrosion resistance. In parallel, the analysis of laser-clad regions involves microstructural examination, surface characterisation through techniques like SEM, and chemical analysis to understand the composition and potential alloying effects.

In the commonly employed potential dynamic assay for electrochemical corrosion testing, the working electrode will be submerged in a corrosive electrolyte solution, usually 3.5% NaCl [72]. The electrode's potential is systematically swept across a range, measuring the resulting current to generate a polarisation curve. Extracted parameters, such as the corrosion potential (E_{corr}) and current (I_{corr}), will provide insights into the material's corrosion behaviour. E_{corr} reflects a material's equilibrium potential in corrosive environments, indicating a balance between oxidation and reduction rates. A more positive E_{corr} suggests enhanced corrosion resistance while monitoring changes provides insights into the material's susceptibility. I_{corr} represents the current linked to surface electrochemical reactions, with lower values indicating superior corrosion resistance, emphasising the effectiveness of the protective oxide layer [73]. Monitoring I_{corr} is critical to assess protective mechanisms, particularly in laser-clad regions with varying composition and microstructure. Tafel slopes from the curve indicate the corrosion kinetics behaviour. The corrosion rate and the polarisation curve results are used to analyse the material's corrosion resistance.

2.5 Applications and Advantages of Laser Cladding

High heating and cooling rates, metallurgical bonding, and a narrow HAZ are why laser-based manufacturing methods are so attractive for surface modification. Turbine blades and aircraft engine components undergo laser cladding to increase their high-temperature oxidation and wear resistance [74]. It also prevents the deterioration of moving engine parts such as cylinder liners, pistons, camshafts, and crankshafts. Laser cladding is used to reinforce the surfaces of components like bucket teeth, conveyor belts, and drilling tools in the mining and heavy equipment industry. Maintenance costs are lowered by reducing wear, and equipment uptime is increased. Laser-cladded coatings on stainless steel exhaust systems improve heat and corrosion resistance, ensuring their durability in high-temperature and corrosive environments [75].

In the medical field, laser cladding is employed to modify the surfaces of titanium-based implants to improve their biocompatibility and osseointegration. By applying bioactive coatings or creating surface textures, laser cladding enhances the interaction between the implant and the surrounding tissue, promoting faster healing and reducing the risk of implant rejection [76].

Laser cladding plays a crucial role in tailoring FGMs for specific applications. Laser-clad FGMs with graded thermal conductivity improve heat dissipation in critical components of aircraft engines, enhancing overall engine efficiency and reliability. In manufacturing, laser-clad FGMs are used for localised strengthening, reducing the weight of components while improving mechanical properties. HEA elements act as matrices reinforced with lightweight alloys (Al, Mg) or ceramic particles (TiN, NbC, TiC), making LC-HEA-based composite coatings particularly effective for ceramic applications (Figure 2.8).

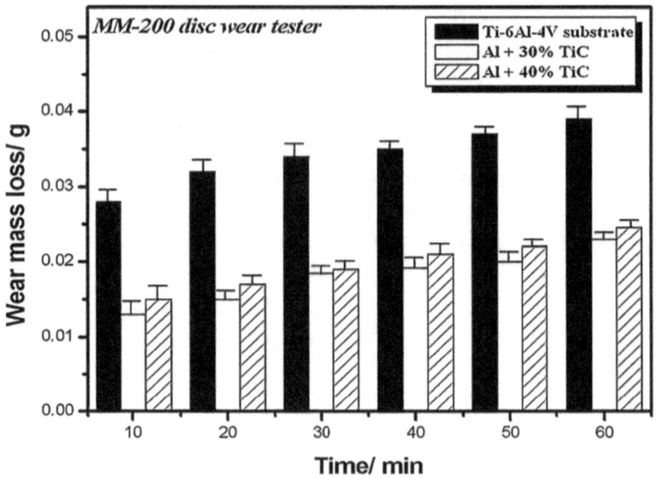

FIGURE 2.8
Improved wear resistance by laser cladding ceramic layers on titanium substrate. (Reproduced with permission from Elsevier.)

2.6 Challenges in Laser Cladding

Even though laser cladding has been used to strengthen and repair mechanical parts' surfaces, a few issues still need to be resolved. In terms of microstructure after laser cladding, porosity formed due to impurities in the powder, excessive protective gas, rapid solidification, and rapid cooling. Residual thermal stress leads to crack formation in the clad region [77]. The macroscopic dimensions and surface quality of laser cladding layers still need to be investigated and improved. Large, tiny, and complex curved parts are frequently strengthened and repaired using cladding in practical applications [78]. Therefore, an urgent issue regarding achieving high-quality cladding and automation for surface strengthening or repair remains.

2.7 Future Trends of Laser Cladding

In the aerospace, petrochemical, mining, and other industries, laser cladding plays a significant role in the surface strengthening or repair of mechanical equipment parts. It has a significant impact on reducing production costs and increasing production efficiency.

Future research should create novel materials like multi-principles HEAs, multi-performance composites, and amorphous and nanocrystalline materials. In addition, developing a gradient of functionally graded coatings is essential, which can reduce internal stresses and crack formation [79]. Research on amorphous materials become more important in recent years, particularly the composition design and control of amorphous coatings based on the characteristics of the cladding process, which decides the hardness and corrosion resistance properties in the clad layer [80]. Future research will focus on the theory of HEA crystallisation and solidification and its relationship with mechanical and physical properties [81].

Monitoring the laser cladding process and control is more important to form the required microstructure and clad quality. The temperature of the clad region is currently being measured; however, in the future, the dilution rate and voids will also be identified, and control measures will be implemented [82]. Regarding the laser cladding equipment and process, in the future, high-power lasers should be developed to cover the large area of mechanical parts. Similarly, small and portable cladding devices should be implemented for small parts, and the cladding nozzles should be optimised to attain cladding quality [83]. Laser technology will be combined with robots to enhance the automation and monitoring of the cladding of complex parts.

2.8 Conclusion

This chapter provides an overview of recent research, developments, and applications of the laser cladding process. It also summarises the current challenges in the cladding process and guides future research direction in the cladding field to its further development. The further chapters in this book will discuss in detail hybrid laser cladding, repair of engineering components through laser cladding, ceramic material modification for cladding, corrosion, high-temperature oxidation behaviours of laser cladding materials, mechanical properties, residual stress of clad materials, effect of alloying elements on wear and fatigue property, and industrial applications of laser cladding.

References

1. L. Zhu et al., "Recent research and development status of laser cladding: A review," *Opt Laser Technol*, vol. 138, p. 106915, 2021, doi:10.1016/j.optlastec.2021.106915.

2. Q. Lai et al., "Influences of depositing materials, processing parameters and heating conditions on material characteristics of laser-cladded hypereutectoid rails," *J Mater Process Technol*, vol. 263, pp. 1–20, 2019, doi:10.1016/j.jmatprotec.2018.07.035.
3. H. Song, J. Lei, J. Xie, S. Wu, L. Wang, and W. Shou, "Laser melting deposition of K403 superalloy: The influence of processing parameters on the microstructure and wear performance," *J Alloys Compd*, vol. 805, pp. 551–564, 2019, doi:10.1016/j.jallcom.2019.07.102.
4. P. Ramiro, M. Ortiz, A. Alberdi, and A. Lamikiz, "Characteristics of Fe-based powder coatings fabricated by laser metal deposition with annular and four stream nozzles," *Proc CIRP*, vol. 74, pp. 201–205, 2018, doi:10.1016/j.procir.2018.08.094.
5. T. M. Yue, G. H. Meng, H. Xie, X. Lin, and H. O. Yang, "Two-step laser cladding of AlCoCrCuFeNi high-entropy alloy on AZ91-Mg," *Int J Metall Mater Eng*, vol. 1, no. 2, p. 1, 2015, doi:10.15344/2455-2372/2015/115.
6. R. M. Volu, K. Zilnyk, S. A. S. Dyer, C. L. dos Santos, J. J. Neto, and G. de Vasconcelos, "A new two-step method for laser cladding of silicon carbide in Wc-Co substrates," *Mater Res*, vol. 26, 2023, doi:10.1590/1980-5373-MR-2022-0195.
7. R. Vilar, "Laser cladding," *J Laser Appl*, vol. 11, no. 2, pp. 64–79, 1999, doi:10.2351/1.521888.
8. H. Qin, Z. H. Cai, P. Zhang, and Z. Yang, "Development status of laser cladding technologies," *Appl Mech Mater*, vol. 584–586, pp. 1500–1503, 2014, doi:10.4028/www.scientific.net/AMM.584-586.1500.
9. S. T. Nyadongo, E. O. Olakanmi, and S. L. Pityana, "Experimental and numerical analyses of geometrical and microstructural features of Tribaloy T-800 composite coating deposited via laser cladding-assisted with pre-heat (LCAP) process," *J Manuf Process*, vol. 69, pp. 84–111, 2021, doi:10.1016/j.jmapro.2021.07.043.
10. M. Dias Da Silva, K. Partes, T. Seefeld, and F. Vollertsen, "Comparison of coaxial and off-axis nozzle configurations in one step process laser cladding on aluminum substrate," *J Mater Process Technol*, vol. 212, no. 11, pp. 2514–2519, 2012, doi:10.1016/j.jmatprotec.2012.06.011.
11. V. M. Weerasinghe and W. M. Steen, "Laser cladding with pneumatic powder delivery," in *Applied Laser Tooling*, O. D. D. Soares and M. Perez-Amor (Eds.), Dordrecht: Springer Netherlands, 1987, pp. 183–211. doi:10.1007/978-94-009-3569-3_10.
12. F. Caiazzo, "Additive manufacturing by means of laser-aided directed metal deposition of titanium wire," *Int J Adv Manuf Technol*, vol. 96, no. 5–8, pp. 2699–2707, 2018, doi:10.1007/s00170-018-1760-0.
13. P. Henri, N. Jonne, T. Sebastian, T. Jari, N. Steffen, and V. Petri, "Laser cladding with coaxial wire feeding," in *Proceedings of the 31st International Congress on Applications of Lasers & Electro-Optics*, Laser Institute of America, Anaheim, California, USA, 2012, pp. 1196–1201. doi:10.2351/1.5062408.
14. J.-D. Kim and Y. Peng, "Plunging method for Nd:YAG laser cladding with wire feeding," *Opt Lasers Eng*, vol. 33, no. 4, pp. 299–309, 2000, doi:10.1016/S0143-8166(00)00046-4.
15. H. C. Man, K. H. Leong, and K. L. Ho, "Process monitoring of powder pre-paste laser surface alloying," *Opt Lasers Eng*, vol. 46, no. 10, pp. 739–745, 2008, doi:10.1016/j.optlaseng.2008.05.006.

16. E. Lugscheider, H. Bolender, and H. Krappitz, "Laser cladding of paste bound hardfacing alloys," *Surf Eng*, vol. 7, no. 4, pp. 341–344, 1991, doi:10.1179/sur.1991.7.4.341.
17. F. Brückner, S. Nowotny, and C. Leyens, "Innovations in laser cladding and direct metal deposition," in *High Power Laser Materials Processing: Lasers, Beam Delivery, Diagnostics, and Applications*, SPIE, San Francisco, California, USA, 2012, p. 823904. doi:10.1117/12.906706.
18. J. I. Arrizubieta, S. Martínez, A. Lamikiz, E. Ukar, K. Arntz, and F. Klocke, "Instantaneous powder flux regulation system for laser metal deposition," *J Manuf Process*, vol. 29, pp. 242–251, 2017, doi:10.1016/j.jmapro.2017.07.018.
19. E. Toyserkani, S. Corbin, and A. Khajepour, *Laser Cladding*. CRC Press, Taylor & Francis Group, Boca Raton, FL, USA, 2005.
20. E. M. Birger, G. V. Moskvitin, A. N. Polyakov, and V. E. Arkhipov, "Industrial laser cladding: Current state and future," *Weld Int*, vol. 25, no. 3, pp. 234–243, 2011, doi:10.1080/09507116.2010.540880.
21. S. Huang, D. Li, L. Zhang, X. Zhang, and W. Zhu, "Tailoring the mechanical properties of laser cladding-deposited ferrous alloys with a mixture of 410L alloy and Fe-Cr-B-Si-Mo alloy powders," *Materials*, vol. 12, no. 3, p. 410, 2019, doi:10.3390/ma12030410.
22. H. He, T. Zhang, M. Ma, and W. Liu, "Microstructure and wear resistance of laser cladding particulate reinforced Fe-based composite coating on railway steel," *J Laser Appl*, vol. 29, no. 2, p. 022503, 2017, doi:10.2351/1.4983232.
23. R. A. Rahman Rashid et al., "Effect of clad orientation on the mechanical properties of laser-clad repaired ultra-high strength 300 M steel," *Surf Coat Technol*, vol. 380, p. 125090, 2019, doi:10.1016/j.surfcoat.2019.125090.
24. L. Meng et al., "A comparison of microstructure and mechanical properties of laser cladding and laser-induction hybrid cladding coatings on full-scale rail," *Mater Sci Eng A*, vol. 748, pp. 1–15, 2019, doi:10.1016/j.msea.2019.01.068.
25. N. Jeyaprakash, C. H. Yang, and S. P. Tseng, "Wear tribo-performances of laser cladding colmonoy-6 and stellite-6 micron layers on stainless steel 304 using Yb:YAG disk laser," *Metals Mater Int*, vol. 27, no. 6, pp. 1540–1553, 2021, doi:10.1007/s12540-019-00526-6.
26. S. P. Kumar and P. D. Babu, "Effect of laser energy density on surface morphology, microstructure and mechanical behaviour of direct metal laser melted 17-4 PH stainless steel," *Int J Mater Res*, vol. 114, pp. 10–11, 2023, doi:10.1515/ijmr-2022-0242.
27. M. Duraiselvam, R. Galun, V. Wesling, B. L. Mordike, R. Reiter, and J. Oligmüller, "Cavitation erosion resistance of AISI 420 martensitic stainless steel laser-clad with nickel aluminide intermetallic composites and matrix composites with TiC reinforcement," *Surf Coat Technol*, vol. 201, no. 3–4, pp. 1289–1295, 2006, doi:10.1016/j.surfcoat.2006.01.054.
28. S. Martin Vinoth and P. Dinesh Babu, "Comparing and analysing the characterisation of laser cladding of CrNiW and CrNiFeAlZr powders on H13 tool steel," *Proc Inst Mech Eng C J Mech Eng Sci*, vol. 237, no. 13, pp. 2969–2989, 2023, doi:10.1177/09544062221141267.
29. G. Telasang, J. Dutta Majumdar, N. Wasekar, G. Padmanabham, and I. Manna, "Microstructure and mechanical properties of laser clad and post-cladding tempered AISI H13 tool steel," *Metall Mater Trans A Phys Metall Mater Sci*, vol. 46, no. 5, pp. 2309–2321, 2015, doi:10.1007/s11661-015-2757-z.

30. S. S. Joshi, D. Choudhuri, S. A. Mantri, R. Banerjee, N. B. Dahotre, and S. Banerjee, "Rationalizing surface hardening of laser glazed grey cast iron via an integrated experimental and computational approach," *Mater Des*, vol. 156, pp. 570–585, 2018, doi:10.1016/j.matdes.2018.07.022.
31. I. Bunaziv, O. M. Akselsen, X. Ren, B. Nyhus, M. Eriksson, and S. Gulbrandsen-Dahl, "A review on laser-assisted joining of aluminium alloys to other metals," *Metals*, vol. 11, no. 11, p. 1680, 2021, doi:10.3390/met11111680.
32. C. M. Grohol, Y. C. Shin, and A. Frank, "Laser cladding of aluminum alloy 6061 via off-axis powder injection," *Surf Coat Technol*, vol. 415, p. 127099, 2021, doi:10.1016/j.surfcoat.2021.127099.
33. N. Jeyaprakash and C. H. Yang, "Comparative study of NiCrFeMoNb/FeCrMoVC laser cladding process on nickel-based superalloy," *Mater Manuf Process*, vol. 35, no. 12, pp. 1383–1391, 2020, doi:10.1080/10426914.2020.1779933.
34. N. Jeyaprakash, C. H. Yang, and S. P. Tseng, "Characterization and tribological evaluation of NiCrMoNb and NiCrBSiC laser cladding on near-α titanium alloy," *Int J Adv Manuf Technol*, vol. 106, no. 5–6, pp. 2347–2361, 2020, doi:10.1007/s00170-019-04755-2.
35. A. K. Das, "Recent trends in laser cladding and alloying on magnesium alloys: A review," *Mater Today: Proc*, vol. 51, pp. 723–727, 2021, doi:10.1016/j.matpr.2021.06.217.
36. J. Liu, H. Yu, C. Chen, F. Weng, and J. Dai, "Research and development status of laser cladding on magnesium alloys: A review," *Opt Lasers Eng*, vol. 93, pp. 195–210, 2017, doi:10.1016/j.optlaseng.2017.02.007.
37. Z. Qi, C. Chen, C. Wang, Z. Zhou, J. Zhou, and Y. Long, "Effect of different laser wavelengths on laser cladding of pure copper," *Surf Coat Technol*, vol. 454, pp. 129181, 2023, doi:10.1016/j.surfcoat.2022.129181.
38. L. Jin et al., "A review of laser cladding on copper and copper alloys," *Int J Electrochem Sci*, vol. 17, p. 220920, 2022, doi:10.20964/2022.09.14.
39. J. Liu, H. Liu, P. Chen, and J. Hao, "Microstructural characterization and corrosion behaviour of AlCoCrFeNiTi x high-entropy alloy coatings fabricated by laser cladding," *Surf Coat Technol*, vol. 361, pp. 63–74, 2019, doi:10.1016/j.surfcoat.2019.01.044.
40. Z. U. Arif, M. Y. Khalid, E. ur Rehman, S. Ullah, M. Atif, and A. Tariq, "A review on laser cladding of high-entropy alloys, their recent trends and potential applications," *J Manuf Process*, vol. 68, pp. 225–273, 01, 2021, doi:10.1016/j.jmapro.2021.06.041.
41. C. Cui et al., "Characteristics of cobalt-based alloy coating on tool steel prepared by powder feeding laser cladding," *Opt Laser Technol*, vol. 39, no. 8, pp. 1544–1550, 2007, doi:10.1016/j.optlastec.2006.12.005.
42. K. Wang, B. Chang, Y. Lei, H. Fu, and Y. Lin, "Effect of cobalt on microstructure and wear resistance of Ni-based alloy coating fabricated by laser cladding," *Metals (Basel)*, vol. 7, no. 12, p. 551, 2017, doi:10.3390/met7120551.
43. N. Jeyaprakash, G. Prabu, C. H. Yang, M. P. Banda, and E. Mohan, "Structural and nanohardness of laser melted CM247LC nickel-based superalloy," *Trans Indian Inst Metals*, vol. 76, no. 2, pp. 287–295, 2023, doi:10.1007/s12666-022-02734-y.
44. N. Jeyaprakash, C. H. Yang, P. Susila, and S. S. Karuppasamy, "Laser cladding of NiCrMoFeNbTa particles on Inconel 625 alloy: Microstructure and corrosion resistance," *Trans Indian Inst Metals*, vol. 76, no. 2, pp. 599–612, 2023, doi:10.1007/s12666-022-02701-7.

45. Y. Lu, G. Huang, Y. Wang, H. Li, Z. Qin, and X. Lu, "Crack-free Fe-based amorphous coating synthesized by laser cladding," *Mater Lett*, vol. 210, pp. 46–50, 2018, doi:10.1016/j.matlet.2017.08.125.
46. Q. Zhu, S. Qu, X. Wang, and Z. Zou, "Synthesis of Fe-based amorphous composite coatings with low purity materials by laser cladding," *Appl Surf Sci*, vol. 253, no. 17, pp. 7060–7064, 2007, doi:10.1016/j.apsusc.2007.02.055.
47. H. Yan, A. Wang, K. Xu, W. Wang, and Z. Huang, "Microstructure and interfacial evaluation of Co-based alloy coating on copper by pulsed Nd:YAG multilayer laser cladding," *J Alloys Compd*, vol. 505, no. 2, pp. 645–653, 2010, doi:10.1016/j.jallcom.2010.06.099.
48. M. M. Quazi, M. A. Fazal, A. S. M. A. Haseeb, F. Yusof, H. H. Masjuki, and A. Arslan, "Laser-based surface modifications of aluminum and its alloys," *Crit Rev Solid State Mater Sci*, vol. 41, no. 2, pp. 106–131, 2016, doi:10.1080/10408436.2015.1076716.
49. R. L. Sun, D. Z. Yang, L. X. Guo, and S. L. Dong, "Laser cladding of Ti-6Al-4V alloy with TiC and TiCq NiCrBSi powders," *Surf Coat Technol*, vol. 135, pp. 307–312, 2001.
50. Z. Zhang and R. Kovacevic, "Laser cladding of iron-based erosion resistant metal matrix composites," *J Manuf Process*, vol. 38, pp. 63–75, 2019, doi:10.1016/j.jmapro.2019.01.001.
51. D. Bartkowski, A. Bartkowska, and P. Jurči, "Laser cladding process of Fe/WC metal matrix composite coatings on low carbon steel using Yb: YAG disk laser," *Opt Laser Technol*, vol. 136, 2021, p. 106784, doi:10.1016/j.optlastec.2020.106784.
52. L. Chen, Y. Chen, X. Chen, T. Yu, and Z. Wang, "Microstructure and properties of in situ TiC/Ni functionally gradient coatings by powder-fed laser cladding," *Ceram Int*, vol. 48, no. 24, pp. 36789–36801, 2022, doi:10.1016/j.ceramint.2022.08.243.
53. T. M. Yue and T. Li, "Laser cladding of Ni/Cu/Al functionally graded coating on magnesium substrate," *Surf Coat Technol*, vol. 202, no. 13, pp. 3043–3049, 2008, doi:10.1016/j.surfcoat.2007.11.007.
54. K. Liu, Y. Li, and J. Wang, "In-situ reactive fabrication and effect of phosphorus on microstructure evolution of Ni/Ni-Al intermetallic composite coating by laser cladding," *Mater Des*, vol. 105, pp. 171–178, 2016, doi:10.1016/j.matdes.2016.05.074.
55. Y. Liu, Y. Ding, L. Yang, R. Sun, T. Zhang, and X. Yang, "Research and progress of laser cladding on engineering alloys: A review," *J Manuf Process*, vol. 66, pp. 341–363, 2021, doi:10.1016/j.jmapro.2021.03.061.
56. P. K. Shanmuganathan, D. B. Purushothaman, and M. Ponnusamy, "Effect of high laser energy density on selective laser melted 316L stainless steel: Analysis on metallurgical and mechanical properties and comparison with wrought 316L stainless steel," *3D Print Addit Manuf*, vol. 10, no. 3, pp. 383–392, 2023, doi:10.1089/3dp.2021.0061.
57. V. Ocelík, U. de Oliveira, M. de Boer, and J. T. M. de Hosson, "Thick Co-based coating on cast iron by side laser cladding: Analysis of processing conditions and coating properties," *Surf Coat Technol*, vol. 201, no. 12, pp. 5875–5883, 2007, doi:10.1016/j.surfcoat.2006.10.044.
58. Vernon-Parry and K.D, "Scanning electron microscopy: an introduction." *III-Vs Review*, vol. 13, pp. 40–44, 2000.
59. L. Shepeleva, B. Medres, W. D. Kaplan, M. Bamberger, and A. Weisheit, "Laser cladding of turbine blades," *Surf Coat Technol*, vol. 125, pp. 45–48, 2000.

60. Z. Mei, W. Y. Wang, and A. H. Wang, "Transmission electron microscopy characterization of laser-clad iron-based alloy on Al-Si alloy," *Mater Charact*, vol. 56, no. 3, pp. 185–191, 2006, doi:10.1016/j.matchar.2005.11.002.
61. S. Wu, Z. Liu, X. Huang, Y. Wu, and Y. Gong, "Process parameter optimization and EBSD analysis of Ni60A-25% WC laser cladding," *Int J Refract Metals Hard Mater*, vol. 101, pp. 105675, 2021, doi:10.1016/j.ijrmhm.2021.105675.
62. J. C. Pereira, J. C. Zambrano, M. J. Tobar, A. Yañez, and V. Amigó, "High temperature oxidation behavior of laser cladding MCrAlY coatings on austenitic stainless steel," *Surf Coat Technol*, vol. 270, pp. 243–248, 2015, doi:10.1016/j.surfcoat.2015.02.050.
63. M. R. Fernández, A. García, J. M. Cuetos, R. González, A. Noriega, and M. Cadenas, "Effect of actual WC content on the reciprocating wear of a laser cladding NiCrBSi alloy reinforced with WC," *Wear*, vol. 324–325, pp. 80–89, 2015, doi:10.1016/j.wear.2014.12.021.
64. S. W. Yang, J. Yoon, H. Lee, and D. S. Shim, "Defect of functionally graded material of inconel 718 and STS 316L fabricated by directed energy deposition and its effect on mechanical properties," *J Mater Res Technol*, vol. 17, pp. 478–497, 2022, doi:10.1016/j.jmrt.2022.01.029.
65. M. Zhong and W. Liu, "Laser surface cladding: The state of the art and challenges," *Proc Inst Mech Eng C J Mech Eng Sci*, vol. 224, no. 5, pp. 1041–1060, 2010, doi:10.1243/09544062JMES1782.
66. P. Dinesh Babu, B. Prasannakumar, P. Marimuthu, R. K. Mishra, and T. Ram Prabhu, "Microstructure, wear and mechanical properties of plasma sprayed TiO 2 coating on Al–SiC metal matrix composite," *Arch Civil Mech Eng*, vol. 19, no. 3, pp. 756–767, 2019, doi:10.1016/j.acme.2019.03.001.
67. S. Sashank, P. Dinesh Babu, and P. Marimuthu, "Experimental studies of laser borided low alloy steel and optimization of parameters using response surface methodology," *Surf Coat Technol*, vol. 363, pp. 255–264, 2019, doi:10.1016/j.surfcoat.2019.02.036.
68. G. Sun, R. Zhou, J. Lu, and J. Mazumder, "Evaluation of defect density, microstructure, residual stress, elastic modulus, hardness and strength of laser-deposited AISI 4340 steel," *Acta Mater*, vol. 84, pp. 172–189, 2015, doi:10.1016/j.actamat.2014.09.028.
69. F. Wang, H. Mao, D. Zhang, X. Zhao, and Y. Shen, "Online study of cracks during laser cladding process based on acoustic emission technique and finite element analysis," *Appl Surf Sci*, vol. 255, no. 5 PART 2, pp. 3267–3275, 2008, doi:10.1016/j.apsusc.2008.09.039.
70. X. L. Yan, S. Y. Dong, B. S. Xu, and Y. Cao, "Progress and challenges of ultrasonic testing for stress in remanufacturing laser cladding coating," *Materials*, vol. 11, no. 2, p. 293, 2018, doi:10.3390/ma11020293.
71. J. M. Lourenço et al., "Fatigue and fracture behavior of laser clad repair of AerMet® 100 ultra-high strength steel," *Int J Fatigue*, vol. 85, pp. 18–30, 2016, doi:10.1016/j.ijfatigue.2015.11.021.
72. Y. Liu, D. Xiang, K. Wang, and T. Yu, "Corrosion of laser cladding high-entropy alloy coatings: A review," *Coatings*, vol. 12, no. 11, p. 1669, 2022, doi:10.3390/coatings12111669.
73. S. P. Kumar et al., "Investigation of crystallographic orientation and mechanical behaviour in laser-welded stainless steel 316L additive components," *Trans Indian Ins Metals*, vol. 76, no. 2, pp. 527–535, 2023, doi:10.1007/s12666-022-02756-6.

74. K. Hareharen, S. Pradeep Kumar, T. Panneerselvam, P. Dinesh Babu, and N. Sriraman, "Investigating the effect of laser shock peening on the wear behaviour of selective laser melted 316L stainless steel," *Opt Laser Technol*, vol. 162, p. 109317, 2023, doi:10.1016/j.optlastec.2023.109317.
75. J. Li, C. Chen, T. Squartini, and Q. He, "A study on wear resistance and microcrack of the Ti 3 Al/TiAl+TiC ceramic layer deposited by laser cladding on Ti-6Al-4V alloy," *Appl Surf Sci*, vol. 257, no. 5, pp. 1550–1555, 2010, doi:10.1016/j.apsusc.2010.08.094.
76. Q. Liu, M. Janardhana, B. Hinton, M. Brandt, and K. Sharp, "Laser cladding as a potential repair technology for damaged aircraft components," *Int J Struct Integr*, vol. 2, no. 3, pp. 314–331, 2011, doi:10.1108/17579861111162914.
77. W. Ya, B. Pathiraj, D. T. A. Matthews, M. Bright, and S. Melzer, "Cladding of tribaloy T400 on steel substrates using a high power Nd:YAG laser," *Surf Coat Technol*, vol. 350, pp. 323–333, 2018, doi:10.1016/j.surfcoat.2018.06.069.
78. N. Jeyaprakash and C. H. Yang, "Improvement of tribo-mechanical properties of directionally solidified CM-247 LC nickel-based super alloy through laser material processing," *Int J Adv Manuf Technol*, vol. 106, no. 11–12, pp. 4805–4814, 2020, doi:10.1007/s00170-020-04931-9.
79. G. Lian, M. Yao, Y. Zhang, and X. Huang, "Analysis and respond surface methodology modeling on property and performance of two-dimensional gradient material laser cladding on die-cutting tool," *Materials*, vol. 11, no. 10, p. 2052, 2018, doi:10.3390/ma11102052.
80. J. Natarajan, C. H. Yang, and S. S. Karuppasamy, "Investigation on microstructure, nanohardness and corrosion response of laser cladded colmonoy-6 particles on 316l steel substrate," *Materials*, vol. 14, no. 20, p. 6183, 2021, doi:10.3390/ma14206183.
81. H. Wang, Q. Liu, Y. Guo, and H. Lan, "MoFe1.5CrTiWAlNbx refractory high-entropy alloy coating fabricated by laser cladding," *Intermetallics (Barking)*, vol. 115, p. 106613, 2019, doi:10.1016/j.intermet.2019.106613.
82. N. Jeyaprakash, C. H. Yang, S. S. Karuppasamy, and M. Duraiselvam, "Stellite 6 cladding on AISI type 316L stainless steel: Microstructure, nanohardness and corrosion resistance," *Trans Indian Inst Metals*, vol. 76, no. 2, pp. 491–503, 2023, doi:10.1007/s12666-022-02731-1.
83. M. Cortina, J. I. Arrizubieta, J. E. Ruiz, A. Lamikiz, and E. Ukar, "Design and manufacturing of a protective nozzle for highly reactive materials processing via laser material deposition," *Proc CIRP*, vol. 68, pp. 387–392, 2018. doi:10.1016/j.procir.2017.12.100.

3

Hybrid Laser Cladding Process

V. Lakshmanan, D. Manikandan, and S. Hari Prasadh

3.1 Laser Cladding

By using filler material, laser cladding, also known as "laser deposition," forms a metallurgical link between a metal substrate and a metal covering. Usually delivered in powder, strip, or wire form, the filler material is exposed to the laser's heat either coaxially or laterally. It is feasible to produce components with material gradients or even components deposited with several materials by utilizing different types of powder and varying the feed rate of each [1]. Numerous industrial operations, such as surface coating application, fast production, and component repair, need the use of laser cladding. The three types of lasers used in this process are CO_2, Nd:YAG (neodymium-doped yttrium aluminum garnet), and fiber lasers. This procedure is advantageous for a number of reasons. Creating a comprehensive, original literature review on the topic of laser cladding, especially tailored for a mechanical engineering master's degree that focuses on composite materials, requires thorough exploration and synthesis of current research, methodologies, applications, benefits, limitations, and future directions. Given the constraints, I'll provide a structured overview that you can expand upon for a more detailed and extensive literature review. Note that the following content is a guideline and should be further developed with more detailed studies, critical analysis, and references to specific research articles and textbooks. Laser cladding is an advanced surface engineering technique that employs a laser to melt material, typically in powder or wire form, onto a substrate, creating a high-quality coating layer. This process is distinguished by its ability to produce coatings with excellent metallurgical bonds to the substrate, minimal dilution, and controlled microstructures. Originating as a method for repairing and enhancing the surface properties of industrial components, laser cladding has evolved into a critical technology for manufacturing new parts, particularly within the realm of composite materials. Its applications span various industries, including aerospace, automotive, and healthcare, driven by its ability to improve wear resistance, corrosion resistance, and mechanical properties of metal surfaces. Laser cladding involves complex physical and chemical processes, such as rapid melting and solidification, which are influenced by various factors, including laser power, beam size,

DOI: 10.1201/9781032693576-3

scanning speed, and the feed rate of the cladding material. The interaction of the laser beam with the material creates a melt pool that fuses the cladding material to the substrate. Subsequently, a rapid cooling procedure ensues, leading to the solidification of the material. The process parameters and material characteristics, including the substrate and cladding material, have a significant impact on the cladding layer's quality, microstructure, and mechanical properties. An essential element of laser cladding is the establishment of a robust metallurgical connection between the cladding layer and the substrate, accomplished by a minor incorporation of the substrate material into the molten pool. This dilution is carefully controlled to ensure the integrity and properties of the cladding layer. The rapid solidification rates can lead to the formation of non-equilibrium microstructures, including metastable phases and fine grain structures, which often result in enhanced mechanical properties. Laser cladding allows for the utilization of both metal and ceramic materials as cladding materials, depending on the specific properties desired for the surface layer. Metals like stainless steel, nickel-based alloys, and cobalt-based alloys are frequently utilized due to their exceptional resistance to wear and corrosion. Ceramics, such as carbides and oxides, are chosen for applications that demand increased hardness and thermal stability. In laser cladding, composite materials have recently attracted attention because of their exceptional properties that combine the advantages of metals and ceramics. Researchers extensively investigate metal matrix composites (MMCs) and ceramic matrix composites (CMCs) for their potential to enhance wear resistance, thermal stability, and mechanical strength. The choice of composite material is determined by the application's specific demands and its compatibility with the substrate material. Laser cladding offers numerous advantages over traditional coating processes, including precise control over the cladding layer's thickness and composition, minimal heat input to the substrate, and the ability to clad complex geometries. The process is also characterized by low dilution between the cladding material and the substrate, which is crucial for maintaining the integrity of the cladded layer. Furthermore, laser cladding can be used to deposit composite materials, enabling the engineering of surface layers with tailored properties. However, the process also has limitations, such as the high cost of equipment and operation, the need for precise control of process parameters to avoid defects such as cracks and porosity, and limitations in cladding materials due to differences in thermal expansion coefficients between the substrate and the cladding material. Additionally, the process requires a clean working environment and careful preparation of the substrate surface to ensure a quality coating. Laser cladding has been successfully applied in various sectors for surface modification, repair, and fabrication of parts with enhanced surface properties. In the aerospace industry, it is used to repair turbine blades and other critical components, extending their service life. In the automotive sector, laser cladding is employed to improve the wear resistance of engine components and transmission systems. The healthcare industry benefits from laser-cladded implants with improved biocompatibility and wear resistance. The future of laser cladding lies in the development of new composite materials with enhanced properties and the integration of laser cladding with additive manufacturing technologies. This integration opens up possibilities for creating complex parts with tailored properties not only on the surface but also throughout the component. Research is also focused on improving process control and understanding to minimize defects and optimize the properties of the cladded layer. Laser cladding represents a versatile and powerful technique for surface engineering, offering significant advantages in terms of material properties and functional performance of coated or repaired components.

3.2 Hybrid Laser Cladding

The prime objective in combining laser cladding with other sources as a hybrid laser cladding process is to attain high rates of deposition with little thermal effect. Possible applications include coatings for enhancing wear resistance or for repairing steel components through welding. The hybrid laser cladding process parameters are shown in Table 3.1. The laser and plasma arc combine to create a joint process zone, which is essential for achieving a stable process at laser power [2].

Hybrid laser cladding, a cutting-edge surface modification technique, merges the benefits of traditional laser cladding with additional energy sources or materials to enhance coating quality, process efficiency, and the resulting material properties. This innovative approach is characterized by its ability to simultaneously utilize multiple energy sources, such as lasers in conjunction with electric arcs (laser-arc hybrid cladding) or lasers combined with induction heating (laser-induction hybrid cladding), to improve the melting and bonding of cladding materials to the substrate. By integrating supplementary energy sources, hybrid laser cladding facilitates better

TABLE 3.1

Hybrid Laser Cladding Process Parameters

Sl. No.	Parameter	Description	Typical Range
1	Laser power (W)	Power output of the laser source	500–4,000 W
2	Scan speed (mm/s)	Speed at which the laser moves	10–100 mm/s
3	Powder feed rate (g/min)	Rate at which cladding material is fed	10–50 g/min
4	Working distance (mm)	Distance between laser and substrate	5–20 mm
5	Shielding gas flow (L/min)	Flow rate of the shielding gas	10–30 L/min

TABLE 3.2

Common Materials Used in Hybrid Laser Cladding

Sl. No.	Material Type	Examples	Desired Properties
1	Metals	Stainless steel, titanium alloys	Corrosion resistance
2	Ceramics	Silicon carbide, alumina	Wear resistance
3	Carbides	Tungsten carbide, titanium carbide	Hardness
4	Composite materials	Metal matrix composites (MMCs)	Customized properties

control over the thermal gradients and solidification rates within the melt pool, leading to the formation of coatings with superior microstructural properties, reduced porosity, and enhanced metallurgical bonds compared to those produced by conventional laser cladding. This process is particularly beneficial in applying composite coatings, where the synergistic effects of multiple energy sources can aid in the more effective distribution and bonding of reinforcement phases within the matrix, thus achieving coatings with tailored thermal, mechanical, and wear-resistant properties. Material considerations in hybrid laser cladding are paramount, with a wide range of metals, ceramics, and composite powders being utilized to cater to specific application requirements as shown in Table 3.2.

The choice of materials and the precise control of process parameters, such as energy input, feed rate, and shielding gas composition, are critical in achieving the desired coating characteristics and performance. The applications of hybrid laser cladding span across various industries, including aerospace, automotive, and oil and gas, where the technique is employed to restore worn-out parts, enhance surface properties, or fabricate components with complex geometries and customized functionalities. Despite its significant advantages, hybrid laser cladding faces challenges such as the need for sophisticated process control systems to manage the interactions between multiple energy sources and the cladding material, the potential for increased operational costs, and the requirement for specialized equipment. Nevertheless, ongoing research and development efforts are focused on overcoming these challenges by optimizing process parameters, exploring new material combinations, and integrating advanced monitoring and control technologies.

The future of hybrid laser cladding lies in the further exploration of its potential to produce high-performance coatings and components with unparalleled material properties, as well as its integration with additive manufacturing processes for the creation of complex, functionally graded materials and structures. The performance outcomes of hybrid laser cladding are shown in Table 3.3.

As the technology matures and becomes more accessible, it is poised to revolutionize the field of surface engineering and material science, offering

TABLE 3.3

Performance Outcomes of Hybrid Laser Cladding

Sl. No.	Outcome	Measurement Technique	Desired Value/Range
1	Microstructure	Microscopy, SEM, TEM	Fine, homogeneous grains
2	Hardness (HV)	Vickers hardness test	>500 HV
3	Wear resistance	Pin-on-disk test, abrasion test	Low wear rate
4	Surface roughness (Ra, μm)	Surface profilometer	<10 μm

TABLE 3.4

Advantages and Limitations of Hybrid Laser Cladding

Sl. No.	Aspect	Advantages	Limitations
1	Efficiency	High deposition rate, reduced material waste	High initial cost
2	Precision	Accurate control over cladding area	Requires skilled operation
3	Material properties	Enhanced wear and corrosion resistance	Material compatibility issues
4	Application flexibility	Suitable for complex geometries, repair, and manufacturing	Limited by equipment size

new possibilities for the design and manufacture of advanced components with enhanced durability, performance, and lifespan as also the advantages and limitations of hybrid laser cladding processes as shown in Table 3.4. This condensed narrative serves as a springboard for more exhaustive research and writing on the topic of hybrid laser cladding. For a thorough and academically robust literature review or dissertation chapter, it is crucial to delve deeper into each mentioned aspect, cite specific studies and experiments, and critically analyze the outcomes and implications of recent advancements in the field.

3.3 Types of Hybrid Laser Cladding Process

3.3.1 GMAW-Based Hybrid Laser Cladding

GMAW (gas metal arc welding) as shown in Figure 3.1 is based on hybrid laser cladding and represents an innovative fusion of traditional welding techniques with high-precision laser cladding, aimed at enhancing the quality and efficiency of surface coatings. This technique synergizes the deep penetration and filler material addition capabilities of GMAW with precise heat control and minimal heat-affected zone of laser cladding. By combining these processes, GMAW-based hybrid laser cladding enables the deposition of metal coatings on substrates with improved metallurgical properties, such as enhanced bond strength, reduced porosity, and optimized microstructural characteristics. The process is highly versatile, allowing for the application of a wide range of materials, including metals and alloys, to enhance surface properties like wear resistance, corrosion resistance, and restoration of dimensional features. Critical to achieving optimal outcomes with this method are the careful control of process parameters such as laser power, arc current, wire feed rate, and travel speed—and the selection of appropriate shielding gases to protect the weld pool from atmospheric contamination.

This hybrid approach finds applications in industries requiring high-performance coatings and precise repair capabilities, including aerospace, automotive, and manufacturing, offering a cost-effective solution for extending the life of components and improving their functional performance. The ongoing development and research into GMAW-based hybrid laser cladding focus on refining the process to achieve even finer microstructural control,

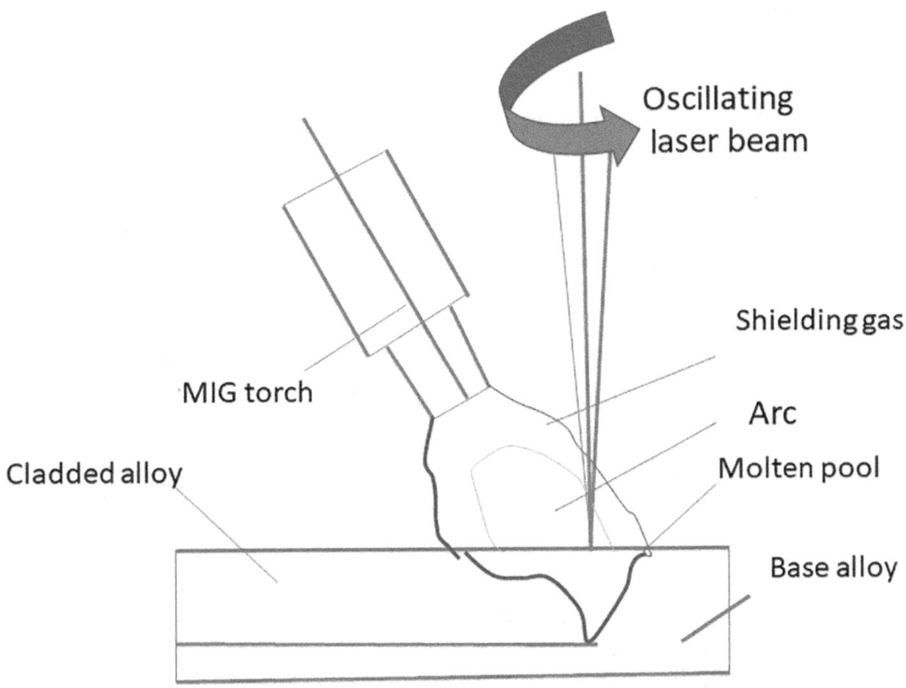

FIGURE 3.1
Diagram of GMAW-based hybrid laser cladding.

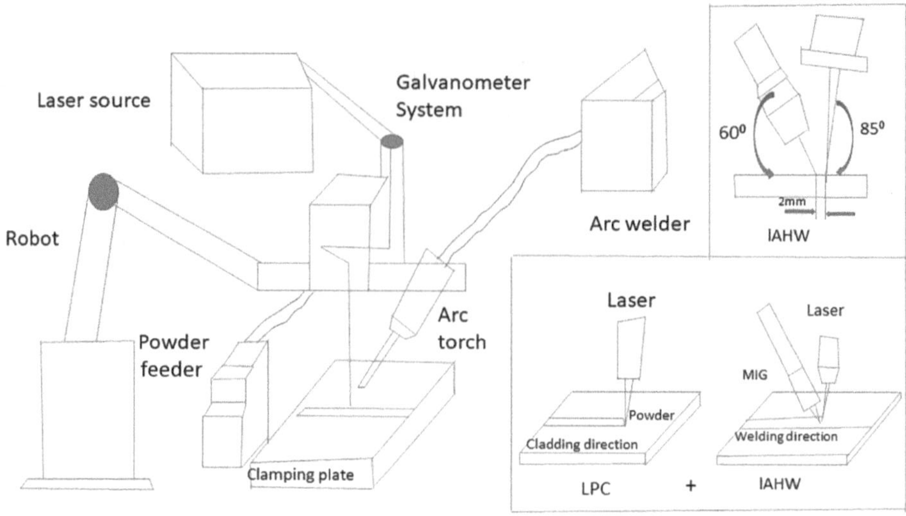

FIGURE 3.2
Dual-energy cladding system.

increase the efficiency of material deposition, and expand the range of applicable materials, promising to open new avenues in advanced manufacturing and repair techniques as shown in Figure 3.2.

3.3.2 GTAW-Based Laser Hybrid Welding Process

The parent metal is produced with a sound weld of excellent integrity using the gas tungsten arc welding (GTAW) method. In addition, there is a lot of wear, corrosion, and abrasion on the components in these sectors. Using the cladding procedure increases their durability at work. GTAW is frequently utilized in the cladding process to provide corrosion protection and hardfacing. By utilizing sophisticated hotwire multi-cathode GTAW procedures, it has a higher rate of deposition. Optimizing weld parameters such as mean voltage, wire feed rate, frequency, hotwire current, and type of current source direct current (DC), DC pulsed, and alternate current (AC) can further increase the deposition rate. The GTAW-based laser hybrid welding process is a cutting-edge technique that amalgamates the precision and efficiency of laser welding with the versatility and filler material addition capability of GTAW as shown in Figure 3.3. This innovative approach harnesses the focused energy of a laser beam to achieve deep penetration and high welding speeds, while the GTAW component allows for the introduction of filler material, enhancing the weld's mechanical properties and enabling the welding of thicker materials with improved control over the weld pool. The combination of these two methods facilitates a more flexible welding process, capable of producing high-quality joints with reduced heat input, minimizing

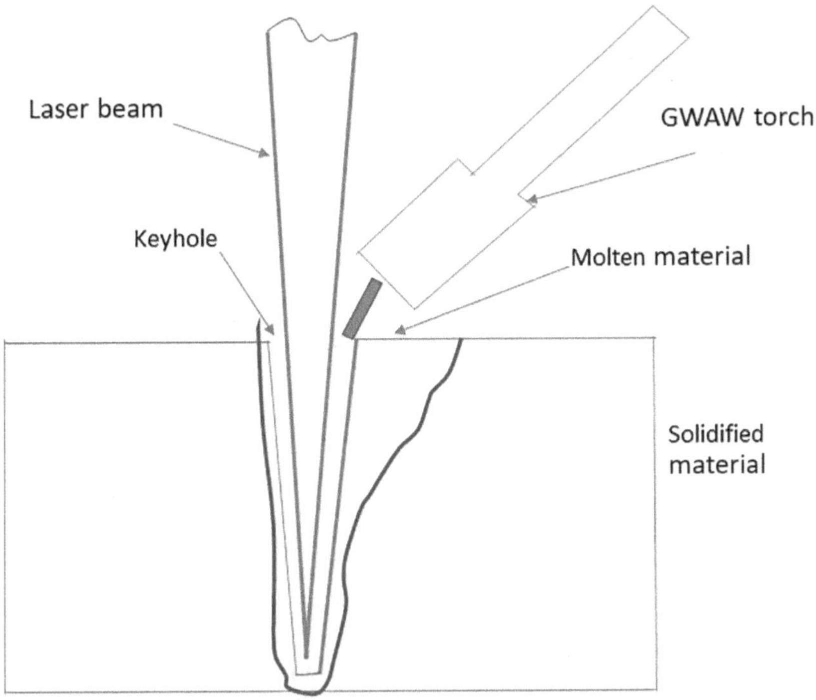

FIGURE 3.3
Diagram of hybrid laser-GMAW cladding.

thermal distortion and the potential for defects such as porosity and cracking. The process is particularly beneficial in applications requiring the welding of dissimilar materials or the precise control of the microstructure of the weld zone, making it highly relevant for industries such as aerospace, automotive, and medical devices where material integrity and performance are critical. Key to the success of the GTAW-based laser hybrid welding process is the optimization of process parameters, including laser power, arc current, welding speed, and the type and flow rate of shielding gas, to ensure a stable arc and adequate protection of the weld area from atmospheric contamination. As research and development in this field continue, the GTAW-based laser hybrid welding process is expected to see further refinements, enhancing its applicability and efficiency for advanced manufacturing needs.

3.3.3 Plasma Arc-Based Hybrid Laser Cladding

A hybrid process configuration refers to a setup where the plasma arc and laser, which are two different energy sources, create a common process zone. Figure 3.4 illustrates the setup of this process variant in the laboratory, both

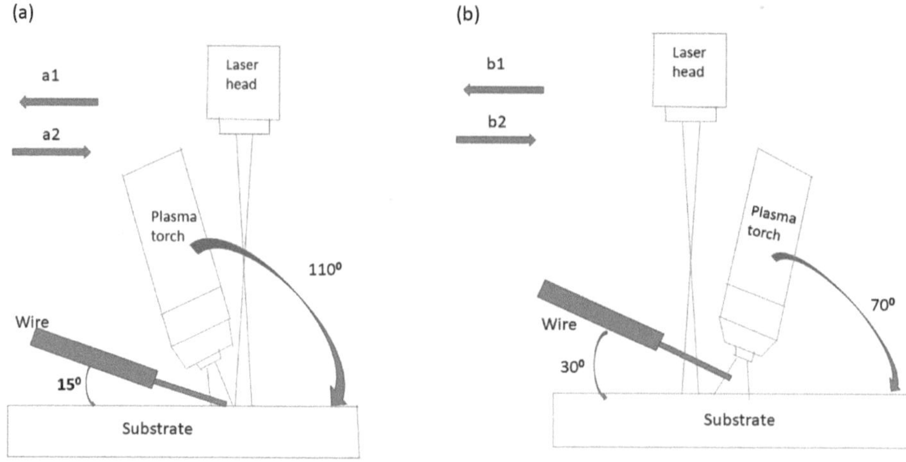

FIGURE 3.4
Diagram of plasma arc-based hybrid laser cladding.

conceptually (b) and experimentally (a). The plasma torch is placed in a position that follows the direction of movement, with a right angle to the workpiece and the laser at a specific angle α. The laser radiation and plasma arc are focused in a single process zone due to the alignment of the two energy sources. In order to establish the laser as the primary component, it is strategically placed a small distance in front of the electrode, aligned perpendicularly with the plasma torch. The commercially available plasma torch has been enhanced by incorporating a laser beam aperture in its front section. Six powder feeds are strategically placed around the arc to facilitate the passage of metal powder through the 4 mm-diameter electrode of the torch. The Plasma Transferred Arc (PTA) system is incorporated into a laser processing workstation to facilitate the execution of the studies. The optics and 2 kW Ytterbium doped Yttrium Aluminum Garnet (Yb:YAG) disk laser on the laser processing workstation are specifically designed to possess a focal length of 200 mm. Argon functions as both the carrier gas and the plasma gas, with flow rates of 2 and 3 L/min, respectively. The wear protection layer experiments employ a filler material [3]. Plasma arc-based hybrid laser cladding is an advanced surface modification technique that integrates the high energy density of a laser beam with the stabilizing and preheating effects of a plasma arc. This innovative approach leverages the laser's capability to achieve precise, localized melting of the cladding material, ensuring minimal thermal distortion and a narrow heat-affected zone, while the plasma arc enhances the process by preheating the substrate and improving the wettability of the surface. This synergy not only facilitates a stronger metallurgical bond between the cladding layer and the substrate but also enables the deposition of coatings with enhanced properties, such as increased hardness, improved wear resistance, and reduced porosity.

The process is particularly advantageous for applying high-performance coatings to critical components in industries requiring exceptional surface characteristics, such as aerospace, automotive, and energy. By optimizing process parameters like laser power, plasma current, and feed rate, this hybrid technique can be tailored to a wide range of materials and applications, offering significant improvements over traditional cladding methods. Continuous advancements in control technologies and process understanding are expanding the potential of plasma arc-based hybrid laser cladding, paving the way for more efficient, reliable, and versatile surface engineering solutions.

3.3.4 Multi-Energy Source Deposition

Multi-energy source deposition represents a pioneering approach in additive manufacturing, blending various energy sources—such as lasers, electron beams, and plasma arcs—to create or repair components with unparalleled precision and material properties. This innovative technique allows for the simultaneous use of multiple energy inputs, optimizing the melting, deposition, and solidification phases of materials in a way that a single energy source cannot achieve. By carefully controlling the interaction of different energy sources, this method can tailor thermal gradients, reduce residual stresses, and achieve finer microstructures, leading to improved mechanical and physical properties of the final product. Multi-energy source deposition is particularly advantageous for manufacturing complex geometries and functionally graded materials, offering the potential for creating components with locally optimized properties that meet specific functional requirements as shown in Figure 3.5. The versatility and efficiency of this process find applications in aerospace, automotive, and

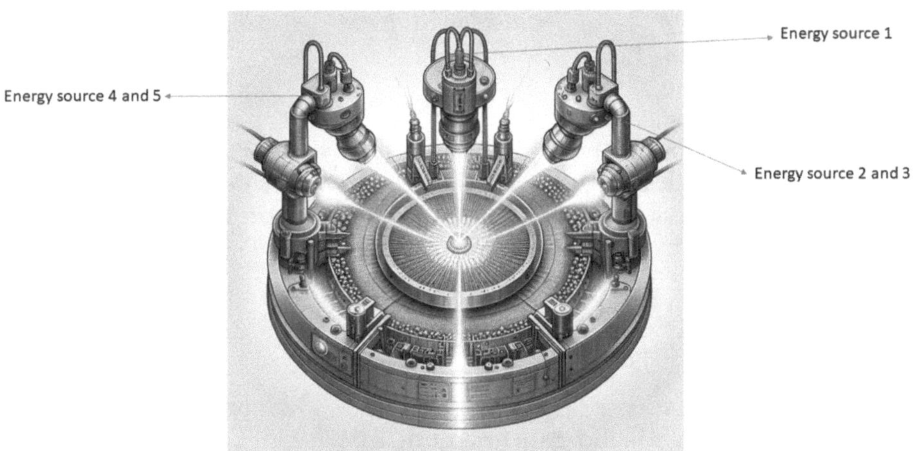

FIGURE 3.5
Multi-energy source deposition hybrid laser cladding processes.

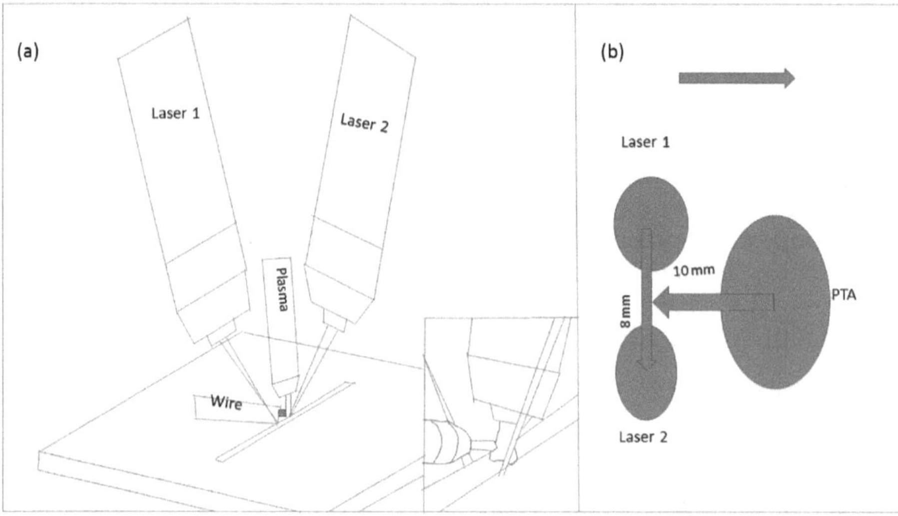

FIGURE 3.6
Two different laser power with plasma arc power sources using hybrid laser cladding.

medical industries, where the demand for high-performance, custom-tailored parts is ever-increasing. As research and development continue to advance in this field, multi-energy source deposition is set to revolutionize the landscape of additive manufacturing, enabling the production of components with previously unattainable levels of complexity and functionality.

For each unit mass of material, the energy required for deposition may be used to determine the efficiency of the deposition process.

$$\varepsilon = P/R \tag{3.1}$$

P – Power
R – Deposition rate
ε – Deposition efficiency

The energy distribution must be changed to direct more energy toward the margins of the melt pool to overcome the low process deposition efficiency and lower the process's remelting depth where an optimized power density profile is produced by combining three or more energy sources as shown in Figure 3.6.

3.3.5 Types of Laser Power Sources

The different types of laser power sources are shown in Table 3.5. A solid-state laser employs a solid-gain medium that is doped with impurities. When this medium is stimulated, it emits photons with a consistent wavelength. Mirrors confine photons to induce further emission, resulting in

TABLE 3.5

Types of Laser Used in Hybrid Laser Cladding and Its Applications

Laser (Wavelength)	Applications
CO_2 laser (10.6 μm) Excimer laser_KrF (0.248 μm)	Fast prototyping, welding and cutting, surface modification, such as alloying and cladding, laser glazing and ablation, stereo lithography via optical means
Excimer laser XeCl (0.312 μm)	Drilling, scribing, and precision micromachining associated with etching, machining, and drilling
Nd:YAG laser (1.06 μm)	Fast prototyping, drilling, welding, machining, and marking CD drives, optical computers, laser printers, scanners and photocopiers, industry alignment in optical communication
Semiconductor laser/diode laser (0.7–1.0 μm) optical communication industrial alignment	Spectroscopy, bar code scanners, biodetectors, pollution detectors, ozone layer detectors, holography, and 3D image scanners. Both laser ablation and cutting, paint stripping and laser cleaning for the preservation of artifacts, and rapid-fire photography
Copper vapor laser (0.51 μm)	Detection of fingerprints, exact drilling and cutting of microholes energy source for isotope separation excitation

amplification and ultimately generating a coherent laser beam via an output mirror. A dye laser is an example of a laser that uses a liquid as its medium. A dye laser is a type of laser that employs an organic dye, which is in the form of a liquid solution, as the laser medium. These lasers generate laser light by utilizing the excited energy states of organic dyes that have been broken down or dissolved in liquid solvents. It emits laser light in the range of wavelengths from near-ultraviolet to near-infrared. A gas laser is a type of laser that utilizes an electric current to pass through a gas within the laser medium to generate laser light. Gas lasers utilize a gaseous condition as the laser medium. Recombination radiation can occur by introducing charge carriers into the specific area defined by the junction. If the current infusion is adequate, it can lead to population inversion and stimulate emissions. Due to these attributes, they are considered one of the primary categories of lasers now in use. They are used in various applications, such as optical data storage, optical fiber networking, and as a power source for solid-state lasers.

These lasers utilize a fiber composed of silicate or occasionally phosphate glass to absorb the first light emitted by the pump laser diodes. The fiber then converts this light into a laser beam with a certain wavelength. In order to do this, the optical fiber is doped. Through the incorporation of different doping materials, it is possible to generate laser beams with a diverse array of wavelengths.

3.4 Laser Pre-Cladding (LPC) and Laser Hybrid Arc Cladding (LHAC)

Laser Pre-cladding (LPC) and laser hybrid arc cladding (LHAC) are two advanced techniques in the field of surface engineering, each with distinct methodologies and advantages for enhancing surface properties as shown in Figure 3.7. LPC involves a preparatory step where a laser is used to preheat the substrate before the deposition of the cladding material. This preheating process is crucial for improving the substrate's surface characteristics, making it more conducive to forming a stronger metallurgical bond with the cladding material. The laser preheating method precisely controls the temperature of the substrate, optimizing the conditions for the subsequent cladding process to minimize thermal distortions and stresses, resulting in a cladding layer with superior adhesion, reduced porosity, and enhanced mechanical properties. LPC is particularly useful in applications requiring precise control over the microstructure of the cladded layer, enabling the production of surfaces with tailored properties for specific operational requirements. LHAC, on the other hand, combines laser cladding with traditional arc welding techniques (such as Metal Inert Gas/Metal Active Gas (MIG/MAG) or Tungsten Inert Gas (TIG) welding) in a single process. This hybrid approach harnesses the deep penetration and filler material addition capabilities of arc welding, alongside the high precision and minimal heat input characteristics of laser cladding. The result is a cladding process that offers enhanced efficiency, improved material properties, and greater versatility in the types of materials that can be deposited. LHAC is particularly effective for applying thick coatings, repairing components, and joining dissimilar

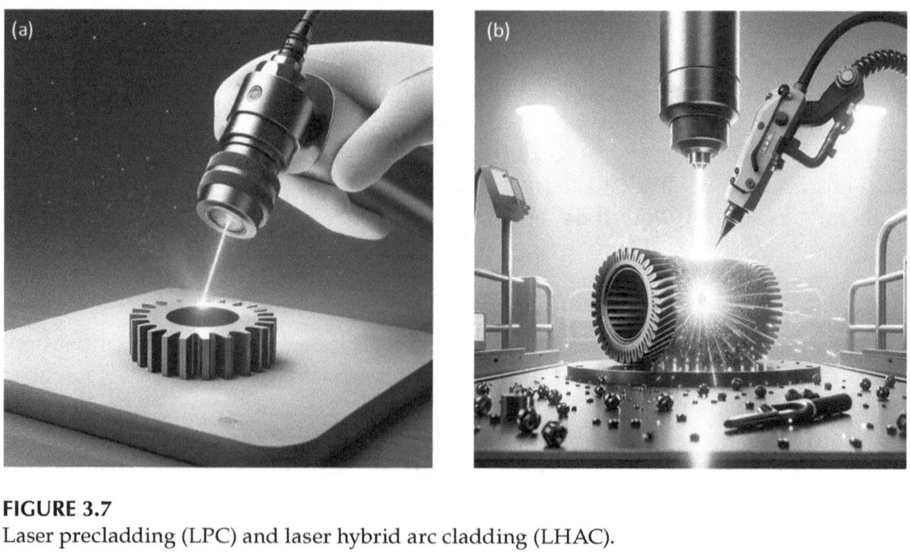

FIGURE 3.7
Laser precladding (LPC) and laser hybrid arc cladding (LHAC).

materials, producing joints with excellent mechanical strength and minimal thermal distortion. The process is adaptable to a wide range of industrial applications, from aerospace to automotive to heavy machinery, where the durability and performance of surface coatings are critical. Both LPC and LHAC represent significant advancements in materials engineering, providing innovative solutions for extending the life of components, enhancing surface properties, and facilitating the development of new materials and composites. While LPC focuses on optimizing substrate conditions for better cladding outcomes, LHAC leverages the strengths of both laser and arc welding to achieve efficient, high-quality coatings. As research and technological developments continue, these methods are set to further expand the possibilities for manufacturing and repairing components across various industries, offering tailored solutions to complex material challenges. In this case, the accumulation of filling wire caused the arc zone to become wider and the laser zone of the LPC and LAHW sample to become smaller, which was in line with the previous observation. The formation of columnar grains at the fusion line was entirely prevented during LAHW. As a result, the equiaxed grains took on the characteristics of columnar grains, causing the average grain size to decrease from 74.7 to 46.4mm. Furthermore, the laser zone's average particle size decreased from 92.0 to 46.9mm, while the arc zone's average grain size decreased from 86.6 to 39.7mm. [4].

3.5 Morphology of Molten Pool

The morphology of a molten pool, particularly in processes like welding, laser cladding, or additive manufacturing, plays a crucial role in determining the quality and characteristics of the final product. This morphology refers to the shape, size, depth, and fluid dynamics within the molten material during the process, which are influenced by various factors including the heat input, material properties, and process parameters. In these thermal processes, the morphology of the molten pool is critical because it directly affects the microstructure, mechanical properties, and potential defects in the solidified material. For instance, the depth and width of the molten pool can influence the penetration of the weld or cladding, affecting the bond strength and integrity of the material. Similarly, the cooling rate, which is partially determined by the pool's morphology, can impact the formation of microstructures within the solidified material, leading to variations in hardness, tensile strength, and ductility. The fluid dynamics within the molten pool, which encompass the movement speed of the fluid and variations in temperature, are also crucial. If not adequately managed, they have the potential to result in the development of welding flaws such as porosity, cracks, and incomplete fusion. Marangoni convection, which is caused by variations in surface tension due to temperature or composition disparities, has a notable impact on the behavior of molten pools. It affects the dispersion of alloying elements and the resulting uniformity of the material.

Furthermore, in laser-based processes, the laser parameters such as power, speed, and focal point, as well as the interaction between the laser beam and the material, have a substantial impact on the shape and structure of the molten pool. The energy density and operational mode (continuous or pulsed) of the laser can modify the distribution and absorption of heat in the material, thereby impacting the shape and stability of the molten pool.

Gaining knowledge and managing the shape and structure of the liquid pool is crucial for maximizing production methods, minimizing flaws, and attaining the intended characteristics of the material. The development of monitoring and control technologies, including real-time imaging and temperature measurement, has enhanced the capacity to observe and regulate the dynamics of the molten pool during the process. This has resulted in improved results in material processing applications. When subjected to infrared laser power, the molten pool creates a surface-level metallurgical connection with the substrate. As the infrared laser becomes more powerful, the slivery cladding track becomes continuous and maintains a consistent shape. Upon analyzing the morphology of the cladding track's cross-section under various laser sources, two significant findings were discovered: firstly, employing a blue laser improves the quality and efficiency of the cladding process by augmenting the absorption rate. Furthermore, the blue laser lacks the ability to improve the quality of the bonding interface between the two high-reflectivity materials in this specific experiment. The study aims to analyze the fluctuation pattern of molten pool size and the distribution of each parameter under hybrid and infrared lasers. The hybrid laser's reduction in infrared laser power significantly decreases the overall dimensions of the molten pool, including its height, breadth, and depth-to-width ratio, similar to the effects observed with a pure infrared laser. In contrast, a reduction in the infrared laser power of the hybrid laser results in an increase in both the proportion of the overall molten pool height and the proportion of the molten pool width. Consequently, the infrared blue hybrid laser can produce a larger molten pool, resulting in a higher cladding efficiency. Furthermore, there is a growing decrease in the ratio of the depth to width of the molten pool. As a result, the hybrid laser improves the cladding stability by decreasing the molten pool's aspect ratio and increasing the shift propensity toward the conduction mode [5].

3.6 Hybrid Laser Cladding Applied in Magnesium Alloys

Hybrid laser cladding applied to magnesium alloys represents a cutting-edge approach in surface engineering, aimed at enhancing the surface properties of these lightweight, yet inherently corrosion-prone materials. Magnesium alloys, known for their high strength-to-weight ratio, are increasingly used in the automotive, aerospace, and electronics industries to reduce weight and improve fuel efficiency. However, their susceptibility to wear and corrosion significantly limits their application. Hybrid laser cladding offers a promising solution to these challenges by depositing a protective layer of material onto the magnesium alloy surface, thereby improving its wear resistance, corrosion resistance, and overall durability. The hybrid laser cladding process combines the precision and minimal heat input of laser cladding with another complementary energy source, such as an arc or plasma, to enhance the cladding material's bonding and coating quality. This dual-energy approach allows for better control over the thermal input, reducing the risk of overheating and damaging the heat-sensitive magnesium substrate. By carefully selecting cladding materials, such as aluminum alloys, ceramics, or MMCs that are compatible with magnesium alloys, the hybrid process can tailor the surface properties of the magnesium alloy to meet specific application requirements. One of the key advantages of applying hybrid laser cladding to magnesium alloys is the ability to create a dense, homogeneous coating with a strong metallurgical bond to the substrate, which significantly enhances the surface's resistance to wear and corrosion without compromising the material's inherent lightweight characteristics. Additionally, this process can be precisely controlled and localized, preserving the integrity of the base material and enabling the treatment of selected areas where enhanced properties are most needed. Moreover, hybrid laser cladding on magnesium alloys opens up new possibilities for the design and manufacturing of components with complex geometries and graded functionalities. By adjusting process parameters and material compositions, engineers can create coatings with tailored thermal, mechanical, and physical properties, enabling the development of advanced components that leverage the unique benefits of magnesium alloys while overcoming their limitations. Despite the potential benefits, the application of hybrid laser cladding to magnesium alloys requires careful consideration of process parameters, material compatibility, and coating characteristics to avoid defects and ensure optimal performance. Ongoing research and technological developments are focused on overcoming these challenges, exploring new material systems, and refining the process to unlock the full potential of magnesium alloys across various industries. When compared to other structural alloys, magnesium alloys are widely known to possess the highest strength-to-weight ratio. Magnesium alloys possess a low density and high specific strength, along with notable damping capacity, advantageous casting properties, machinability, and recyclability. Magnesium alloys offer significant benefits in the automobile and aerospace industries due to their ability to decrease weight and enhance energy efficiency. Regrettably, magnesium alloys' limited hardness and insufficient resistance to wear and corrosion have impeded their extensive use in engineering, especially in settings with higher levels of wear, friction, and corrosive substances [6].

Cladding and alloying techniques are employed to enhance the surface characteristics of magnesium alloy, thereby circumventing these issues. Aluminum alloys were commonly applied as protective coatings on the surface of magnesium alloy components due to their significant enhancement of the strength and corrosion resistance of magnesium alloys. Moreover, aluminum alloys exhibit exceptional metallurgical compatibility with magnesium. Various melting techniques, including selective laser melting, cold-arc melting, and electron-beam melting, have been developed to deposit alloy layers onto components for applications such as cladding, dissimilar junctions, and additive manufacturing. These high-energy melting techniques are sophisticated methods employed to rapidly generate non-equilibrium conditions. They are frequently employed for surface engineering and remanufacturing components made from magnesium alloy [7]. There is currently a lot of interest in using the LAHW technique to apply a protective layer to magnesium alloys. This method synergistically utilizes the thermal energy generated by both an arc and a laser, leading to multiple benefits.

The hybridization effect leads to a decrease in porosity, along with several other advantages, including improved surface appearance of the weld bead, faster welding speeds, increased ability to accommodate larger gaps, deeper penetration, and a lower occurrence of welding errors. The hybrid laser-arc welding technique synergistically utilizes both the electric arc and laser beam to create a unified weld pool, leading to enhanced weld quality and productivity. The process of LAHW is intricate, involving numerous variables and challenges. Porosity is the most prevalent flaw in LAHW. When the speed of solidification of the molten metal surpasses the rate is filling. Creating and reducing porosity in LAHW is challenging due to the introduction of two sources during welding. Extensive research has been carried out to optimize welding parameters, including torch angle, welding speed, arc location, laser, and arc power, to reduce porosity. Furthermore, novel methods such as ultrasonic vibration, magnetic fields, and oscillating laser beams have been devised to mitigate porosity, along with the optimization of welding parameters. Among these innovative techniques, laser beam oscillating welding stands out because it can enhance mechanical properties by reducing macro-segregation, preventing welding fractures, and mitigating porosity. Wang et al. utilize circular beam oscillations. In relation to the LAHW of 8 mm AA6082, the level of porosity was decreased from 5% to below 1% [3]. It has been found that beam oscillation welding can significantly improve the mechanical properties of a welded joint. A welded junction made of AA6061-T6 can endure stress levels of up to 277 MPa, which is approximately 89% of the stress resistance of A6061-T6 sheets. Moreover, the extension of the joint is enhanced by approximately 65% when compared to a sample that was welded using non-oscillation techniques [7]. It is asserted that the use of an oscillating laser beam decreases the presence of small holes in the Al-6Mg alloy lock butt weld, resulting in a 36.3% increase in its ability to withstand stretching forces compared to a weld that does not use oscillation. Wang et al.

examined three distinct oscillation paths for laser beam oscillating welding of 5A06 aluminum alloy sheets: linear [8].

The LAHW process was employed to clad AZ80 magnesium alloy plates with 5356 aluminum alloy. We investigated the influence of the oscillating amplitude on the microstructure, internal flaws, and weld shape. There was a discussion of the connection between the cladding layer's wear resistance and oscillation amplitude, ocular, and infinite [9].

3.7 Hybrid Laser Cladding on Microstructure, Hardness, Wear Resistance, and Roughness

Hybrid laser cladding is a sophisticated surface engineering technique that amalgamates the precision and control of laser technology with additional energy sources to deposit materials on substrates, significantly enhancing their surface characteristics. This process impacts various properties of the treated surface, including microstructure, hardness, wear resistance, and roughness, each of which plays a critical role in the performance and application of the materials.

3.7.1 Microstructure

The microstructure of materials modified through hybrid laser cladding is notably refined and homogeneous due to the rapid cooling and solidification that occurs. This process can lead to the formation of fine grains and a dense structure, which are beneficial for improving mechanical properties. The precise control over the laser and secondary energy source allows for tailoring the thermal gradients and solidification rates, influencing the distribution and size of phases within the microstructure, thereby enhancing the material's overall performance [10].

3.7.2 Hardness

An immediate benefit of the altered microstructure through hybrid laser cladding is the noticeable improvement in hardness. The fine-grained microstructure, along with the potential formation of hard phases or compounds within the clad layer, contributes to a significant increase in surface hardness. This enhancement is crucial for applications requiring resistance to deformation under load, such as in wear-prone components.

3.7.3 Wear

The wear resistance of materials treated with hybrid laser cladding is substantially improved, a direct consequence of the increased hardness and optimized microstructure. The process allows for the addition of wear-resistant materials, such as carbides and ceramics, into the substrate's surface, further enhancing its resistance to abrasion, erosion, and fatigue. This makes hybrid laser cladding an invaluable technique for extending the life of mechanical components subjected to harsh operational conditions.

3.7.4 Roughness

While hybrid laser cladding generally aims to improve surface characteristics, the impact on roughness can be variable and highly dependent on the specific process parameters and materials used. In some cases, the rapid solidification can lead to a smooth finish, especially when fine particles are used as the feedstock material. However, adjustments in process parameters are often required to minimize roughness and achieve a surface finish that meets specific application requirements. The ability to control the energy input and the interaction between the laser and the material allows for the optimization of surface roughness, balancing the need for enhanced mechanical properties with the requirements for a functional surface finish. Hybrid laser cladding represents a versatile and powerful technique for improving the surface properties of materials, offering significant advantages in terms of microstructure, hardness, wear resistance, and roughness. Its ability to tailor these properties makes it an essential tool for extending the durability and functionality of components across various industries.

3.8 Conclusions

The LAHW employs varying beam oscillation amplitudes. The look, profile, distribution of components, and performance of the weld surface are all significantly impacted by the amplitude of the laser oscillation.

1. An increase in the laser beam's oscillation area causes the welds to grow, while penetration depth dramatically falls with an increase in oscillating amplitude.
2. The laser beam oscillation can inhibit the result of alloy segregation and keyhole-induced porosity, which are made possible by the stirring action that promotes liquid melt flow and diffusion.
3. The laser beam stirring can dilute the concentration in the upper portion of the seam with an increase in oscillation amplitude. The brittle phase cannot develop due to the cladded alloy's diluting, which therefore weakens the cladded layer's hardness and wear resistance.
4. An appropriately hardened weld seam with low segregation and porosity, good wear resistance, and a pleasing look may be produced with an optimized oscillation amplitude of 1 mm.

References

1. C. Brunner-Schwera, R. Kerstinga, B. Grafa, M. Rethmeie, Laser-plasma-cladding as a hybrid metal deposition-technology applying a SLM-produced copper plasma nozzle, *Procedia CIRP* 74 (2018) 738–742.
2. N. Jeyaprakash, C.-H. Yang, P. Susila, S.-S. Karuppasamy, Laser cladding of NiCrMoFeNbTa particles on Inconel 625 alloy: Microstructure and corrosion resistance, *Transaction of the Indian Institute of Metals* 76 (2022) 1–14.
3. N. Jeyaprakash, C.-H. Yang, S.-S. Karuppasamy, M. Duraiselvam, Stellite 6 cladding on AISI type 316L stainless steel: Microstructure, nanohardness and corrosion resistance, *Transaction of the Indian Institute of Metals* 76 (2022) 491–503.
4. J. Huang, K. Hao, L. Xu, Y. Han, L. Zhao, W. Ren, Grain refinement of laser-arc hybrid welded 2219 aluminum alloy by introducing TiB_2 particles employing laser pre-cladding, *Journal of Material Research and Technology* 127 (2023) 194–199.
5. H. Yang, J. Wu, Q. Wei, Z. Tang, A. Wang, X. Jin, X. Li, Y. Wu, G. Lu, H. Wang, H. Wang, Stable cladding of high reflectivity pure copper on the aluminum alloy substrate by an infrared-blue hybrid laser, *Additive Manufacturing Letters* 3 (2022) 100040.
6. N. Jeyaprakash, C.-H. Yang, Comparative study of NiCrFeMoNb/FeCrMoVC laser cladding process on nickel based super-alloy, *Materials and Manufacturing Processes* 35(12) (2020) 1383–1391.
7. N. Jeyaprakash, C.-H. Yang, K.R. Ramkumar, G.-Z. Sui, Comparison of micro-structure, mechanical and wear behavior of laser cladded SS 410 substrate using SS 420 and Colmonoy 5 particles, *Journal of Iron and Steel Research International* 27 (2020) 1446–1455.
8. N. Jeyaprakash, C.-H. Yang, S.-P. Tseng, Optimization of tribological parameters over WC-12%Co laser alloyed pearlitic ductile iron using Taguchi based Grey relational analysis, *Journal Central South University* 27 (2020) 736–751.
9. Z. Ren, Y. Zhao, G. Han, W. Wang, K. Zhou, T. He, Y. Sun, Laser–arc hybrid cladding of Al-Mg alloy coating on AZ80 Mg alloy: Effect of laser beam oscillations amplitude. *Materials* 15 (2022) 7272. https://doi.org/10.3390/ma15207272.
10. D. Manikandan, V.C. Sathish Gandhi, R. Kumaravelan, V. Vignesh, Effect of hybrid filler loading (Polyalthia longifolia seed and graphite) on the mechanical and thermal properties of vinyl ester composites, *Polymer Composites* 45 (2024) 6318–6331. https://doi.org/10.1002/pc.28199.

4

Repair of Engineering Components through Laser Cladding

Natarajan Jeyaprakash, Sundara Subramanian Karuppasamy, and Che-Hua Yang

4.1 Introduction

Along with the Industrial Revolution, the usage of pure metals and metallic alloys has a tremendous impact on every industrial sector, such as aerospace, automobile, nuclear, energy, marine and so on. The metallic alloys made from metals, such as iron, copper, aluminium, magnesium, titanium, nickel and so on, provide a significant contribution in the abovementioned industries [1,2]. Each alloy has significant characteristics, such as mechanical, chemical, electrochemical, thermal, physical, electrical and magnetic behaviours. Based on these behaviours and the area of application, metallic alloys are chosen and implemented in respective industries to promote growth and accomplish techWnological advancements. As reported, these metallic alloys work in highly aggressive environments, and the mechanical characteristics offered by the components made from these metallic alloys are reported to reduce due to prevailing service environmental conditions [3–5].

The mechanical properties of the metallic alloy components, such as tensile strength, toughness, elasticity, wear resistance, corrosion resistance, fatigue behaviours and hardness, are greatly influenced by exposure to highly aggressive work environments for a prolonged interval. The aggressive environments include variations in temperature, pressure, dissolved salt concentration, dissolved oxygen concentration and so on. The variation in temperature, especially high temperature, will lead to minimising the tensile strength of the components. In the case of boilers, the high temperature leads to the destruction of the metallic surfaces, which in turn promotes crack growth on the surface [6,7]. After prolonged exposure to high temperature, cracks grow and propagate into the material, thereby resulting in the catastrophic failure of the material followed by property loss. In the presence of residual stress and high temperature, the material structure will deform in terms of fatigue due to numerous amounts of pores formation in

 DOI: 10.1201/9781032693576-4

the material. Additionally, this combination of stress and high temperature induces the stress corrosion cracking phenomenon due to the reduction in the corrosion resistance behaviour offered by the components. On the other hand, the higher pressure exerts significant amount of stress, causing the material to deform drastically. This variation in pressure also renders the mechanical characteristics offered by the components [8–10].

The components that work in marine environments are exposed to dissolved oxygen. After a long service life, the metallic surface reacts with the dissolved oxygen, which leads to an electrochemical phenomenon called corrosion. Corrosion is unavoidable in most industries, and according to failure survey, it is reported that more than 65% of industrial accidents and failures are caused by corrosion. The chloride ion in seawater tends to be more aggressive, which combines with dissolved oxygen to form corrosive sites on the metallic surface. These corrosive sites develop into pores and voids, thereby reducing the corrosion resistance offered by the components. Other than high temperature, high pressure and dissolved oxygen, dissolved salts influence the mechanical behaviours offered by the components [11,12]. Particularly, in the case of fuel tanks of automobiles and aerospace engines, the petrochemical fuels contain fluoride, chloride, nitrides and sulphides. After long-term usage, the Cl^-, F^-, N^- and S^- ions react with the metallic anions to form metallic salts that are deposited in the inner walls of the engine. Over time, the deposited metallic sulphides react with the metallic surface in the inner walls of the engines, thereby inducing pores and cavity formation. These pores constantly develop into major leak holes and cracks, which decrease the tensile, corrosion resistance, wear resistance and hardness offered by the metallic alloys (aerospace engines). Thus, it can be concluded that the service environments containing high pressure, high temperature, dissolved salts and dissolved oxygen will alter the mechanical characteristics offered by the components made from metallic alloys. Hence, the researchers are working on several techniques to improve the mechanical characteristics and service life of the metallic components implemented in highly aggressive environments [13–15].

Surface engineering and its methods play a vital role in improving the mechanical properties of the parent material and its service life [16,17]. It is reported that a wide range of mechanical characteristics provided by the parent material can be proliferated by making alterations at the surface level. Some of the most used surface engineering methods are as follows [18–20]. Vapour deposition is a kind of surface engineering technique where the materials are deposited on the parent materials in a vapour form, and at the end, a thin coating is obtained on the parent material, which is reported to enhance the mechanical characteristics offered by the parent material. Vapour deposition can be done in two ways namely physical and chemical vapour deposition [21,22]. The main difference between these two deposition techniques is that in physical vapour deposition, the coating material's vapour is obtained in terms of physical manner, whereas chemical methods are used to produce vapour film on the substrate (parent material) while considering the chemical vapour deposition technique. Both vapour deposition techniques have limitations, such as the toxicity of the precursor used, not being suitable for complex geometries, high initial and maintenance costs of the equipment and so on [23]. Another technique called thermal spraying could be adapted for a wide range of substrates. In this method, with the aid of heat-assisted arc gun, the coating material is deposited on the substrate. However, this method encounters several drawbacks, such as the inability to produce thick layered coatings, poor metallurgical bonding between the substrate and the coated layer, holds good only for thin substrates, production of toxic fumes, need for skilled labour and so on [24,25].

Laser-based surface treatment techniques play a significant role in enhancing the mechanical, physical, electrochemical, thermal and tribological characteristics offered by the parent material (substrates) when compared with traditional surface treatment techniques such as vapour deposition, thermal spraying, hardfacing, etc. [26,27]. As the name implies, these techniques repair the worn-out or damaged surfaces with the aid of high-energy laser beam and efficiently alter the damaged surfaces when compared with the other techniques. The laser-based surface treatments are broadly classified into three different categories based on the surface temperature, melting temperature and vapourisation temperature of the substrates used in the repairing process, as shown in Figure 4.1. Among the three categories, the laser cladding process belongs to the second category, where the surface temperature of the substrate lies between the melting temperature and vapourisation temperature of the substrate. This process is considered one of the most prominent surface treatment techniques, which employ a high-energy laser beam to irradiate the sample's surface in such a way that the substrate's surface is locally melted but not evaporated [28]. In this localised melt pool, the coating material is deposited to obtain a thick coating on the substrate, thereby improving the mechanical characteristics and service life of the components. Moreover, this technique is best suited for a wide range of substrates, including metallic alloys, polymers and composites, and a broad range of coating materials benefit from this process. It is reported that the coated layer produced via the cladding process is thick and has better metallurgical bonding between the coated surface and the coating layer. Thus, the clad layer proliferates the mechanical characteristics such as tensile strength, toughness, wear resistance, corrosion resistance, fatigue resistance, hardness and so on [29,30]. Many researchers suggested that the laser cladding process can be widely adapted to repair or recreate the worn-out surfaces of the engineering components used in various industrial sectors. This chapter focuses on addressing the laser cladding principle, process parameters and how this cladding process is adapted in various industrial sectors, such as aerospace, automobiles, nuclear, biomedical, marine and other sectors, to repair the industrial (engineering) components along with relevant literature.

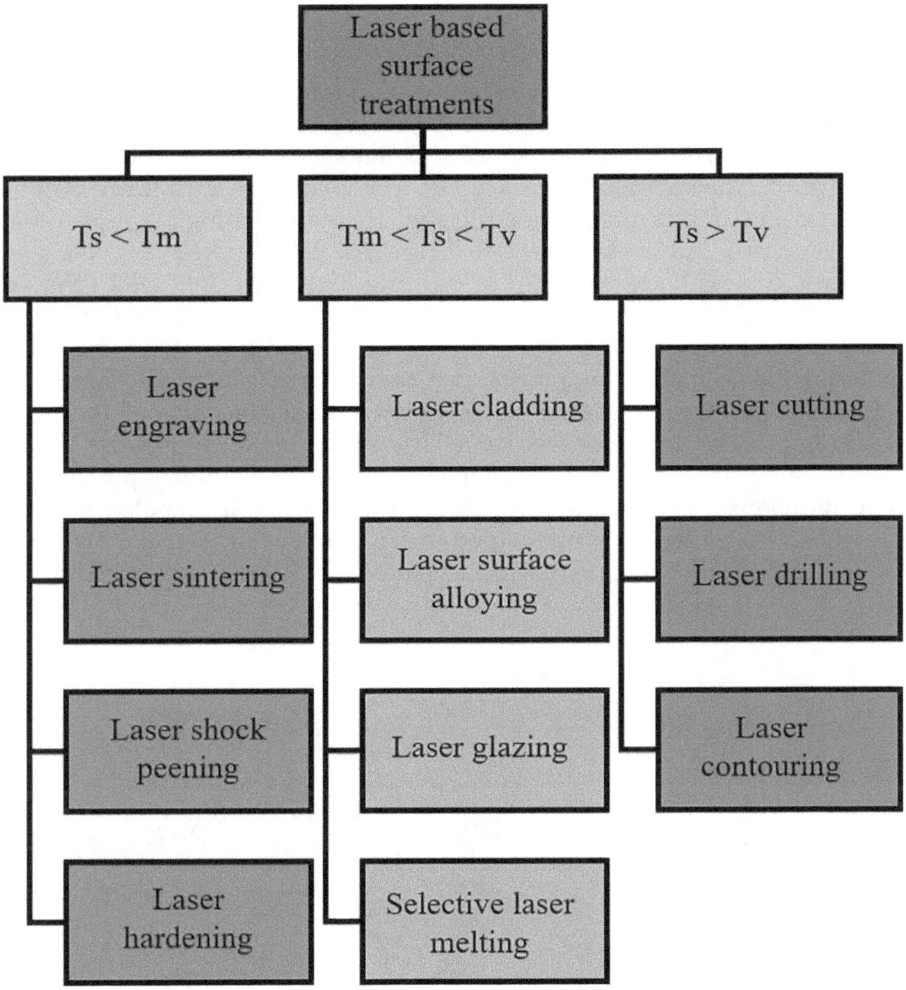

FIGURE 4.1
Classification of laser-based surface treatments where Ts is the substrate's surface temperature, Tm is the substrate's melting temperature and Tv is the substrate's vapourising temperature.

4.2 Laser Cladding

This section briefly discusses the working principle in the laser cladding process for creating clad layers in parent materials and explains the influence of process parameters to obtain highly efficient clad layers on the substrates.

4.2.1 Laser Cladding: Principle

As mentioned earlier, the laser cladding process alters the repaired or worn-out surfaces to a greater extent when compared with other surface treatment techniques. This process employs a high-energy laser beam, which irradiates the sample's surface and melts it in order to create a molten melt pool on the substrate's surface [31]. The coating materials are eventually deposited in this molten metallic pool, thereby forming a compact coating layer on the substrate's surface (Figure 4.2). It is reported that the laser cladding process holds good for fabricating highly efficient clad layers in terms of mechanical strength, anti-wear, anti-corrosive and fatigue resistance characteristics. Moreover, based on the condition of the feedstock material, the laser cladding process can be further classified into two major categories as shown in Figure 4.3. They are powder feedstock-based laser cladding process and wire feedstock-based laser cladding process [32,33]. As the name implies, in the powder feedstock laser cladding process, the coating material is deposited in powder form to obtain the clad layer on the substrate, whereas wire-shaped coating material is used, which is melted using the laser source, and the melted particles get deposited on the

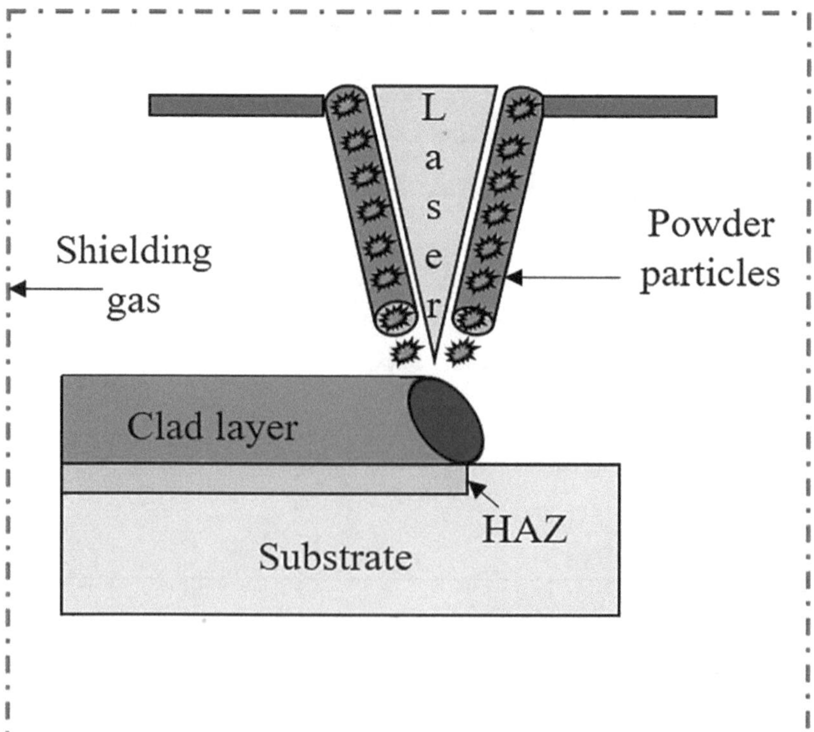

FIGURE 4.2
Laser cladding process, where HAZ represents the heat-affected zone.

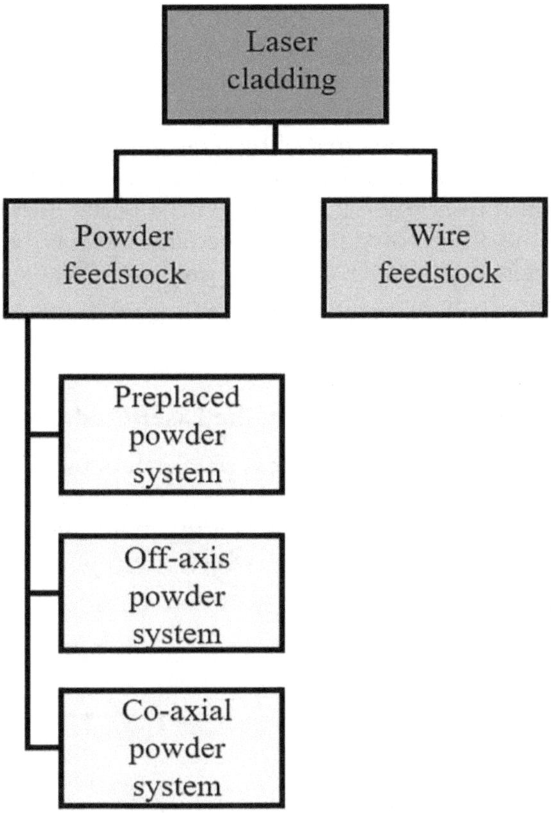

FIGURE 4.3
Classification of the laser cladding process based on feedstock.

substrate's surface to obtain a uniform clad layer on the substrate. Further, the powder feedstock-based cladding system is classified into three categories based on the orientation of the feedstock nozzle, which directs the powder particles on the molten metallic pool. The three classifications of the powder-based feedstock system are as follows: (i) preplaced powder-based laser cladding [34], (ii) off-axis powder feedstock laser cladding [35] and (iii) co-axial powder feedstock laser cladding system [36]. In the preplaced system, the coating material, which exists in powder form, is mixed with some liquid additives, such as acetone, in proper ratio in order to form a gel or paste-like substance. This paste-like substance (coating powder particles + acetone) is then applied on the substrate's surface and the laser is directed on it. During cladding, the heat from the laser source will melt the powder paste where the acetone is evaporated from the powder particles in the paste are eventually deposited on the substrate's surface. On the other hand, the powder particles will reach the molten melt pool in a separate path, which is termed the off-axis powder feed system. The co-axial powder feed system is mostly employed in many industrial sectors. This system consists of powder pathways, which are aligned in such a way that the powder particles are directed simultaneously onto the molten metallic pool, which is reported to be created by irradiating the substrate's surface with the laser source. The clad layer obtained via the co-axial system is reported to contain finer microstructure with a better surface finish when compared with the other powder-based feedstock and wire feedstock cladding systems. Hence, the laser cladding process serves as the dominant technique for repairing worn-out engineering components in almost all industrial sectors.

4.2.2 Process Parameters Influencing the Laser Cladding Process

To obtain a better clad layer, the process parameters used to create the clad layer need to be optimised. The optimisation of the process parameters is entirely dependent on the substrate and coating materials. However, the four major process parameters (Figure 4.4), which are reported to influence the laser cladding process, are as follows.

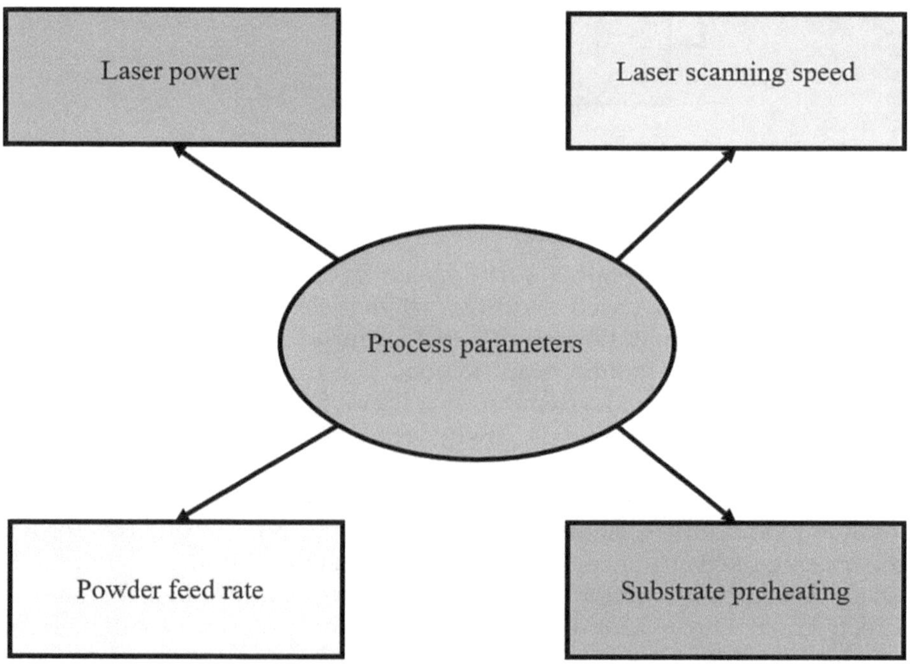

FIGURE 4.4
Laser cladding process parameters.

4.2.2.1 Laser Power

To obtain a fine cladding layer, the laser cladding process parameters need to be optimised. The optimisation process depends on the melt pool temperature of every substrate material used in the cladding process. Since the clad layer is formed by the localised melting of the substrate's surface, there is a need to make sure that sufficient energy transfers occur between the laser beam and the substrate surface. In general, for most of the metals and metallic alloy substrates, a higher level of laser power is required so that the high-energy laser beam provides sufficient irradiation on the substrate's surface, thereby creating an adequate metallic molten pool [37,38]. It is reported that the coating particles that are deposited in the molten melt pool, which is created by the high-energy laser beam, tend to be uniform with a better surface finish. Also, the surface defects, such as pores, voids and cracks, were said to be either absent or less when a high-energy laser beam is used. On the other hand, a low-energy laser beam will result in improper melting of the substrate's surface, which will lead to an uneven distribution of the coating powder particles on the substrate surface [39,40]. This uneven distribution will pave the way for the formation of surface defects such as pores, cracks, voids and other defects on the clad layer. Also, the metallurgical bonding strength that exists between the clad layer and the substrate's surface tends to be low at low laser power. Hence, after a prolonged time, the clad layer obtained by using the low-energy laser beam experiences a reduction in the bonding strength, and in due course, this reduction will lead to the detachment of the clad layer from the mating (substrate) surface. Thus, in most cases, sufficient laser power is required to obtain a better clad layer in terms of mechanical aspects.

4.2.2.2 Laser Scanning Speed

In the laser cladding process, there is a need for localised melting of the substrate's surface to fabricate the clad layer on the parent material. Though the laser cladding process takes place with sufficient laser power, to obtain a uniform and continuous clad layer, the scanning speed of the laser beam needs to be optimised. The laser scanning speed is defined as the rate at which the laser beam moves along the axial (cladding direction) direction so that a continuous clad layer is obtained after the cladding process. The laser scanning speed is usually represented in mm/second, where the laser beam is precisely moved to fabricate a continuous clad layer. The control over the laser scanning speed is decided by the material type, which is preferred as the substrate. In general, scanning speed decides the temperature gradient and its distribution over the surface of the substrate. Higher scanning speed will lead to subsequent heating and cooling cycles, thereby obtaining a higher solidification rate of the clad layer [41,42]. This higher solidification rate tends to the formation of finer grains in the clad layer, which will proliferate the mechanical characteristics of the clad layer. Also, it is reported that the precipitation rate of secondary and intermetallic precipitates is higher at the increased solidification rate, which is obtained at the higher laser scanning speed. On the other hand, lower laser scanning speed results in improper localised melting, which tends to disturb the clad layer. Also, the solidification rate is reported to be low, thereby causing discontinuity in the clad layer. These discontinuities will act as the nucleation sites for wear and fatigue phenomenon, thereby reducing the mechanical properties of the clad layer. Thus, to obtain a better uniform and continuous clad layer, the laser scanning speed should be high to achieve a higher solidification rate, which in turn influences the service life of the clad layer [43,44].

4.2.2.3 Powder Feeding Rate

The third most influential process parameter in the laser cladding process is the powder feeding rate. This process parameter is only for the powder-based laser cladding systems, namely off-axis and co-axial powder feedstock laser cladding systems, and not for the wire feedstock-based laser cladding system. For obtaining a better clad layer, other than the laser power and laser scanning speed, the feedstock rate needs to be optimised. The powder feeding rate is defined as the rate at which the powder (coating) particles are directed (deposited) on the molten metallic pool obtained on the substrate's surface. Many researchers reported that this powder feeding rate is more important since the majority of the mechanical properties exhibited by the clad layer are entirely dependent on the powder particles. In general, a higher powder feeding rate results in a better distribution of powder particles in the molten metallic pool [45,46]. This better distribution leads to fine-grain formation at the clad layer along with a higher segregation rate of intermetallic precipitates in the clad region. Also, this higher powder feeding rate commences with a higher solidification rate, thereby attaining better grain refinement at the clad layer. Further, some researchers reported that the grain size and shape obtained at the clad layer are dependent on the powder particle's size and shape distribution. The size of the powder particles used is directly proportional to the grain size obtained at the clad layer. The larger the particle size, the greater the grain size in the clad layer. On the other hand, a lower powder feeding rate results in improper distribution of the powder particles, resulting in the formation of voids on the clad layer [47,48]. Thus, an adequate powder feeding rate is required to obtain a clad layer with a better surface finish.

4.2.2.4 Preheating the Substrate Material

Many researchers reported that the mechanical characteristics offered by the pure metal and metallic alloys could be enhanced by performing heat treatments. Heat treatment is the process where the material is heated to a

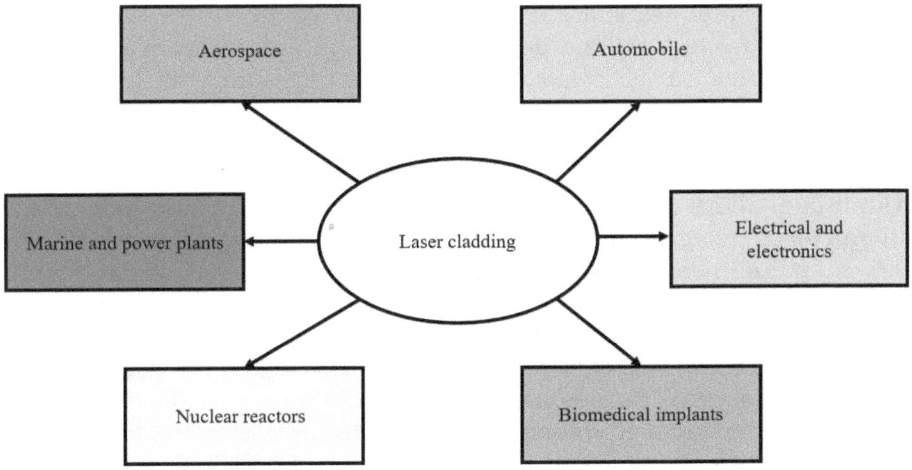

FIGURE 4.5
Major industries that employ thelaser cladding process.

particular temperature and cooled via various pathways such as annealing, quenching, furnace cooling, etc. This process promotes the diffusion and distribution of phases in the metallic alloys, thereby attaining better grain refinement than the non-heat-treated samples [49,50]. This better grain refinement proliferates the mechanical, thermal, electrical, electrochemical and other characteristics offered by pure metals and metallic alloys. In the laser cladding process, the substrate material plays a vital role in deciding the bonding strength, surface roughness, tensile strength and hardness of the clad layer. It is reported that via preheating the substrate, the bonding strength and other mechanical characteristics of the clad layer are greatly improved when compared with the clad layer obtained on the non-heat-treated substrate. Through preheating, the grain refinement at the clad layer is uniform, and the clad layer exhibits better microstructure with better distribution of intermetallic precipitates. Thus, these four process parameters are reported to influence the mechanical characteristics offered by the clad layer [51,52]. The following sections will elaborate on the laser cladding process adopted in major industrial sectors (Figure 4.5).

4.3 Repairing of Engineering Components in the Aerospace Industry

Aluminium is a metal whose mechanical and physical properties can be changed by combining it with other elements such as silicon, magnesium, copper, zinc and manganese. To obtain the required qualities for a particular application, these metals are combined in precise ratios. For example, an alloy that is strong, light and perfect for usage in the automotive and aerospace industries can be created by mixing magnesium with aluminium. Aluminium alloys are widely used in a wide range of sectors due to their low density, corrosion resistance and thermal conductivity. One of the major industrial beneficiaries of aluminium alloys is the aerospace industry [53,54]. Aluminium alloys contribute nearly 65% to the weight composition of any aircraft since the high strength-to-weight ratio is the major key aspect. The service environment of this alloy renders its mechanical characteristics, such as hardness, tensile strength, toughness, wear resistance, fatigue behaviours and corrosion resistance, thereby minimising the service life span of the aluminium alloy components. Hence, the researchers employ the laser cladding process and fabricate various clad layers on the aluminium alloy substrates. Ni et al. [55] investigated the microstructure and microhardness behaviour of the high-entropy alloy clad layer obtained on the 5083 aluminium substrate fabricated by the laser cladding process. By keeping the laser power at a constant value (1.1 kW) and by varying the laser scanning speed within the range of 270–630 mm/min, the clad layers were obtained on the aluminium substrate. Their outcomes are as follows. The microstructure tends to be finer by increasing the laser scanning speed. However, at extreme scanning speed (630 mm/min), the authors reported the formation of fragile patches over the clad layer, which might render the mechanical properties offered by the clad layer. Moreover, the hardness property is also influenced by increasing the laser scanning speed. The hardness continues to increase when increasing the laser scanning speed, but at extreme speed (630 mm/min), due to the fragile patches, the hardness behaviour tends to diminish thereby decreasing the wear resistance property [56]. Song et al. [57] developed a numerical model for improving the anti-fatigue characteristics of the aircraft skin by laser cladding it with aluminium alloy. By varying the cladding layer thickness and length, the optimal thickness was characterised, and the major outcomes are as follows. The fatigue life of the clad layer is directly proportional to the clad layer's thickness and length. The higher the thickness and length, the greater the fatigue resistance offered by the clad layer. The fatigue life of samples and stress intensity variables are affected differently by cladding coating length, patterns and angles, with the longest cladding coating having the most impact. In order to balance the tension load, a longer cladding coating length can increase the length of the laser action time on samples. This will result in a larger residual compressive stress. As a result, the fatigue life is increased and the stress intensity components are decreased. Liang et al. [58] examined the strengthening mechanism of the Ni60 clad layer fabricated on the 6063Al substrate via the laser cladding process. Also, they compared the microstructure and strengthening mechanism of the Ni 60 clad layer and Ni60 alloy+oxides of rare elements such as lanthanum, cerium and yttrium. The thick columnar crystal structure of the Ni60 cladding layer's upper strengthening layer has some pores that are around 5 μm in size and severe segregation. With finely dispersed dendrites and no visible pore flaws, the upper strengthening layer of the cladding layer enhanced with rare earth has a refined structure, and structural segregation has been much enhanced. The base structure of Al; the solid solution of Cr, Fe and C; and the sparse, long strip-like dendritic structure of the Al-rich phase Al_3Ni, which builds on the base structure, make up all of the interface transition layers of the cladding layer. The rare earth element was also dissolved by the rare earth cladding layer. Li et al. [59] examined the microstructure and mechanical properties of Ti clad layer obtained on AA7075 alloy for improving the tribological behaviours offered by the AA7075 alloy for aerospace applications. The mechanical properties, such as hardness and corrosion resistance, were compared between the clad layer and the AA7075 substrate. Their results were concluded as follows. The clad layer was rich in the α-Ti, α-Al, TiBCN, TiN and Al_3Ti phases, which were distributed in the clad layer matrix. Further, from the hardness test, it is revealed that the Ti alloy clad layer exhibited excellent hardness, which is nearly five times the hardness offered by the AA 7075 substrate. The sample's cross-sectional morphology is separated into four sections: the coating zone, transition zone, substrate and heat-affected zone. The coating zone's microstructure is primarily composed of equiaxed grains and white lath-like crystals, while the transition zone's microstructure, at a TiBCN content of 15%, is primarily composed of elongated dendrite crystals and white small particles. When the TiBCN concentration in the coating zone increases, the average hardness values steadily rise. At 15 wt.% the maximum microhardness is experienced within 0.2 mm of the surface. The corrosion resistance offered by the clad layer is higher than that of the AA 7075 substrate, which is decided by the Ecorr and Icorr values. Also, the Ti alloy clad layer exhibits better wear resistance characteristics due to the Ti-rich dissolved phases in the clad layer. Further, the mechanical and tribological behaviours offered by the aluminium alloys used to make aerospace structures could be improved by the clad layer made from either metal matrix composites or combining metal matrix composites with other metallic alloy powders.

4.4 Repairing of Engineering Components in the Electrical and Electronics Industries

Copper alloys belong to the category of rare earth metal alloys that are implemented in a wide range of applications. This alloy contains a significant amount of copper alloying with other metallic elements such as aluminium, silicon, chromium, zirconium, tungsten and so on. The combined effect of alloying elements with copper paved the way for copper alloys to exhibit several significant mechanical, physical, electrical and thermal characteristics. This alloy holds good in applications where electrical conductance along with thermal resistance are the key factors. This alloy serves as a good conductor of electricity than any other metallic alloys and dissipates the heat to a greater extent, which is produced during the charge or ion transfers without affecting the working medium. Other than the electrical conductivity, these alloys are well known for their corrosion resistance, hardness and tensile characteristics [60,61]. These characteristics made this alloy to get implemented while manufacturing the components of power transmission systems, electrical rail cables (wires), heat exchangers, electrical circuits, power junction boxes and so on. Most copper alloys are reported to work in severe harsh environments containing dissolved salts and oxygen, high temperatures and high pressure. After a long-term exposure to these environments, the copper alloys are said to render their mechanical and electrical characteristics, thereby causing failure in electrical components and sometimes leading to property loss. Hence, the researchers adapted the laser cladding (surface enhancement technique) to repair the worn-out copper alloy components to extend their service life. Qi et al. [62] investigated the influence of various laser wavelengths on the microstructure and mechanical properties offered by the pure copper clad layer obtained on the 316L stainless steel substrate. They have adapted two laser beams of 1,080 and 455 nm wavelengths, whereas the beam size and the laser power tend to be constant at 1.2 mm and 400 W, respectively. By varying the laser scanning speed in the range of 10–30 mm/s with different laser wavelengths, the pure copper clad layer was obtained on the stainless steel 316L substrate. Their major findings are as follows. At lower laser scanning speed (10 mm/s), the pure copper clad layer obtained at two wavelengths has an uneven microstructure. These clad layers are reported to contain many pores and voids, which might have formed due to the lower laser scanning speed. The uniformity in the microstructure of the clad layer increases with an increase in the laser scanning speed. At 50 mm/s, the clad layer obtained at the two various wavelengths is said to be uniform and contains fine-grains in its microstructure. The precipitation of copper-rich and iron-rich phases was distributed through the clad layer. Also, considering the hardness aspects, the pure copper clad layer obtained by the 1,080 nm has higher hardness than the pure copper clad layer obtained by the 455 nm wavelength. Prabu Balu et al. [63] analysed the microstructure and hardness distributions of the nickel-based clad layer obtained on the copper substrate for electrical applications. By varying the laser scanning speed and the powder feed rate within the range of 2.5–25 mm/s and 19–55 g/min, the nickel alloy clad layer was obtained on the copper substrate, which was subjected to preheat treatment at 300°C. Their results are concluded as follows. The microstructure of the clad layer contains Ni-rich columnar dendrites followed by the segregation of the secondary precipitates, which are distributed in the grain boundaries of the columnar dendrites. Further, the Ni-based clad layer obtained on the preheated copper substrate has higher hardness than the clad layer obtained on the non-heat-treated copper substrates. By using the pulsed Nd:YAG laser, Yan et al. [64] examined the wear behaviours and phases present in the cobalt-based clad layer obtained on the copper substrate via the multilayer laser cladding process. Their research outcomes are as follows. The cobalt clad layer contains chromium carbides, nickel-tungsten and chromium-nickel-tungsten intermetallic precipitates which are distributed in the α-Co matrix. When compared to the copper substrate, the cobalt-based clad layer exhibited excellent hardness, which is nearly five times the hardness offered by the copper substrate. This increase in hardness at the clad layer is due to the precipitation of the Ni-, Cr-, W-rich intermetallic precipitates, which were absent in the copper substrate. Moreover, the Co-based clad layer has significant wear resistance, which is almost 15 times the wear resistance offered by the copper substrate. Ng et al. [65] interrogated the molybdenum clad layer obtained on the copper substrate to improve the tribological characteristics such as hardness and wear resistance of the copper substrate for electrical and electronic applications. With the aid of a microhardness tester and pin-on-disc setup, the hardness and wear resistance offered by the copper substrate and molybdenum clad layer were compared, and the results were concluded as follows. By introducing a Ni intermediary layer, cladding challenges caused by the thermal and metallurgical mismatch between Mo and Cu can be resolved. The powder preplacement approach can be used to improve laser energy absorption and overcome the challenges associated with laser treatment of Cu due to its high reflectivity. Strong interfacial bonding was achieved in the fabrication of a superior duplex coating on Cu substrate, consisting of an intermediate layer of Ni–Cu solution and a top Mo layer, through an appropriate selection of laser settings. The Mo-clad surface exhibited a hardness approximately seven times higher than the Cu substrate. Additionally, wear resistance regularly showed a considerable improvement according to pin-on-disc wear testing. Thus, it can be concluded that by performing laser cladding on copper and copper-based alloy substrates, the mechanical, electrical, electromechanical, thermal and other physical characteristics of the copper substrate were improved, thereby attaining a higher service life span in the copper components.

4.5 Repairing of Engineering Components in the Automobile Industry

One of the most prominent benefiters of the laser cladding surface enhancement technique is the automobile industry. In most automobiles, iron-based alloys (ferrous alloys), such as carbon steel, mild steel and stainless steel, were used for fabricating the automobile structures. These alloys are reported to exhibit several dominant mechanical, thermal, chemical and physical characteristics, including tensile strength, toughness, malleability, rigidity, hardness, wear resistance, fatigue resistance, anti-corrosive behaviours, ductility, heat resistance and so on. These characteristics have paved the way for iron-based alloys such as mild steel, carbon steel, maraging steel and stainless steel to get implemented in fabricating most of the components of automobiles. The prevailing service environments of these alloys greatly influence the mechanical and tribological behaviours of these alloys, thereby rendering their service life span [66,67]. The prevailing high temperature and high pressure will degrade the tensile and fatigue behaviours of the steels, whereas the dissolved oxygen and salts concentration will minimise the wear- and corrosion-resistant characteristics due to the formation of nucleation zones on the surface of mild, carbon and stainless steels, respectively. In due course, these nucleation sites will develop into pores, voids and cracks, which will propagate into the inner surface thereby causing catastrophic failures and accidents. Thus, there is a need to improve the mechanical and tribological characteristics of the steel and stainless steel components to enhance the service life of the engineering components in the automobile industry. Many researchers work on the laser cladding technique for improving the mechanical and tribological characteristics of automobile components, which work in highly aggressive environments. Zhang et al. [68] performed the laser cladding process on mild steel A36 substrate by coating the substrate with a combination of stainless steel 420 and molybdenum powder particles. The authors established a numerical model to analyse the thermal gradients and their distribution in the molten metallic pool formed on the mild steel substrate. With the constant laser power of 4,100 W and powder feeding rate of 0.43 g/s, the clad layer was obtained by varying the laser scanning speed from 3 to 7 mm/s. Their observations are as follows. Under the chosen conditions, the laser power attenuation model that was employed indicated a 1% decrease in laser power. The laser beam warmed the powder before it got to the molten pool. The maximum temperature of the powder was 172.1°C, although the temperature at the molten pool's edge was only estimated to be 62.3°C. Clad geometrical properties were achieved by constantly stacking thin layers of melting powder in the molten pool. There was good agreement between the expected and experimental clad geometries. From the bottom to the top surface of the clad, the microstructure morphology of the cladded layer altered from cellular to columnar and finer dendrites. The percentage of Mo in the clad layer declined as scanning speed increased, but there were more locations with higher concentrations of Mo. At room temperature, the clad layer's microstructures were martensite and ferrite. Ferrite was dispersed at the borders of the grains and subgrains. As the scanning speed increased, the fraction of ferrite decreased. Mahmoud et al. [69] investigated the microstructure and hardness distribution of the stainless steel 316L clad layers obtained on the mild steel substrate at various laser powers ranging from 2,400 to 2,800 W and the laser scanning speed was kept constant at 6 mm/s. At greater laser power, the stainless steel-coated layers showed no macro-defects and were firmly adhered to the substrate. Crack-related flaws started to show up at 2,000 W of power. With all of the processed laser characteristics, the dilution process at the interface was severely constrained. The highly refined mixed columnar and cellular dendritic austenite structure that was stretched perpendicular to the substrate surface made up the majority of the stainless steel clad layer. There is a direct correlation between the laser power and the depth of the carbon steel impacted zone. Large ferrite grains encased in feathery bainite made up the majority of the carbon steel fusion zone. Throughout the clad zone, the hardness values were nearly uniform and increased as the laser power was reduced. Zhang et al. [70] aimed to improve the anti-erosive, anti-corrosive and wear characteristics of the stainless steel substrate by laser cladding it with two nickel-based alloy powders, namely Ni-chromium carbide and Ni-tungsten carbide, respectively. The segregated phases in both clad layers were analysed in terms of SEM and XRD analysis, whereas the hardness offered by both clad layers was examined by the microhardness tester, and their results were concluded as follows. Metal carbides were formed when the chromium carbide (Cr_3C_2) particles in the laser-melted pool completely dissolved. The biphase structure of the metal carbide makes up the microstructure of the laser-clad Ni+chromium carbide composite coating, whereas the Ni+tungsten carbide composite coating features a high melting point where the tungsten carbide (WC) particles have partially dissolved in the molten metallic pool of the stainless steel substrate. On examining the hardness behaviour, the stainless steel substrate has the lowest hardness of the two clad layers. Among the two clad layers, the Ni-tungsten carbide alloy clad layer exhibits higher hardness than the Ni-chromium carbide clad layer. The primary cause of the variation in microhardness between the two coatings is their distinct solidified structures. On the other hand, the erosive corrosive wear phenomenon was very high in the stainless steel substrate than in the other two clad layers. Among the two clad layers, the Ni-chromium carbide clad layer experiences a lower erosive corrosive wear phenomenon than the Ni-tungsten carbide layers. Thus, via the cladding process, the mechanical and tribological characteristics of the iron-based substrates were reported to be improved to sustain the service environmental conditions.

4.6 Repairing of Engineering Components in Nuclear Reactors

These days, superalloys based on nickel have become more significant due to their wide range of uses in the chemical, nuclear reactor, marine and aerospace industries. Their extensive use can be attributed to many qualities, including hardness, remarkable resistance to thermal shocks and good mechanical and chemical capabilities at high temperatures. These are high-temperature, corrosion-resistant alloys that are normally utilised at temperatures above

500°C. Typically, they have notable concentrations of up to ten alloying elements, which can include heavy refractory metals like tantalum, tungsten or rhenium, as well as light elements like carbon or boron. Even at temperatures that are almost at their melting points, these alloys show outstanding resistance to creep, sulfidation and oxidation [71–73]. Because nickel-based superalloys have excellent characteristics at high temperatures, they are frequently utilised to fabricate core components for nuclear reactors. Since the core components are employed in hostile environments, prolonged exposure of these components in these environments will lead to failures and accidents. Hence, to avoid such failures and accidents, researchers used the laser cladding technique for altering the surfaces of the nickel alloy-based reactor core components. Feng et al. [74] investigated the microstructure, hardness and wear resistance characteristics of the nickel-based clad layer (Inconel 625) obtained on the martensitic steel for nuclear power plant applications. Through optimal process parameters, the Inconel 625 clad layer was obtained and compared with the arc-welded Inconel 625 layer. Their results showed that the microstructure of the Inconel 625 clad consists of columnar dendrites along with intermetallic Ni-Fe precipitates as the grain boundary. The hardness profile of the Inconel 625 coating that is arc-welded shows a modest decrease in hardness in the coating zone and a significant decrease in hardness in the iron dilution zone. The loss in hardness in arc-welded Inconel 625 coating is observed to be associated with coarser microstructure and worsened dilution of iron, as determined by SEM and EDS measurements. At high temperatures, the Inconel 625 coating's hardness reduces for both laser-cladded and arc-welded materials. In contrast, the Inconel 625 coating that has been laser clad has a harder texture than one that has been arc welded. At ambient temperature, the Inconel 625 coatings show abrasive wear with a high friction coefficient of around 0.6, while adhesive wear with a lower friction coefficient of 0.32–0.36 predominates at higher temperatures. Due to various wear mechanisms and the enhanced lubricating action of the debris or particles worn off of the mating surface, the wear rate lowers as the temperature rises. In contrast, the laser-cladded coating has a reduced wear rate due to its increased hardness and lower iron dilution. Tabernero et al. [75] investigated the mechanical characteristics offered by the Inconel 718 clad layers obtained by various patterns, namely zigzag, spiral, transversal and longitudinal directions, respectively. The clad layer obtained via longitudinal direction showed higher yield and ultimate tensile strength than the other patterns. This is due to the fine-grains formed on the longitudinal side than on the other sides. Additionally, because the deposited material's microstructure is homogenised, precipitation hardening enhances the mechanical qualities. Higher tensile strength and ductility increments of up to 100% were seen in the tests conducted on the treated specimens. On the other hand, it can also be stated that the specimens' mechanical qualities have decreased, and in the best of situations, they are only roughly 75% as strong as wrought Inconel 718. Further, the mechanical characteristics of the nickel-based clad layer could be improved by doping the nickel-based powder particles with metal carbides of chromium, tungsten and titanium. Thus, the laser cladding process is used to repair the worn-out surfaces of the engineering components used in nuclear reactors.

4.7 Repairing of Engineering Components in the Biomedical Industry

The biomedical industry has evolved along with the usage of non-invasive metallic alloys for fabricating biomedical implants. In real time, the biomedical implants are exposed to highly aggressive environmental conditions, such as variations in the pH level of blood, acidity nature of the body fluids, alkaline nature of the body pH levels and so on. These prevailing service environmental conditions of biomedical implants lead to the initiation of failures that are reported to be very dangerous to mankind. Over prolonged usage, the biomedical implants undergo a corrosion process, and as a result, the corrosion by-products are released into the blood serum, which might cause serious issues in the human body. In most cases, the biomedical implants are fabricated from pure titanium or titanium alloys [76–78]. Titanium alloys have made a tremendous impact in the biomedical industry due to their extreme level of biocompatibility and corrosion-resistant characteristics. Owing to its biocompatibility, this alloy has excellent mechanical and tribological features such as tensile strength, toughness, wear resistance, hardness and fatigue resistance, which made them to get implemented while fabricating the biomedical implants. It is reported that prolonged usage in pH variation environments, such as the human body, could initiate the corrosion process, thereby reducing the service life of the biomedical implants. In order to improve the mechanical characteristics and biocompatibility of the biomedical implants, researchers employed the laser cladding process for fabricating the clad layer on the biomedical implants. Nabhani et al. [79] investigated the microstructure and corrosive phenomenon of the Ti-6Al-4V clad layer obtained on the Ti-6Al-4V substrate via the laser cladding process. They adopted various corrosive media, such as solutions containing NaCl, HCl and H_2SO_4, respectively. With the aid of SEM micrographs, the microstructure has been analysed, and the corrosion performance was evaluated in terms of polarisation curves plotted in different electrolytic media as mentioned above. Their results are concluded as follows. The microstructure of the Ti-6Al-4V clad layer contains needle-like Ti phases, which are distributed throughout its microstructure. Moreover, by performing the XRD analysis, these needle-like phases are rich in α and α' Ti phases, which contribute to the biocompatibility behaviours offered by the clad layer. When comparing the corrosion resistance offered by the Ti-6Al-4V substrate and Ti-6Al-4V clad layers, the clad layers treated with all three electrolytic solutions offered higher resistance against corrosion when compared with the substrate samples. Among the different electrolytic solutions, the clad layer offered significant resistance against corrosion in both NaCl and H_2SO_4 media when compared with the HCl medium. Diao et al. [80] examined the corrosion resistance, hardness and biocompatibleness of the Ti alloy substrate by cladding it via composite coating performed using pure titanium, titanium carbide and titanium boride powders. Their results are summarised as follows. The XRD reveals the composition of the TiC and TiB_2 phases in the cladding layer following the laser cladding treatment. The intermetallic phases and grain refinement cause the clad layer's microhardness to increase dramatically, reaching a level three to four times greater than that of the substrate. It is evident that the laser cladding specimens' corrosion resistance is improving over that of the substrate. This could be explained by the quick solidification followed by grain refining and the segregation of the secondary phases. The decreasing scanning rate improves the corrosion property and further increases the microhardness.

4.8 Repairing of Engineering Components in Marine Industry and Power Plants

The components of the marine and power plants are made from iron alloys, nickel alloys, magnesium alloys and aluminium alloys. The variation in temperature and pressure (power plants) and the variation in dissolved oxygen and dissolved salt concentration (marine equipment) will initiate the failure process in those components. The high temperature and pressure will lead to stress-induced cracks, whereas the dissolved salts and oxygen concentration will start the corrosion in these components, thereby rendering the mechanical, thermal and physical behaviours offered by the components. It is reported that the service life of the components used in marine and power plant sectors is improved by the laser cladding technique. Cao et al. [81] optimised the process parameters for coating magnesium alloy for power plant applications. They employed the Taguchi method and optimised the laser power, feeding rate and laser cladding speed for performing laser cladding on the magnesium alloy. Their results were concluded as follows. From their observations, it is found that laser power plays a crucial role in obtaining a better clad layer with fine-grain structures. Lowering the laser power leads to an uneven distribution of the clad particles in the clad layer, thereby decreasing the reliability of the clad layer. Under high feed rate and high laser power, the clad layer exhibited a lower dilution effect with a higher solidification rate. Yang et al. [82] investigated the Ai-Si clad layer obtained on the magnesium alloy for enhancing the anti-wear characteristics of the magnesium substrate.

The clad layer was found to be free from pores and voids due to the optimal process parameters followed during fabrication. The hardness offered by the Al-Si clad layer was quite higher than the magnesium alloy substrate, and this same trend is observed while comparing the wear resistance characteristics offered by the clad layer and substrate. Jun et al. [83] performed laser cladding on the magnesium alloy to improve the hardness offered by the substrate via metal matrix composite coating. They concluded that the laser cladding process holds good for improving the hardness and wear resistance behaviours offered by the substrate. Thus, the laser cladding process aids in improving the mechanical, thermal, physical and electrochemical characteristics offered by the marine and power plant components.

4.9 Conclusion

Surface treatments play a major role in uplifting the mechanical characteristics offered by the parent or base materials. These techniques efficiently repair the worn-out surfaces, thereby improving the surface characteristics and offering greater resistance towards corrosion, wear and fatigue failures. Among the surface treatment techniques, the laser cladding process is considered a predominant technique, which is implemented in most of the industrial sectors. This technique employs a high-energy laser beam that irradiates the test sample's surface, thereby creating a molten melt pool on it. The coating powders were dropped on this melt pool, and after solidification, the clad layer with a better surface finish was obtained. This chapter discussed the laser cladding principle, followed by the process parameters that influence the clad layers and how the laser cladding principle is adapted in nuclear, aerospace, automobile, electrical, marine, power plants and biomedical sectors with relevant literature.

Acknowledgement

The authors would like to acknowledge the editors of this book for providing an opportunity to write this chapter.

Conflicts of Interest

The authors declare that there are no conflicts of interest.

References

1. Xu M, David JM, Kim SH. The fourth industrial revolution: Opportunities and challenges. *International Journal of Financial Research*. 2018;9(2):90–5.
2. Berg M, Hudson P. Rehabilitating the industrial revolution 1. *The Economic History Review*. 1992;45(1):24–50.
3. Karuppasamy SS, Jeyaprakash N, Yang CH. Microstructure, nanoindentation and corrosion behavior of colmonoy-5 deposition on SS410 substrate using laser cladding process. *Arabian Journal for Science and Engineering*. 2022;47:8751–67.
4. Niinomi M, Nakai M, Hieda J. Development of new metallic alloys for biomedical applications. *Acta Biomaterialia*. 2012;8(11):3888–903.
5. Urban K. *Complex Metallic Alloys: Fundamentals and Applications*. John Wiley & Sons; 2010.
6. Basu P, Kefa C, Jestin L. *Boilers and Burners: Design and Theory*. Springer Science & Business Media; 2012.
7. Barma MC, Saidur R, Rahman SM, Allouhi A, Akash BA, Sait SM. A review on boilers energy use, energy savings, and emissions reductions. *Renewable and Sustainable Energy Reviews*. 2017;79:970–83.
8. Arjunwadkar A, Basu P, Acharya B. A review of some operation and maintenance issues of CFBC boilers. *Applied Thermal Engineering*. 2016;102:672–94.
9. Saidur R, Ahamed JU, Masjuki HH. Energy, exergy and economic analysis of industrial boilers. *Energy Policy*. 2010;38(5):2188–97.
10. Natarajan J, Yang CH, Karuppasamy SS. Investigation on microstructure, nanohardness and corrosion response of laser cladded colmonoy-6 particles on 316l steel substrate. *Materials*. 2021;14(20):6183.
11. Bhandari J, Khan F, Abbassi R, Garaniya V, Ojeda R. Modelling of pitting corrosion in marine and offshore steel structures–A technical review. *Journal of Loss Prevention in the Process Industries*. 2015;37:39–62.
12. Natesan M, Selvaraj S, Manickam T, Venkatachari G. Corrosion behavior of metals and alloys in marine-industrial environment. *Science and Technology of Advanced Materials*. 2008;9:045002.
13. Jeyaprakash N, Yang CH, Karuppasamy SS, Dhineshkumar SR. Evaluation of microstructure, nanoindentation and corrosion behavior of laser cladded Stellite-6 alloy on Inconel-625 substrate. *Materials Today Communications*. 2022;31:103370.
14. Brondel D, Edwards R, Hayman A, Hill D, Mehta S, Semerad T. Corrosion in the oil industry. *Journal of Petroleum Technology*. 1987;39:756–62.
15. Chandler KA. *Marine and Offshore Corrosion: Marine Engineering Series*. Elsevier; 2014.
16. Jeyaprakash N, Yang CH, Karuppasamy SS, Rajendran DK. Correlation of microstructural with corrosion behaviour of Ti-6Al-4V specimens developed through selective laser melting technique. *Proceedings of the Institution of Mechanical Engineers, Part E: Journal of Process Mechanical Engineering*. 2022;236(5):2240–51.
17. Molitor P, Barron V, Young T. Surface treatment of titanium for adhesive bonding to polymer composites: A review. *International Journal of Adhesion and Adhesives*. 2001;21(2):129–36.
18. Pan X, Shi Z, Shi C, Ling TC, Li N. A review on surface treatment for concrete–Part 2: Performance. *Construction and Building Materials*. 2017;133:81–90.

19. Jeyaprakash N, Yang CH, Susila P, Karuppasamy SS. Laser cladding of NiCrMoFeNbTa particles on Inconel 625 alloy: Microstructure and corrosion resistance. *Transactions of the Indian Institute of Metals*. 2023;76(2):599–612.
20. Le Guéhennec L, Soueidan A, Layrolle P, Amouriq Y. Surface treatments of titanium dental implants for rapid osseointegration. *Dental Materials*. 2007;23(7):844–54.
21. Rossnagel SM. Thin film deposition with physical vapor deposition and related technologies. *Journal of Vacuum Science & Technology A: Vacuum, Surfaces, and Films*. 2003;21(5):S74–87.
22. Carlsson JO, Martin PM. Chemical vapor deposition. In *Handbook of Deposition Technologies for Films and Coatings*, 2010 (pp. 314–363). William Andrew Publishing.
23. Zhang YI, Zhang L, Zhou C. Review of chemical vapor deposition of graphene and related applications. *Accounts of Chemical Research*. 2013;46(10):2329–39.
24. Herman H, Sampath S, McCune R. Thermal spray: Current status and future trends. *MRS Bulletin*. 2000;25(7):17–25.
25. Pawlowski L. *The Science and Engineering of Thermal Spray Coatings*. John Wiley & Sons; 2008.
26. Draper CW, Mazzoldi P, editors. *Laser Surface Treatment of Metals*. Springer Science & Business Media; 2012.
27. Lawrence JR, Waugh D, editors. *Laser Surface Engineering: Processes and Applications*. Elsevier; 2014.
28. Jeyaprakash N, Yang CH, Karuppasamy SS, Duraiselvam M. Stellite 6 cladding on AISI Type 316L stainless steel: microstructure, nanohardness and corrosion resistance. *Transactions of the Indian Institute of Metals*. 2023;76(2):491–503.
29. Toyserkani E, Khajepour A, Corbin SF. *Laser Cladding*. CRC Press; 2004.
30. Jeyaprakash N, Yang CH, Karuppasamy SS. Laser cladding of Colmonoy 6 particles on Inconel 625 substrate: Microstructure and corrosion resistance. *Surface Review and Letters*. 2022;29(08):2250102.
31. Shepeleva L, Medres B, Kaplan WD, Bamberger M, Weisheit A. Laser cladding of turbine blades. *Surface and Coatings Technology*. 2000;125(1–3):45–8.
32. Reddy L, Preston SP, Shipway PH, Davis C, Hussain T. Process parameter optimisation of laser clad iron based alloy: Predictive models of deposition efficiency, porosity and dilution. *Surface and Coatings Technology*. 2018;349:198–207.
33. Heigel JC, Gouge MF, Michaleris P, Palmer TA. Selection of powder or wire feedstock material for the laser cladding of Inconel® 625. *Journal of Materials Processing Technology*. 2016;231:357–65.
34. Zhang H, Lian G, Zhang Y, Pan Y, Cao Q, Yang J, Ke D. The influence of powder size on the microstructure and properties of Mo2FeB2 coating fabricated via laser cladding with pre-placed powder. *The International Journal of Advanced Manufacturing Technology*. 2022;120(9–10):6041–52.
35. Kumar SP, Elangovan S, Mohanraj R, Srihari B. Critical review of off-axial nozzle and coaxial nozzle for powder metal deposition. *Materials Today: Proceedings*. 2021;46:8066–79.
36. Jeyaprakash N, Karuppasamy SS, Yang CH. 1 Application of wear-resistant laser claddings. In Hitesh Vasudev, Chander Prakash (Eds), *Handbook of Laser-Based Sustainable Surface Modification and Manufacturing Techniques*, 2023 (pp. 1–26). CRC Press.
37. Barnes S, Timms N, Bryden B, Pashby I. High power diode laser cladding. *Journal of Materials Processing Technology*. 2003;138(1–3):411–6.

38. Pekkarinen J, Kujanpää V, Salminen A. Laser cladding with scanning optics: Effect of power adjustment. *Journal of Laser Applications*. 2012;24(3):032003.
39. Karuppasamy SS, Jeyaprakash N, Yang CH. 27 Application of corrosion-resistant laser claddings. *Handbook of Laser-Based Sustainable Surface Modification and Manufacturing Techniques*. 2023;51(51):27.
40. Nie MH, Zhang S, Wang ZY, Zhang CH, Chen HT, Chen J. Effect of laser power on microstructure and interfacial bonding strength of laser cladding 17-4PH stainless steel coatings. *Materials Chemistry and Physics*. 2022;275:125236.
41. Jiao X, Wang J, Wang C, Gong Z, Pang X, Xiong SM. Effect of laser scanning speed on microstructure and wear properties of T15M cladding coating fabricated by laser cladding technology. *Optics and Lasers in Engineering*. 2018;110:163–71.
42. Qiao Y, Huang J, Huang D, Chen J, Liu W, Wang Z, Zhibin Z. Effects of laser scanning speed on microstructure, microhardness, and corrosion behavior of laser cladding Ni45 coatings. *Journal of Chemistry*. 2020;2020:1–11.
43. Karuppasamy SS, Jeyaprakash N, Yang CH. The sensors in future vehicles. In Jeyaprakash Natarajan, Mahendra Babu Kantipudi, Che-Hua Yang, Yaojung Shiao (Eds), *The Future of Road Transportation*, 2023 (pp. 261–283). CRC Press.
44. Wang H, Cheng Y, Wan Y, Jeyaprakash N, Wang Y, Ma K, Yang J. Influence of scanning speed on microstructure and corrosion resistance of Fe-based amorphous coatings by high-speed laser cladding. *Surface and Coatings Technology*. 2024;479:130449.
45. Lee HK. Effects of the cladding parameters on the deposition efficiency in pulsed Nd: YAG laser cladding. *Journal of Materials Processing Technology*. 2008;202(1–3):321–7.
46. Karuppasamy SS, Jeyaprakash N, Yang CH. Alternative propulsion systems. In Jeyaprakash Natarajan, Mahendra Babu Kantipudi, Che-Hua Yang, Yaojung Shiao (Eds), *The Future of Road Transportation*, 2023 (pp. 47–68). CRC Press.
47. Aghili SE, Shamanian M. Investigation of powder fed laser cladding of NiCr-chromium carbides single-tracks on titanium aluminide substrate. *Optics & Laser Technology*. 2019;119:105652.
48. Huang YL, Liu J, Ma NH, Li JG. Three-dimensional analytical model on laser-powder interaction during laser cladding. *Journal of Laser Applications*. 2006;18(1):42–6.
49. Ding C, Cui X, Jiao J, Zhu P. Effects of substrate preheating temperatures on the microstructure, properties, and residual stress of 12CrNi2 prepared by laser cladding deposition technique. *Materials*. 2018;11(12):2401.
50. Murugesan P, Satheeshkumar V, Jeyaprakash N, Yang CH, Karuppasamy SS. Effect of α-Al and Si precipitates on microstructural evaluation and corrosion behavior of laser powder bed fusion printed AlSi10Mg plates in seawater environment. *Metals and Materials International*. 2023;29:2515–2532.
51. Jin Z, Kong X, Ma L. Investigating the impact of substrate preheating on the thermal flow and microstructure of laser cladding of nickel-based superalloy. *Materials*. 2024;17(2):399.
52. Liu H, Du X, Guo H, Liu J, Chen P, Yang H, Hao J. Finite element analysis of effects of dynamic preheating on thermal behavior of multi-track and multi-layer laser cladding. *Optik*. 2021;228:166194.
53. Karuppasamy SS, Yang CH. Adapting the time-domain synthetic aperture focusing technique (T-SAFT) to laser ultrasonics for imaging the subsurface defects. *Sensors*. 2023;23(19):8036.

54. Riquelme A, Rodrigo P, Escalera-Rodriguez MD, Rams J. Wear resistance of aluminum matrix composites' coatings added on AA6082 aluminum alloy by laser cladding. *Coatings*. 2021;12(1):41.
55. Ni C, Shi Y, Liu J, Huang G. Characterization of Al0. 5FeCu0.7NiCoCr high-entropy alloy coating on aluminum alloy by laser cladding. *Optics & Laser Technology*. 2018;105:257–63.
56. Karuppasamy SS, Jeyaprakash N, Yang CH. Battery technology for hybrid and battery vehicles. In Jeyaprakash Natarajan, Mahendra Babu Kantipudi, Che-Hua Yang, Yaojung Shiao (Eds), *The Future of Road Transportation*, 2023 (pp. 136–158). CRC Press.
57. Song M, Wu L, Liu J, Hu Y. Effects of laser cladding on crack resistance improvement for aluminum alloy used in aircraft skin. *Optics & Laser Technology*. 2021;133:106531.
58. Liang CJ, Wang CL, Zhang KX, Liang ML, Xie YG, Liu WJ, Yang JJ, Zhou SF. Nucleation and strengthening mechanism of laser cladding aluminum alloy by Ni-Cr-B-Si alloy powder based on rare earth control. *Journal of Materials Processing Technology*. 2021;294:117145.
59. Li Y, Zhang P, Bai P, Wu L, Liu B, Zhao Z. Microstructure and properties of Ti/TiBCN coating on 7075 aluminum alloy by laser cladding. *Surface and Coatings Technology*. 2018;334:142–9.
60. Prabu G, Jeyaprakash N, Yang CH, Alnaser IA. Investigation on deformation of nano-twins of LPBF produced Cu alloy through Triboindenter. *Tribology International*. 2024;191:109117.
61. Ivanov AD, Nikolaev AK, Kalinin GM, Rodin ME. Effect of heat treatments on the properties of CuCrZr alloys. *Journal of Nuclear Materials*. 2002;307:673–6.
62. Qi Z, Chen C, Wang C, Zhou Z, Zhou J, Long Y. Effect of different laser wavelengths on laser cladding of pure copper. *Surface and Coatings Technology*. 2023;454:129181.
63. Balu P, Rea E, Deng J. Laser cladding of nickel-based alloy coatings on copper substrates. In Mike Green, Cath Rose (Eds), *Industrial Laser Applications Symposium (ILAS 2015)*, 2015 (Vol. 9657, pp. 8–17). SPIE.
64. Yan H, Wang A, Xu K, Wang W, Huang Z. Microstructure and interfacial evaluation of Co-based alloy coating on copper by pulsed Nd: YAG multilayer laser cladding. *Journal of Alloys and Compounds*. 2010;505(2):645–53.
65. Ng KW, Man HC, Cheng FT, Yue TM. Laser cladding of copper with molybdenum for wear resistance enhancement in electrical contacts. *Applied Surface Science*. 2007;253(14):6236–41.
66. Dhanasekar R, Kumar TS, Kumar KK, Jeyaprakash N. Physical simulation of the heat-affected zone in nickel-free high-nitrogen stainless steel weld at various peak temperatures and cooling rates. *Materials Today Communications*. 2024;38:107996.
67. Aghasibeig M, Fredriksson H. Laser cladding of a featureless iron-based alloy. *Surface and Coatings Technology*. 2012;209:32–7.
68. Zhang Z, Farahmand P, Kovacevic R. Laser cladding of 420 stainless steel with molybdenum on mild steel A36 by a high power direct diode laser. *Materials & Design*. 2016;109:686–99.
69. Mahmoud ER, Khan SZ, Ejaz M. Laser surface cladding of mild steel with 316L stainless steel for anti-corrosion applications. *Materials Today: Proceedings*. 2021;39:1029–33.

70. Zhang D, Zhang X. Laser cladding of stainless steel with Ni–Cr_3C_2 and Ni–WC for improving erosive–corrosive wear performance. *Surface and Coatings Technology*. 2005;190(2–3):212–7.
71. Jeyaprakash N, Yang CH, Karuppasamy SS, Radhika N. Microstructural evolution and anti-corrosion characteristics of as-cast and LPBF printed Inconel plates in seawater environment. *Surface Review and Letters*. 2023;31:2450032.
72. Sanchez S, Smith P, Xu Z, Gaspard G, Hyde CJ, Wits WW, Ashcroft IA, Chen H, Clare AT. Powder bed fusion of nickel-based superalloys: A review. *International Journal of Machine Tools and Manufacture*. 2021;165:103729.
73. Jeyaprakash N, Yang CH. Comparative study of NiCrFeMoNb/FeCrMoVC laser cladding process on nickel-based superalloy. *Materials and Manufacturing Processes*. 2020;35(12):1383–91.
74. Feng K, Chen Y, Deng P, Li Y, Zhao H, Lu F, Li R, Huang J, Li Z. Improved high-temperature hardness and wear resistance of Inconel 625 coatings fabricated by laser cladding. *Journal of Materials Processing Technology*. 2017;243:82–91.
75. Tabernero I, Lamikiz A, Martínez S, Ukar E, Figueras J. Evaluation of the mechanical properties of Inconel 718 components built by laser cladding. *International Journal of Machine Tools and Manufacture*. 2011;51(6):465–70.
76. Jeyaprakash N, Kantipudi MB, Yang CH. Study on microstructure and anti-corrosion property of stainless steel particles deposition on Ti–6Al–4V substrate using laser cladding technique. *Lasers in Manufacturing and Materials Processing*. 2022;9(2):214–27.
77. Rack HJ, Qazi JI. Titanium alloys for biomedical applications. *Materials Science and Engineering: C*. 2006;26(8):1269–77.
78. Jeyaprakash N, Yang CH. Microstructure and wear behaviour of SS420 micron layers on Ti–6Al–4V substrate using laser cladding process. *Transactions of the Indian Institute of Metals*. 2020;73:1527–33.
79. Nabhani M, Razavi RS, Barekat M. Corrosion study of laser cladded Ti-6Al-4V alloy in different corrosive environments. *Engineering Failure Analysis*. 2019; 97:234–41.
80. Diao Y, Zhang K. Microstructure and corrosion resistance of TC2 Ti alloy by laser cladding with Ti/TiC/TiB_2 powders. *Applied Surface Science*. 2015;352:163–8.
81. Cao X, Xiao M, Jahazi M, Fournier J, Alain M. Optimization of processing parameters during laser cladding of ZE41A-T5 magnesium alloy castings using Taguchi method. *Materials and Manufacturing Processes*. 2008;23(4):413–8.
82. Yang Y, Wu H. Improving the wear resistance of AZ91D magnesium alloys by laser cladding with Al–Si powders. *Materials Letters*. 2009;63(1):19–21.
83. Jun Y, Sun GP, Wang HY, Jia SQ, Jia SS. Laser (Nd: YAG) cladding of AZ91D magnesium alloys with Al+ Si+ Al_2O_3. *Journal of Alloys and Compounds*. 2006;407(1–2):201–7.

5

Modification of Ceramic Material through Laser Cladding Process

Govindarajan Prabu, Ibrahim A. Alnaser, Natarajan Jeyaprakash, and Che-Hua Yang

5.1 Introduction

Laser cladding was one of the melting processes by which the coated material was fused with the ceramic materials and form a metallurgical bonding. The coating material was selected based on the demand of ceramic material. An alloying element was introduced to the ceramic material through the laser beam during the laser cladding process [1]. The alloying material was applied either before or during the laser cladding process. A schematic representation of laser cladding process is shown in Figure 5.1. In the pre-placed coating method, the laser energy was passed on the pre-deposited coating layers on the substrate material. In another method, the powder of the alloying elements was blown into the interaction region of laser beam and the ceramic material. As a result, the molten alloying material was deposited on the substrate material. A small amount of the substrate material was melted to form a metallurgical bonding between the ceramic cladding layer and the

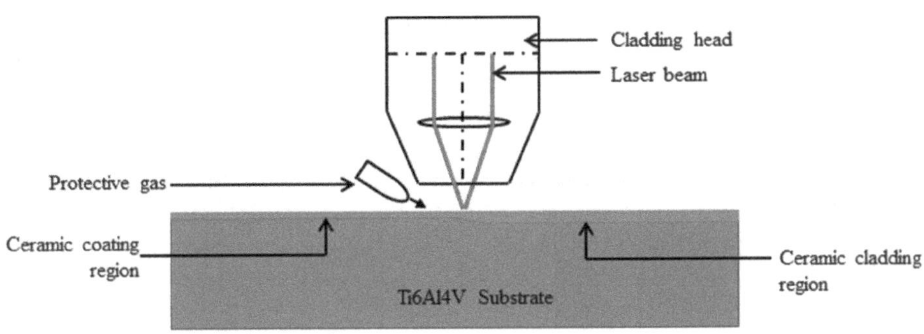

FIGURE 5.1
Schematic representation of laser cladding process.

DOI: 10.1201/9781032693576-5

substrate material [2]. This rapid solidification during the laser cladding process led to significant changes in the microstructure of the cladding region. The surface modification of the substrate material enhanced its wear resistance, decreased the erosion, and diminished the corrosion and oxidation behavior [3,4]. A variety of ceramic alloying elements are suitable to clad on any substrate material to enhance their properties.

Substrate materials include functional glasses, oxides, porous bodies, composite structures, and nonmetallic materials. The laser cladding process was done on both ceramic and nonceramic-based substrate materials. The surface modification of substrate material was achieved using different surface treatment processes [5,6].

These fabricated materials were used in different fields such as biomedical, manufacturing, aerospace industries, and so on. The chemical and the physical properties of the surface were changed, which led to the enhancement of other properties of the ceramic cladding region along with the substrate material. Carbon-based ceramic materials were used to improve the properties of the substrate through the laser cladding process. The laser type was also one of the important factors affecting the cladding region as the surface coverage area, cladding layer thickness, and the chemical properties of the alloy powder and substrate. The large thickness of the cladding region was fabricated using the CO_2 laser, which was in the range of several millimeter. The Nd: YAG laser can be used for fabricating the precise micrometer thickness of cladding region. The precision of the cladding thickness was improved with the use of robot control. The laser cladding process was mainly considered to obtain the appropriate properties, strong metallurgical bonding between the cladding region and the substrate, highest completion rate, usage of minimum quantity of alloying elements, and elimination of defects such as distortion, cracks, pores, and so on.

The energy used in the laser cladding process was slightly below than that of the other surface treatment processes due to precise control. In laser cladding, more energy was used to heat the ceramic substrate material than the alloying elements. This energy value was approximately more than ten times when compared to the alloying elements [7]. The melting of alloying elements in the powder, foil, or wire form was easy than that of the flat solid surface [8]. This is because the ratio between the surface area of alloying elements to its volume was higher than that of the substrate surface in the cladding process. The minimum power required to fabricate the cladding region was around 2kW under economic conditions. Insufficient power limited the melting of the cladding material, whereas excessive power resulted in undesirable cladding regions with dilution [9]. The laser beam interacted with the powder material for 0.1s, which was enough to form the homogeneous cladding region. Then rapid cooling process led to the solidification of the molten alloying elements, resulting in the formation of fine microstructure that was significantly different from the conventionally fabricated material. The clad geometry was formed with respect to the ceramic material surface property.

If a small-diameter laser beam was used, only a discrete region was cladded with a single pass. A large region was cladded using a wider beam diameter with a maximum scanning rate. The power density and the interaction time were appropriately selected to obtain the widest cladding region.

5.2 Importance of Ceramic Coating

The ceramic coating enhanced different properties of the material with respect to the formation of the chemical composition of the cladding layer. The steel substrate was severely affected by wear, which can be modified with the mixture of TiC ceramic particles and AlCoCrFeNi high-entropy alloy (HEA) [10]. The laser cladding process was used to fabricate the ceramic coating on the steel substrate. The high-melting-point ceramic particles such as TiC and Mo were mixed with the AlCoCrFeNi HEA [11]. These mixed particles formed a metallurgical bonding with the substrate using the laser cladding process [12]. A schematic representation of ceramic coating through laser cladding process is shown in Figure 5.2. The result revealed that the ceramic-material-reinforced HEA coating on the steel substrate significantly improved wear resistance. In addition, the wear resistance of TiC-reinforced AlCoCrFeNi HEA cladded region was 2.6 times higher than that of the TiC-only cladded region.

The hydroxyapatite (HA)-based ceramic cladding improved the wettability and microhardness of Ti alloy and enhanced the life of implant application [13]. The chemical composition of hydroxyapatite is shown in Table 5.1. The Ti material was generally used as the implant material as its

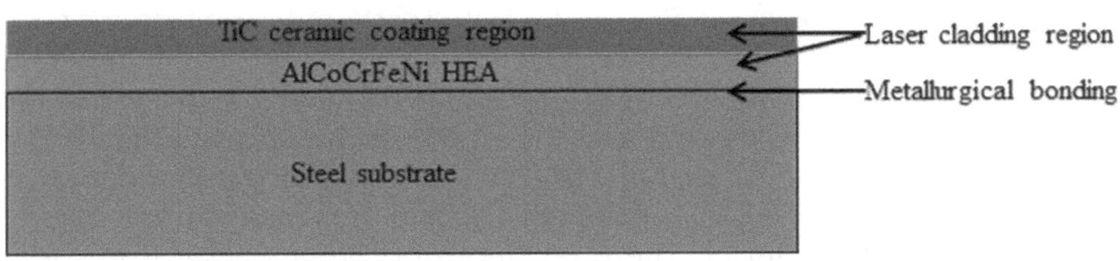

FIGURE 5.2
Schematic representation of ceramic coating through laser cladding process.

TABLE 5.1

Chemical Composition of Hydroxyapatite

Elements	Ca	P	Co	Cu	Ni	Cr	Cd	Others
Wt. (%)	Bal.	17.8	7.8	6.1	10.7	9.5	1.0	<0.3

density was similar to that of the natural bones. In addition, this material possessed higher tribological and mechanical properties [14]. However, the laser-based ceramic coating on the Ti-based implant materials exhibited higher mechanical strength and wettability. The alloying elements selected were 100% HA or a mixture of 50% HA and 50% Ti powder. The prepared blending powders were mixed well with the adhesive material, polyvinyl alcohol, and coated onto the surface of Ti alloy material. Further, the laser energy was passed onto the coated material. Moreover, the HA ceramic has superior biocompatibility and inertness, which was favorable to use it as the implant material. The calcium titanate was formed and arrested the HA in the cladded region, which was the main factor in the improvement of wettability and hardness of the implant. The HA-based ceramic coating on the Ti implant enhanced the overall behavior of bovine-bone biomedical features.

The mixture of TiN ceramic with graphene oxide played an important role in enhancing the strength of the bone joint of Ti6Al4V biomaterial [15]. The tribological property of laser-cladded TiN ceramic with graphene oxide on Ti6Al4V biomaterial was evaluated using sliding balls. The TiN ceramic laser cladding region acted as a self-lubricating element that reduced the wear of bone joint implants [16]. The graphene oxide did not affect the phases of TiN ceramics and improved the quality of the laser cladding region. Hence, the wear and friction coefficient of the implant were significantly reduced. The coefficient of friction of polyethylene, Si_3N_4, and Ti6Al4V was reduced to 25%, 22%, and 17%, respectively, with 3 wt.% of the graphene oxide. The abrasive and oxidation wear occurred in the cladding region. A lubricating film was formed when the graphene-oxide-based cladding region slide with the counter parts. It resulted in the feasibility of using graphene oxide with ceramic material during the laser cladding process for bone joint implants.

The WC ceramics were mixed with Ni50 and the laser cladding process was carried out on the steel substrate [17]. Two different compositions such as Ni50-10WC and Ni50-20WC were coated onto the substrate. Different compounds such as $M_{23}C_6$, WC, M_7C_3, W_2C, and Ni_3Fe were formed in the laser cladding region. The percentage of C and W elements in the Ni50-20WC was higher than that in the Ni50-10WC. Corresponding microhardness were 817.9 HV and 655.2 HV, respectively, which was 3–4 times higher than the substrate material. The wear depth of the laser cladding specimens was 8.3 and 13.1 mm for Ni50-20WC and Ni50-10WC, respectively. The increase of the WC ceramic percentage in the laser cladding region improved the wear resistance of the steel specimen. The W_2C ceramic was mixed with the carbon nanotubes, which enhanced the mechanical properties of the Inconel 625 substrate during the laser cladding process [18], because the carbon nanotubes alone cannot enhance the mechanical properties of Inconel 625 due to their weak Van der Waals' force.

5.3 Different Methods of Ceramic Coating for the Laser Treatment Process

The ceramic coatings were deposited onto different materials to improve the surface properties such as wear resistance, reduced friction, enhanced thermal resistance, and anticorrosion improvement. These coatings were the additional layer that was formed on the substrate material. There are different types of coating methods used to apply the ceramic materials to substrates. The deposition of ceramic coatings depends on pressure, properties of substrate material, composition, temperature, electric current, electrolyte, etc., all of which decide the resultant properties of the coating material such as mechanical, tribological, metallurgical, and so on. The thickness of the ceramic coating deposited on the substrate material was varied in the order of micrometer to nanometer scale. Ceramic powders were available in the range of nanosize. Hence, it was possible to fabricate high thermal stability with superior mechanical features. As a result, the durability of the ceramic coating much improved even under severe environmental conditions. Sol-gel was one type of coating method that was cheapest and easiest for ceramic material deposition. The coating was bonded to the substrate through chemical bonds at low temperature. Micro-arc oxidation coating method was also used to deposit the ceramic material onto the substrate. This method was superior to the traditional oxidation method and modified the substrate surface, resulting in higher hardness and thickness than the traditional oxidation method.

The coating fabrication method named as liquid deposition was used to improve the wear resistance and heat resistance of the substrate material. This method coated the ceramic particles in the form of nanofibers. The atomic layer coating method was used to deposit the ceramic inorganic material on the polymeric aerogel substrate. In addition, it deposited the inorganic material with metal oxide which had homogeneous structures. Moreover, this method modified both the inner and outer regions of the polymer substrate. The electrochemical method was also used to coat the ceramic material onto the metal substrate. Most of the oxide ceramics were coated using this electrochemical method. The efficiency of coating was improved when the electrochemical anodization process was combined with the grafting method. It fabricated a superhydrophobic property on the aluminum substrate with higher durability of coating. Moreover, it was an effective and eco-friendly method that can be used at any circumstances. The Mg, Al, and Ti materials were coated with the plasma electrolytic method in the form of ceramic oxidation. Hence, this method was used to coat the oxide form of ceramics in an eco-friendly manner. This plasma method has forms a thick coating with high hardness and dense structure [19].

Another method used to fabricate large thickness of the ceramic coating was magnetron sputter. This method used the ceramic material effectively to increase the thickness of the coating. It was considered the superior technique owing to its high deposition rate, better purity, and good adhesion strength. The superhydrophobic property of the substrate was fabricated by the solution immersion coating method. It was one of the convenient methods, using the raw material efficiently and the fabrication cost was less compared to other methods. Hydroxyapatite ceramics were coated onto the magnesium substrate metal and its alloy. Moreover, the fabricated hydroxyapatite coating possessed anticorrosive, thermally stable, biocompatible and eco-friendly properties. Laser cladding was effectively used to fabricate the ceramic coating on the substrate material. It was a high-precision method to fabricate the nano-lever thickness coating. The energy used in this fabrication method was significantly higher than the other methods. The high laser energy was easily used to fabricate the oxide ceramics on the substrate. The antithrombotic surfaces were formed by using the laser cladding process. Chemical vapor deposition was another useful method for fabricating ceramic coatings. This method produces a chemical reaction between the ceramic material and the substrate. A uniform coating thickness can be made using the dip coating method in which the substrate material was dipped into liquid form of a low-viscosity material. The coating thickness depended on the process conditions and the fabrication process was very simple.

5.4 Changes in Microstructure of Ceramic through Laser Cladding

Generally, the laser cladding process improved the bonding and surface composition of the cladding region through homogeneous deposition. This led to producing a high-density cladding and eliminating the lack of fusion defects. Post process was used to study the microstructure of the cladding region and other tests to evaluate the properties such as tribology, corrosion, and. The ceramic coating fabricated on the substrate surface was continuous and defect-free based on the process parameters. The parameters were optimized by changing the laser power, scanning speed, overlap ratio, etc. to obtain a defect-free surface. The thickness of the ceramic cladding region was obtained up to a millimeter level. Metallurgical bonding was formed between the substrate and the ceramic cladding region. The columnar structures were formed at the cladding region with respect to the temperature gradient [20]. Hence, significantly different microstructures were formed at the cladding region.

Based on the analysis carried out by Abioyea et al. [21], the ceramic cladding layer on Inconel 625 material reduced the corrosion and eliminated the pores in the laser cladding region. In addition, Liu et al. [22] found that the laser cladding region possessed dendritic microstructure when 42% wt. of

WC ceramic was used. This cladding region improved the wear and corrosion properties of the substrate. The alloying elements such as Co, Cr, Mn, Fe, and Ni were uniformly distributed in the laser cladding process. Three different regions were formed in the laser cladding specimen, namely cladding region, transition region, and heat-affected region. The cladding region was the top of the coating and the transition region was the intersection region of the coating and substrate. The bottom of the region was the heat-affected region which was affected by thermal energy during the laser cladding process but contained only the substrate elements. During the cladding process, some of the elements such as Fe diffused from the substrate into the cladding region. As a result, the percentage of any particular alloying element in the cladding region was higher than that of the raw powder material.

Hence, the cladding region was dominated by the newly added alloying elements, whereas the transition region possessed both the alloying elements and the substrate materials. It produced fine grains in the laser cladding region, which is shown in Figure 5.3. Ceramic materials were classified into nitrides, carbides, and oxides. The laser power was selected in such a way that the laser energy was spread to overcome the surface tension of powder particles [23]. The presence of porosity in the ceramic cladding region affected the bonding with the substrate. The laser energy induced the

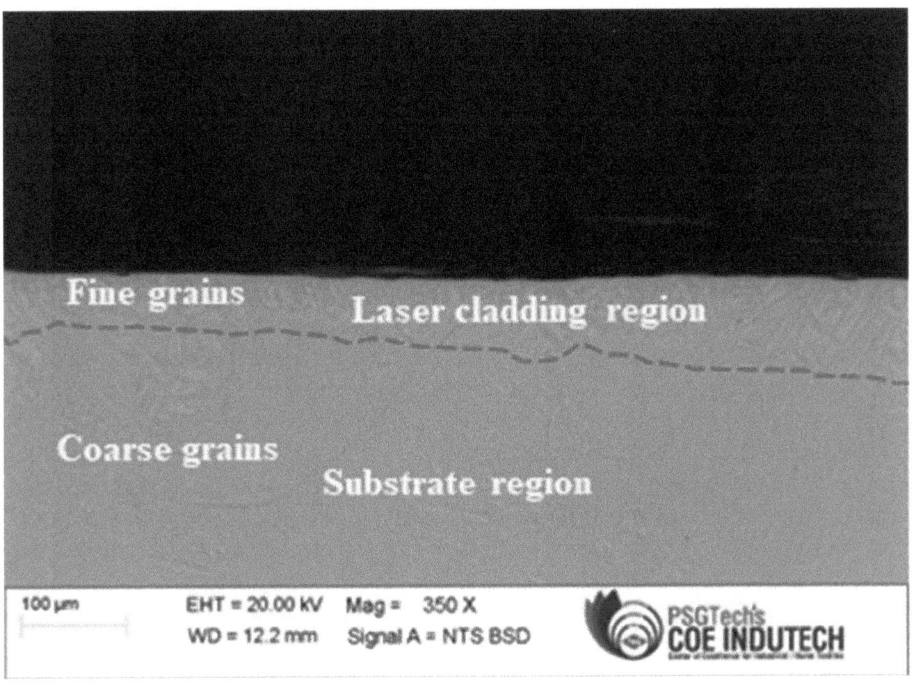

FIGURE 5.3
FESEM image of microstructure of laser cladding specimen.

reaction within the pre-placed powder before it reached the substrate surface. Hence, residual thermal energy induced the chemical reaction that led to chemical bonding with the substrate surface. The La_2O_3 severely affected the microstructure of the ceramic coating of Ni superalloy. The 0.6wt.% of La_2O_3 produced the defect-free microstructure on the ceramic coating [24]. In addition, the metallurgical bonding of ceramic coating occurred with the Ni superalloy substrate. In case of La_2O_3 inclusion in the cladding process, the Fe and Ni powders were melted first in the laser cladding region compared to other alloying elements. On further movement of the laser beam, the molten pool started to solidify.

This microstructure formation in the ceramic laser cladding region was determined with the help of temperature gradient and the rate of solidification [25]. Based on the ratio of temperature gradient to solidification rate, the planar grains formed at the bottom region of the molten pool region because the high-temperature gradient ratio has occurred at the bottom region. The dendrite microstructure was formed away from the bottom of the molten pool region due to the reduction of the temperature gradient ratio. The microstructure of Ni-based laser cladding region significantly differed from that of the cladding region with La_2O_3 inclusion. The La_2O_3 ceramic component refined the microstructure in the dendrite direction.

5.5 Laser Cladding on Ceramic Coating for Wear Resistance

Ceramic materials were used in different fields to enhance their properties through the laser cladding process. The wear was one of the important issues for two matting parts as they degraded after a long time of service. The wear resistance was increased to improve the service life of the component by incorporating the ceramic particles at the surface region [26]. The tribological properties of different types of materials were improved through alloying and laser cladding processes [27]. The coating of ceramic material through the laser cladding process was applied on titanium alloys, nickel superalloys, steel, cast iron, and so on. The tribological property of titanium alloys was increased by the laser cladding of ceramic particles for different applications. The titanium alloy was used in the biomedical field for healing joint pain by implanting artificial joints. The laser cladding process improved the bonding strength of the ceramic-based cladding region with the titanium alloy for enhancing the wear resistance [28,29]. In addition, this process reduced the friction of implant material with respect to the improvement of the wear resistance. TiN is one type of ceramic material that was used for implanting due to its high corrosion resistance and good biocompatibility properties. Specifically, the ceramics were used as artificial joints in femoral heads to overcome the bio-tribological properties. However, the large frictional coefficient of TiN ceramic material led to a reduction in its usage in the biomedical field.

The large frictional coefficient promoted the formation of abrasive particles in the TiN ceramic material. As a result, severe wear occurred on the TiN-coated implant material. It led to inducing the aseptic loosening of the artificial joints. It resulted in the reduction of the life of the artificial joints. The uses of the ceramic material in the human implants were improved by using laser-based modification. The laser cladding process was used to incorporate the self-lubricating material with the TiN ceramic to reduce the friction coefficient [30]. The MoS_2, Ag, WS_2, and Cu materials have a self-lubrication property, which were coated on the nickel-based superalloys using the laser cladding process [31]. The cross-sectional view of the laser cladding specimen is shown in Figure 5.4. The Ag element reduced the wear at the ambient temperature, whereas the sulfide had the property of self-lubricating at high temperatures. Hence, the incorporation of self-lubricating material with ceramic material led to reduced friction and increased wear resistance [32] because the mechanism involved in the antifriction behavior of self-lubricating cladding material was dependent on its characteristics.

The selection of the lubricating material to reduce the friction of ceramic material mainly depended on the application, temperature condition, load,

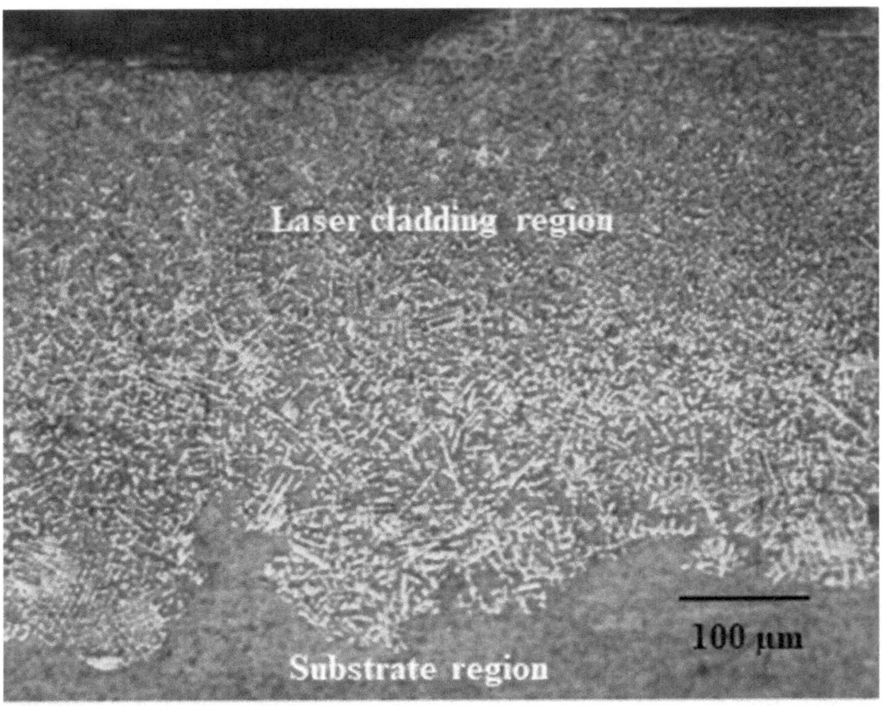

FIGURE 5.4
Cross-sectional view of laser cladding specimen.

and other working conditions. The self-lubricating elements were added with TiN ceramic material to enhance the tribological property. Graphene oxide worked as the solid lubricant that was added with different materials to reduce the frictional coefficient [33,34]. The graphene oxide was added to TiN ceramic powder and the laser cladding process was used to coat on the titanium alloy. The oxide film was formed on the titanium alloy that led to a reduction in the frictional coefficient of ceramic cladding region. The cladding region was rich in oxygen due to the occurrence of tribo-chemical reaction at the wear track region [35]. As a result, the oxide film was formed on the wear track surfaces. The laser cladding region consisting of graphene oxide and TiN ceramic particles exhibited smoother surface than the TiN ceramic cladding region. The wear debris from the matting surface diffused on the oxide tribo-film due to smooth surfaces. The wear track was not formed on the graphene-oxide-doped TiN ceramic-coated specimen. This cladding process stabilized the wear due to the improvement in the quality of the laser coating. Generally, the oxide film was removed during the wear process due to shear and pressure [36]. Hence, the adhesive wear predominantly occurred on the laser cladding region that contained ceramic particles with graphene oxide. The wear mechanism of ceramic particles with oxide coating was adhesion wear and oxidation wear.

5.6 Corrosion Improvement by Laser Cladding on Ceramic Material

Ceramic materials have exceptional characteristics that led to meet the customer demand. For example, nickel superalloy, titanium alloy, etc. have been used in aerospace, automobile, biomedical, chemical process fields and so on due to their excellent properties. However, few of the drawbacks of material properties affected their applications. These drawbacks were eliminated by adding superior property elements on the surface through the coating process. One such example was the corrosion behavior of titanium alloys. The titanium alloys have superior biomechanical biocompatibility, same density as the human bone, etc [37–39]. It was used as an implant material in the biomedical field by adding hydroxyapatite ceramic material on their surface. The surface-modified titanium alloys were used as the artificial joints such as hip joints and knee joints [40]. The hydroxyapatite ceramic coating improved the bioactivity of titanium alloy along with osseointegration strength. The bonding strength of hydroxyapatite ceramic coating fabricated through the laser cladding techniques was higher compared to other methods. The coating fabricated through the laser cladding process consisted of three different layers: top, middle, and bottom layers, which have the characteristics of bioactive, transition, and barrier behavior, respectively [41].

The presence of the Ag element formed a metallurgical bonding with the titanium alloy substrate material. It prevented the chemical reaction of hydroxyapatite ceramic material with the titanium substrate [42]. Hence, the laser cladding technique promoted the hydroxyapatite ceramic material to its osseointegration with matting bone tissue. A schematic representation of the corrosion behavior of substrate and laser cladding specimens is shown in Figure 5.5.

The hydroxyapatite ceramic material interacted with body fluid and produced calcium, phosphate, and other ions. Further, the deposition of the above chemical increased on the surface of hydroxyapatite ceramic material. It resulted in the repaid growth of new tissues and bonds at the interface region between the implant and the bone. Some of the impurities were formed at the laser cladding top surface due to the decomposition of $Ca_3(PO_4)_2$ [43]. These impurities led to induced corrosion on the coated region [44]. However, the increase of hydroxyapatite ceramic material on the coating resulted in crack formation [45]. Improving the corrosion resistance led to stabilizing the hydroxyapatite material and enhanced the service life.

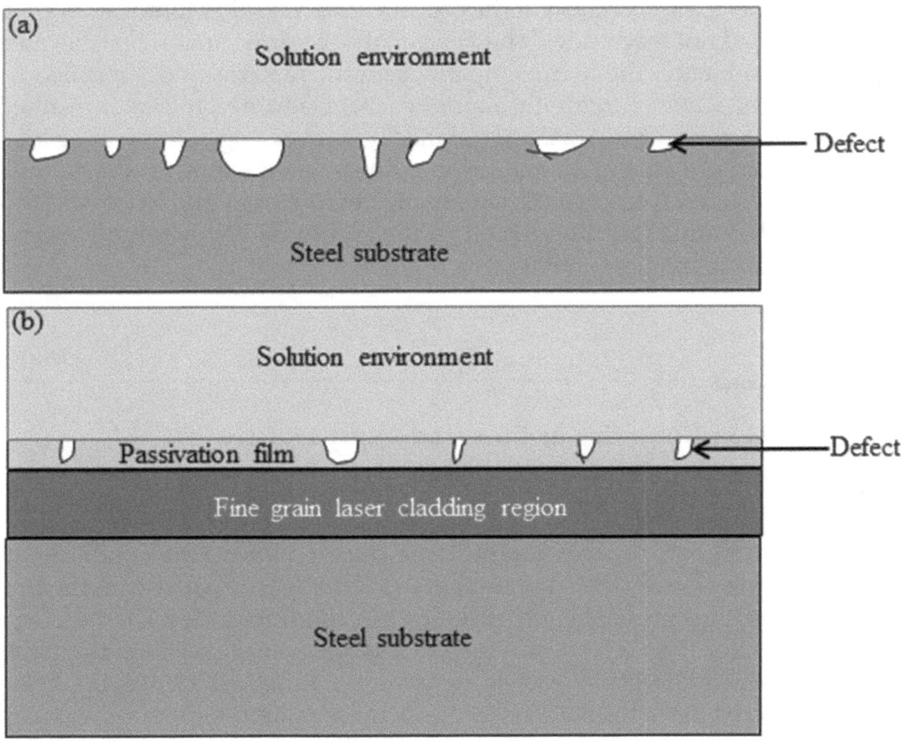

FIGURE 5.5
Corrosion behavior of (a) substrate and (b) laser-cladded specimen.

The increasing coating toughness led to a reduction of the crack propagation due to an increase in energy observation. Hence, the toughness of the hydroxyapatite material through the laser cladding process was improved by reinforcing ceramic material in hydroxyapatite material [46].

The reinforcing ceramic particles were selected in such a way that they noticed the crack initiation energy and prevented further extension of the crack. In addition, the reinforcing particles pull toward the crack and bridge it [47]. Some of the reinforcing ceramic materials for the improvement of corrosion resistance were ZrO_2, TiO_2, etc. The laser cladding process induced the phase changes by transforming the lattice structure of ZrO_2 ceramic material because the ZrO_2 ceramic increased its volume when the operating temperature was 1170°C. As a result, it induced the compressive stress in the nearby area that led to reduced corrosion crack [48]. However, the toughness at a few locations of the cladding region was affected due to the rapid volume change of ZrO_2 ceramic particles. Hence, the ZrO_2 ceramic particles were used to reduce the corrosion when the operating temperature was low. Silicon carbide was one of the ceramic bioinert materials that had higher elastic modulus and toughness. In addition, it possessed superior oxidation resistance with a low frictional coefficient. As a result, it was effectively used in biomedical applications to reduce the corrosion of the coating material [49]. Moreover, the silicon carbide had not reacted with hydroxyapatite ceramic material at elevated temperatures. Hence, the hydroxyapatite ceramic material was modified by reinforcing the silicon carbide through the laser cladding process. It resulted in the enhancement of corrosion resistance by preventing crack propagation through the improvement of coating toughness. As a result, the body fluid had not penetrated into the hydroxyapatite ceramic region. Hence, the corrosion of the ceramic cladding region on the titanium alloy was reduced and maintained the biological activity.

5.7 Oxidation Behavior at Elevated Temperature of Laser-Cladded Ceramic Material

High-temperature oxidation behavior of material was needed when the material was subjected to elevated temperature during its working condition. In the gas turbine engine, the high-temperature fuel-gas mixture met the turbine blade at high pressure and temperature conditions [50] The heat was transferred from the fuel-gas mixture to the blade surface. Here, the blade material working condition was elevated temperature [51]. Hence, the material was selected in such a way that the resistance to high-temperature oxidation was high. In addition, the material should have high erosion resistance as a large number of foreign particles in the combustion product hit the turbine blade surface. Moreover, the material must have superior wear resistance as solid combustion particles slide on the surface of the turbine blade [52]. Using only a single material was not superior in multiple properties. Laser cladding filled this gap to obtain the required property on the single material through the coating process. The ceramic material was used as the coating material in the high-temperature resistance application because this material possessed higher temperature stability when compared to other types of materials. The laser cladding process was applicable in a wide range of materials with less duration of fabrication. The strength of the cladding material increased due to rapid heating and cooling techniques [53]. Zhu et al. [54] coated the F FeCoCrNiSi material through the laser cladding process and found a higher wear resistance of $0.677\times10^{-4}mm^3$ with a lower frictional coefficient of 0.19 at elevated temperature of nearly 600°C. It revealed that the material had withstood high temperatures due to improvement in strength through the laser cladding process.

The Ni60-Cu matrix also exhibited superior tribological characteristics at 600°C temperature due to the laser cladding process. The chemical composition of Ni60 is shown in Table 5.2. This behavior of the material revealed that the Ni60-Cu had the ability to survive at high-temperature applications. The behavior of the material at the elevated temperature was improved by reinforcing the ceramic materials into the matrix materials. The ceramic-reinformed CoCrNiMoCB coating exhibited higher material through laser cladding. Titanium nitride was one of the ceramic materials that had good abrasion resistance, higher hardness, and higher oxidization resistance at elevated temperatures [55]. It was also used to refine the phases that led to the enhancement of the coating performances [56]. This titanium nitride ceramic material was used in various fields such as ornamentation, microelectronics, and material processing units. This titanium nitride ceramic material and Ni-W were fabricated as the composite coating. This composite ceramic material was used to resist wear at the temperature of 600°C–700°C [57]. This type of coating effectively improved the oxidation resistance of the Inconel 718 turbine blade application. Stellite3-Ti_3SiC_2-Cu was one form of ceramic composite coatings that has an oxidation rate of $2.7065\,mg^2/cm^4h$ with a wear rate of $6.09\times10^{-5}mm^3$. It revealed that the oxidation and wear resistance were significantly higher than the Inconel 718 substrate material [58]. In addition, the microhardness of the ceramic coating was 2.5 times of microhardness of the substrate.

The material such as Inconel 718 was subjected to a high-temperature atmosphere, turbine blade, the frictional heat was easily generated on the surface of the blade. It induced the oxidation at the surface region and damaged it.

TABLE 5.2

Chemical Composition of Ni60

Elements	Ni	B	Si	Fe	Cr	C
Wt. (%)	Bal.	2.9	4.5	6.1	10.7	9.5

The material had gained a significant amount of mass after the particular duration of the oxidation process such as 50 or 100 hours at the temperature of 800°C or above. The mass gain during the oxidation process was calculated in mg/cm^2 and the mass values were in the range of two to three digits. The mass gain values increased with the increase of the oxidation testing time. Generally, the materials obeyed the parabolic law of oxidation-related kinetic curve. Hence, the oxidation resistance of the material followed the parabolic curve shape between the operating temperature and its oxidation resistance. The mass gain during the oxidation process was slightly higher than the corresponding substrate material when the laser-cladded ceramic coating was involved. The mass-gaining process on the ceramic coating surface had a rapid and stable oxidation process. The rapid process of oxidation occurred at the initial stage, whereas the stable process occurred when the oxides were not able to contact the coating surface. Hence, the oxidation process was saturated on the laser cladding region and further oxidation of the surface was reduced.

5.8 Laser-Cladded Ceramic Material in Biomedical Field

Biocompatibility materials were preferred to be used as implant materials in the biomedical field. The contribution of ceramic material was vast when compared to other materials. In addition to biocompatibility, other properties such as mechanical, tribological, feasibility of manufacturing, cost-effectiveness, and so on were also considered to select the material for implant applications [59]. Some of the materials were available easily and at low cost for fabrication, but their biocompatibility and bioactivity with the human tissues were less [60]. Such materials were not directly used as the implant materials even though the cost of the implant was less compared to other implant materials because these low-cost implant materials were worn out easily and corroded with less service time. The ceramic material helped to use this low-cost material as the implant through the coating process. The low bioactive material had not contacted the tissues directly as the high biocompatibility ceramic material was deposited on this low bioactive material. Hence, the ceramic material only directly reacted with the tissues and made cell growth on it. Here, the bonding between the ceramic material and the low-cost substrate material was an important factor to be considered because the loose bonding led to encouraging the reaction between the human body fluid and the low-cost substrate material. In addition, some of the elements from the low-cost substrate material such as chromium diffused with the human tissues [61]. It caused inflammation in the surrounding tissues leading to reduction in the healing process. Hence, the failure of the implant occurred, which was overcome by using a high laser cladding process.

The laser cladding process improved the bonding strength between the ceramic coating and the substrate material. In addition, it reduced the induction of cracks at the ceramic cladding region, which frequently occurred in other manufacturing processes [62]. The porous form of ceramic coating was fabricated on the low bioactive substrate material through a laser cladding process. The porous structure was formed due to high temperature and rapid cooling process [63]. The laser cladding process was environmentally friendly, which was suitable for the fabrication of biomedical implant components [64]. The bioactive elements such as hydroxyapatite, Ca, P, and Ti were added to the coating material to fabricate the implant material.

5.9 Defects and Its Remedies in Laser Cladding Treatment on Ceramic Material

The laser cladding process to coat the ceramic material became popular in different industries. The circular- and rectangular-shaped beams were used to fabricate the ceramic coating. The defects such as pores, cracks, and nonuniform distribution of alloying elements were generally present in the cladding region. The crack was induced in the cladding region during the solidification process due to the rapid heating and cooling processes because the elastic strength of the cladding region exceeded the fracture strength. The thermal gradient was the main reason for the increased stress at the time of the cooling process. Different alloying elements have various degrees of thermal contraction with the neighboring alloying elements.

The Ni-based superalloy; the crack was induced on the ceramic cladding region which was eliminated using the different oxides. The La_2O_3 was one type of oxide that was used in the cladding region. The La_2O_3 content value was varied on the cladding region and studied the sensitivity with respect to crack induction. It was found that the 0.6–0.8 wt.% of La_2O_3 content increased the resistance of the Ni-based ceramic cladding region. The cracks induced in the laser ceramic cladding region when the La_2O_3wt.% was increased to 1. Hence, based on the crack initiation, the optimal value of La_2O_3 usage was 0.6–0.8 wt.% when the Ni superalloy was coated with ceramic coating through the laser cladding process [65]. The thermal stress was induced in the cladding region due to the brittleness of ceramic material with a high melting point and low thermal conductivity. La_2O_3 ceramic compound with the Ni elements from the substrate led to significant increase in the thermal gradient. It turned into a thermal shock that initiated the cracks at the laser cladding region, specifically in the centerline and edges of the laser track. This type of crack in the cladding region reduced the service life of the coating. In addition, the crack was initiated at the indentation cavity when the hardness was tested [66]. It shows that the fracture toughness of the cladding region decreased and induced the creak. Based on the presence of the creak at the tip of the indentation cavity, the fracture toughness of the coating was calculated by the following formula [67].

$$\text{Fracture toughness} = \frac{0.035 \times \text{Indentation diagonal length} \times \text{Microhardness}^{0.6}}{\text{Constrain factor}^{0.6} \times \text{Length of the crack at the tip}^{0.5}} \tag{5.1}$$

where the details of units are listed below:

Fracture toughness in MPa, indentation diagonal length and crack length in μm, microhardness in GPa.

The crack in the laser cladding bioactive layer decreased with the increase of the SiC ceramic particles. The hydroxyapatite laser cladding layer has a crack when the percentage of SiC was 0. The crack started to decrease when the hydroxyapatite +5%SiC, hydroxyapatite +15%SiC, and hydroxyapatite +25%SiC were fabricated. It revealed that the increasing content of SiC ceramic particles led to increase in the pinned SiC at the bioactive cladding layer. As a result, less number of crack formations obviously decreased [68]. The incomplete melting of the SiC particles did not participate in minimizing the crack propagation because it was unable to interlock the SiC with the substrate material that led to propagates as the crack. The 25%SiC bridging and pinning well with the substrate impeded the cracks effectively. The fracture toughness increased with the increase of SiC. The crack was initiated where stress concentration was high in the cladding region. This crack propagated and passed through the cladding region to dissipate the stress. The crack tried to pass through the SiC ceramic particles, but it was deflected by the ceramic particles. In some situations, the cracks were initiated at the heat-affected region or the interface of the laser cladding and substrate regions [69]. The SiC particles observed the crack initiation energy and inhibited the further extension of the crack [70]. In addition, the SiC particles bridged with the matrix led to interlock the propagation of the crack [71]. Hence, the SiC ceramic particles consumed the energy that resulted in the enhancement of fracture toughness of the laser cladding region.

5.10 Conclusion

The laser cladding process on the modification of ceramic material was clearly discussed in this chapter. Different ceramic materials such as TiC, titanium nitride, Ni-W composite coating, WC, borides, oxides, nitrides, and Cr_3C_2 were analyzed. This analysis has shown that the laser cladding process induced different phases of the ceramic materials in the cladding region.

As a result, it also enhanced the material properties such as wear resistance, corrosion resistance, biocompatibility, and so on. Moreover, the use of TiN ceramic material reduced the losses of hydroxyapatite material from the Ti-6Al-4V substrate. This Ti-6Al-4V was used as a bioimplant material for replacing the hip and other bone joints. In addition, the laser cladding process modified the AlCoCrFeNi HEA by incorporating the TiC ceramic material resulting in the improvement of wear resistance. The titanium nitride and Ni-W ceramic materials were fabricated as the composite coating using a laser cladding process. This composite laser cladding coating was used to resist the wear at temperatures of 600°C–700°C. Moreover, the oxidation resistance at the elevated temperature was improved using ceramic coating in the laser cladding process.

Acknowledgments

We are grateful for the support from the Ministry of Science and Technology, Taiwan for providing the financial support.

References

1. J. X. Fang, J. X.Wang, Y. J. Wang, H. T. He, D. B. Zhang, Y. Cao, Microstructure evolution and deformation behavior during stretching of a compositionally inhomogeneous TWIP-TRIP cantor-like alloy by laser powder deposition. *Mater. Sci. Eng. A* 847 (2022) 143319. https://doi.org/10.1016/j.msea.2022.143319.
2. N. Jeyaprakash, C.-H. Yang, M. Duraiselvam, G. Prabu, Microstructure and tribological evolution during laser alloying WC-12% Co and Cr_3C_2- 25% NiCr powders on nodular iron surface. *Results Phys.* 12 (2019) 1610–1620.
3. C. Yuhua, M. Yuqing, L. Weiwei, H. Peng, Investigation of welding crack in micro laser welded NiTiNb shape memory alloy and Ti6Al4V alloy dissimilar metals joints. *Optics Laser Technol.* 91 (2017) 197–202.
4. G. Prabu, M. Duraiselvam, N. Jeyaprakash, C.-H. Yang, Microstructural evolution and wear behavior of AlCoCrCuFeNi high entropy alloy on Ti-6Al-4V through laser surface alloying. *Metals Mater. Int.* 27 (2021) 2328–2340.
5. Y. Chen, S. Sun, T. Zhang, X. Zhou, S. Li, Effects of post-weld heat treatment on the microstructure and mechanical properties of laser-welded NiTi/304SS joint with Ni filler. *Mater. Sci. Eng. A* 771 (2020) 138545. https://doi.org/10.1016/j.msea.2019.138545.
6. Z. H. Fu, B. J. Yang, M. L. Shan, T. Li, Z. Y. Zhu, C. P. Ma, W. Gao, Hydrogen embrittlement behavior of SUS301L-MT stainless steel laser-arc hybrid welded joint localized zones. *Corrosion Sci.* 164 (2020) 108337. https://doi.org/10.1016/j.corsci.2019.108337.

7. N. Jeyaprakash, C.-H. Yang, M. Duraiselvam, G. Prabu, S.-P. Tseng, D. Raj Kumar, Investigation of high temperature wear performance on laser processed nodular iron using optimization technique. *Results Phys.* 15 (2019) 102585.
8. F. Peter Prakash, N. Jeyaprakash, M. Duraiselvam, G. Prabu, C.-H. Yang, Droplet spreading and wettability of laser textured C-263 based Nickel superalloy. *Surf. Coat. Technol.* 397 (2020) 1260552.
9. Z. Y. Zhu, Y. L. Liu, G. Q. Gou, W. Gao, J. Chen, Effect of heat input on interfacial characterization of the butter joint of hot-rolling CP-Ti/Q235 bimetallic sheets by laser+CMT. *Sci. Rep.* 11 (1) (2021) 10020.
10. W. Zhao, K. Yu, Q. Ma, C. Song, G. Xiao, H. Zhang, Y. Lv, N. Guo, Z. Li, Synergistic effects of Mo and in-situ TiC on the microstructure and wear resistance of AlCoCrFeNi high entropy alloy fabricated by laser cladding. *Tribol. Int.* 188 (2023) 108827.
11. Q. Guo, H. Hou, Y. Pan, X. Pei, Z. Song, P. K. Liaw, Y. Zhao, Hardening-softening of Al0.3CoCrFeNi high-entropy alloy under nanoindentation. *Mater. Des.* 231 (2023) 112050. https://doi.org/10.1016/j.matdes.2023.112050.
12. G. Ma, P. He, H. Wang, H. Tian, L. Zhou, Q. Yong, D. He, Promoting bonding strength between internal Al-Si based gradient coating and aluminum alloy cylinder bore by forming homo-epitaxial growth interface. *Mater. Des.* 227 (2023) 111764. https://doi.org/10.1016/j.matdes.2023.111764.
13. A.-C. Mocanu, F. Miculescu, G. E. Stan, T. Tite, M. Miculescu, M. H. Ţierean, A. Pas, R.-C. Ciocoiu, T. M. Butte, L.-T. Ciocan, Development of ceramic coatings on titanium alloy substrate by laser cladding with pre-placed natural derived-slurry: Influence of hydroxyapatite ratio and beam power. *Ceram. Int.* 49 (2023) 10445–10454.
14. P. Murugesan, V. Satheeshkumar, N. Jeyaprakash, G. Prabu, C.-H. Yang, Nano-level mechanical and tribological behavior of additively manufactured AlSi10Mg plates. *Arch. Civil Mech. Eng.* 23 (2023) 62.
15. J. Liu, D. Liu, S. Li, Z. Deng, Z. Pan, C. Li, T. Chen, The effects of graphene oxide doping on the friction and wear properties of TiN bioinert ceramic coatings prepared using wide-band laser cladding. *Surf. Coat. Technol.* 458 (15) (2023) 129354.
16. G. Prabu, N. Jeyaprakash, C.-H. Yang, N. Radhika, Analysis on the transformation of melt wear to self-healing crack of wire arc additive manufactured Al 5356 alloy. *Tribol. Int.* 188 (2023) 108802.
17. Y. Liu, X. Gu, C. Lou, L. Kang, Q. Hou, C. Ma, Influence of WC ceramic particles on structures and properties of laser cladding Ni50-WC coatings. *J. Mater. Res. Technol.* 26 (2023) 14–21.
18. S. Zhou, T. Xu, C. Hu, H. Wu, H. Liu, X. Ma, Utilizing carbon nanotubes in ceramic particle reinforced MMC coatings deposited by laser cladding with Inconel 625 wire. *J. Mater. Res. Technol.* 13 (2021) 2026–2042.
19. N. Jeyaprakash, G. Prabu, C.-H. Yang, M. P. Banda, E. Mohan, Structural and nanohardness of laser melted CM247LC nickel-based superalloy. *Trans. Indian Inst. Metals* 76 (2) (2022) 287–295.
20. S. S. Liu, M. Zhang, G. L. Zhao, X. H. Wang, J. F. Wang, Microstructure and properties of ceramic particle reinforced FeCoNiCrMnTi high entropy alloy laser cladding coating. *Intermetallics* 140 (2022) 107402.
21. T. E. Abioyea, D. G. McCartney, A. T. Clarea, Laser cladding of Inconel 625 wire for corrosion protection. *J. Mater. Process. Technol.* 217 (2015) 232–240.

22. J. Liu, D. Liu, S. Li, Z. Deng, Z. Pan, C. Li, T. Chen, The effects of graphene oxide doping on the friction and wear properties of TiN bioinert ceramic coatings prepared using wide-band laser cladding, *Surf. Coat. Technol.* 458 (2023) 129354.
23. W. Zhao, H. Suo, S. Wang, L. Ma, L. Wang, Q. Wang, Z. Zhang, Mg gas infiltration for the fabrication of MgB_2 pellets using nanosized and microsized B powders. *J. Eur. Ceram. Soc.* 42 (2022) 7036–7048. https://doi.org/10.1016/j.jeurceramsoc.2022.08.029.
24. M. Li, B. Han, Y. Wang, K. Pu, Effects of La2O3 on the microstructure and property of laser cladding Ni-based ceramic coating. *Optik* 130 (2017) 1032–1037.
25. M. Gäumann, S. Henry, F. Cléton, J. D. Wagnière, W. Kurz, Epitaxial laser metal forming: Analysis of microstructure formation. *Mater. Sci. Eng. A* 271 (1999) 232–241.
26. P. Zhao, J. Zhu, K. Yang, M. Li, G. Shao, H. Lu, J. He, Outstanding wear resistance of plasma sprayed high-entropy monoboride composite coating by inducing phase structural cooperative mechanism. *Appl. Surf. Sci.* 616 (2023) 156516. https://doi.org/10.1016/j.apsusc.2023.156516.
27. N. Jeyaprakash, G. Prabu, C.-H. Yang, Microstructure and tribological behaviour of FeCoCrNiAlCuMn particles alloyed on Inconel-718 substrate using plasma transferred arc technique. *Intermetallics* 139 (2021) 107360.
28. J. Liu, D. Liu, S. Li, Z. Deng, Z. Pan, C. Li, T. Chen, The effects of graphene oxide doping on the friction and wear properties of TiN bioinert ceramic coatings prepared using wide-band laser cladding. *Surface and Coatings Technology* 458 (2023) 129354.
29. T. Chen, W. Fan, H. Wang, D. Liu, Laser cladding in-situ Ti(C, N) particles reinforced ni-based composite coatings modified with CeO2 nanoparticles. *Metals* 8 (8) (2018) 601.
30. S. S. Akhtar, A critical review on self-lubricating ceramic-composite cutting tools. *Ceram. Int.* 47 (15) (2021) 20745–20767.
31. H. Torres, T. Vuchkov, S. Slawik, C. Gachot, B. Prakash, M. Rodríguez Ripoll, Selflubricating laser claddings for reducing friction and wear from room temperature to 600 °C. *Wear* 408–409 (2018) 22–33.
32. N. Jeyaprakash, C.-H. Yang, G. Prabu, K. Ganesa Balamurugan, Surface alloying of FeCoCrNiMn particles on Inconel-718 using plasma-transferred arc technique: Microstructure and wear characteristics. *RSC Adv.* 11 (2021) 28271–28285.
33. Y. Liu, X. Ge, J. Li, Graphene lubrication. *Appl. Mater. Today* 20 (2020) 100662. https://doi.org/10.1016/j.apmt.2020.100662.
34. J. Zhang, X. Gao, Q. Xu, T. Ma, Y. Hu, J. Luo, Atomistic insights into friction and wear mechanisms of graphene oxide. *Appl. Surf. Sci.* 546 (2021), 149130.
35. N. Jeyaprakash, S. Sivasankaran, G. Prabu, C.-H. Yang, A. Alaboodi, Enhancing the tribological properties of nodular cast iron using multi wall carbon nano-tubes (MWCNTs) as lubricant additives. *Mater. Res. Exp.* 6 (2019) 045038.
36. G. Prabu, M. Duraiselvam, Tribological studies on AlCrFeCuCoNi high entropy alloy surface coated on Ti-6Al-4V using plasma transferred arc technique. *Arch. Metall. Mater.* 67 (2022) 409–420.
37. S. Gautam, D. Bhatnagar, D. Bansal, H. Batra, N. Goyal, Recent advancements in nanomaterials for biomedical implants. *Biomed. Eng. Adv.* 3 (2022) 100029.
38. L. Guo, S. A. Naghavi, Z. Wang, S. N. Varma, Z. Han, Z. Yao, L. Wang, L. Wang, C. Liu, On the design evolution of hip implants: A review. *Mater. Des.* 216 (2022) 110552.

39. N. Jeyaprakash, C.-H. Yang, G. Prabu, N. Radhika, Mechanism correlating microstructure and wear behaviour of Ti-6Al-4V plate produced using selective laser melting. *Metals* 13 (2023) 575.
40. M. Montazerian, F. Hosseinzadeh, C. Migneco, F. M. Fook, Baino, Bioceramic coatings on metallic implants: An overview. *Ceram. Int.* 7 (2022) 48.
41. Z. Deng, D. Liu, Y. Xiong, X. Zhu, H. Mi, J. Liu, T. Chen, Laser cladding preparation of HA-ag gradient bioactive ceramic coating: A feasibility study. *Surf. Coat. Technol.* 427 (2021) 127848.
42. G. Prabu, M. Duraiselvam, M. Matheswaran, P. Amrish, T. Jayabalan, M. Varatharajulu, Laser surface texturing for enhancing microbial fuel cell-based electricity generation from wastewater. *Proc. Inst. Mech. Eng. A: J. Power Energy* 236 (5) (2022) 937–948.
43. M. Bohner, Calcium orthophosphates in medicine: From ceramics to calcium phosphate cements. *Injury* 31 (2000) D37–D47.
44. N. Jeyaprakash, G. Prabu, C.-H. Yang, Inclusion of nano silicon particle in SS316L through LPBF and its responses on corrosion behaviour. *Ceram. Int.* 49 (16) (2023) 26450–26468.
45. C. Han, Y. Li, Q. Wang, D. Cai, Q. Wei, L. Yang, S. Wen, J. Liu, Y. Shi, Titanium/hydroxyapatite (Ti/HA) gradient materials with quasi-continuous ratios fabricated by SLM: Material interface and fracture toughness. *Mater. Des.* 141 (2018) 256–266.
46. D. Shekhawat, M. K. Amit Singh, T. S. Banerjee, A. Patnaik, Bioceramic composites for orthopaedic applications: A comprehensive review of mechanical, biological, and microstructural properties. *Ceram. Int.* 47 (3) (2021) 3013–3030.
47. W. Fan, H. Yang, A. C. Taylor, Numerical analysis of fracture in interpenetrating phase composites based on crack phase field model. *Compos. Sci. Technol.* 232 (2023), 109873.
48. D. Wenya, Y. Ai, W. He, W. Chen, B. Liang, C. Lv, Formation and control of "intragranular" ZrO_2 strengthened and toughened Al_2O_3 ceramics. *Ceram. Int.* 46 (6) (2020) 8452–8461.
49. H. Xiong, L. Zhao, H. Chen, H. Luo, X. Yuan, K. Zhou, D. Zhang, Building SiC-based composites from polycarbosilane-derived 3D-SiC scaffolds via polymer impregnation and pyrolysis (PIP). *J. Eur. Ceram. Soc.* 41 (2) (2021) 1121–1131.
50. L. Gonz'alez-Ferna'ndez, L. del Campo, R. B. P'erez-Sa'ez, M. J. Tello, Normal spectral emittance of Inconel718 aeronautical alloy coated with yttria stabilized zirconia films. *J. Alloy Compd.* 513 (2012) 101–106.
51. A. Khorram, A. D. Jamaloei, A. Jafari, M. Paidar, X. Cao, Microstructural evolution of laser-clad 75Cr3C2+25 (80Ni20Cr) powder on Inconel718 superalloy. *J. Mater. Process. Technol.* 284 (2020) 116735.
52. N. Jeyaprakash, C.-H. Yang, G. Prabu, R. Clinktan, Microstructure and tribological behaviour of Inconel-625 superalloy produced by selective laser melting. *Metals Mater. Int.* 28 (2022) 2997–3015.
53. M. Y. Li, B. Han, L. X. Song, Enhanced surface layers by laser cladding and ion sulfurization processing towards improved wear-resistance and self-lubrication performances. *Appl. Surf. Sci.* 503 (2020) 144226.
54. Z. X. Zhu, X. B. Liu, Y. F. Liu, S. Y. Zhang, Y. Meng, H. B. Zhou, S.-H. Zhang, Effects of Cu/Si on the microstructure and tribological properties of FeCoCrNi high entropy alloy coating by laser cladding. *Wear* 512–513 (2023) 204533.
55. C. Vipin, R. Jayaganthan, C. Ramesh, Structural characterizations of magnetron sputtered nanocrystalline TiN thin films. *Mater. Charact.* 59 (2008) 1015–1020.

56. N. Jeyaprakash, G. Prabu, C.-H. Yang, Formation of different phases and their influences on the mechanical and tribological properties of surface alloyed FeCoCrNiAl particles using PTA technique. *Intermetallics* 142 (2021) 107457.
57. S. B. Hu, J. P. Tu, Z. Mei, Adhesion strength and high temperature wear behaviour of ion plating TiN composite coating with electric brush plating Ni-W interlayer. *Surf. Coat. Technol.* 141 (2–3) (2001) 174–181.
58. S. Y. Zhang, X. B. Liu, Y. Zhu, Y. F. Liu, Y. Meng, J. Liang, S.-H. Zhang, Stellite3-Ti_3SiC_2-Cu composite coatings on IN718 by laser cladding towards improved wear and oxidation resistance. *Surf. Coat. Technol.* 446 (2022) 128766.
59. Z. Hua-Juan, J. Meng, X. Yun-Chuan, S. Guang-Yao, L. Shi-Wei, B. Shan-Hu, C. Xun, J. Shi-Dong, J. Ping, Surface deposition of graphene layer for bioactivity improvement of biomedical 316 stainless steel. *Mater. Lett.* 192 (2017) 123–127.
60. Y. Zhi-Lu, T. Qiu-Fen, M. F. Maitz, Z. Shuo, W. Jin, H. Nan, Direct thrombin inhibitor bivalirudin functionalized plasma polymerized allylamine coating for improved biocompatibility of vascular devices. *Biomaterials* 33 (2012) 7959–7971.
61. M. A. Garcia-Lobato, A. I. Mtz-Enriquez, C. R. Garcia, M. Velazquez-Manzanares, F. Avalosbelmontes, R. Ramos-Gonzalez, L. A. Garciacerda, Corrosion resistance and in vitro bioactivity of dense and porous titania coatings deposited on 316L SS by spraying method. *Appl. Surf. Sci.* 484 (2019) 975–980.
62. R. Kumari, J. D. Majumdar, Microstructure and surface mechanical properties of plasma spray deposited and post spray heat treated hydroxyapatite (HA) based composite coating on titanium alloy (Ti–6Al–4V) substrate. *Mater. Charact.* 131 (2017) 12–20.
63. Z. Q. Yao, Y. Ivanisenko, T. Diemant, A. Caron, A. Chuvilin, J. Z. Jiang, R. Z. Valiev, M. Qi, H. J. Fecht, Synthesis and properties of hydroxyapatite-containing porous titania coating on ultrafine-grained titanium by micro-arc oxidation. *Acta Biomaterialia* 6 (2010) 2816–2825.
64. O. B. Kovalev, D. V. Bedenko, A. V. Zaitsev. Development and application of laser cladding modeling technique: From coaxial powder feeding to surface deposition and bead formation. *Appl. Math. Model.* 57 (2018) 339–359.
65. M. Li, B. Han, Y. Wang, K. Pu, Effects of La_2O_3 on the microstructure and property of laser cladding Ni-based ceramic coating. *Optik* 130 (2017) 1032–1037.
66. Z. Zhang, J. Chen, J. Wang, Y. Han, Z. Yu, Q. Wang, S. Yang, Effects of solder thickness on interface behavior and nanoindentation characteristics in Cu/Sn/Cu microbumps. *Weld. World* 66 (5) (2022) 973–983. https://doi.org.10.1007/s40194-022-01261-0.
67. Z. Zhang, B. Yang, G. Wei, Y. Haiyan, L. Yang, X. Shang, T. Zhou, J. Guo, Significantly toughened PZT-based piezoelectric ceramics simultaneously keeping excellent electrical properties via incorporating trace amount of yttria-partially-stabilized zirconia. *Ceram. Int.* 48 (23) (2022) 35614–35620.
68. Z. Deng, D. Liu, G. Liu, Y. Xiong, S. Xin, S. Li, C. Li, T. Cheu, Study on the corrosion resistance of SiC particle-reinforced hydroxyapatite-silver gradient bioactive ceramic coatings prepared by laser cladding. *Surf. Coat. Technol.* 467 (2023) 129734.
69. A.-C. Alina Mocanu, F. Miculescu, G. E. Stan, T. Tite, M. Miculescu, M. H. Ţierean, A. Pascu, R.-C. Alin Ciocoiu, T. M. Butte, L.-T. Ciocan, Development of ceramic coatings on titanium alloy substrate by laser cladding with pre-placed natural derived-slurry: Influence of hydroxyapatite ratio and beam power. *Ceram. Int.* 49 (2023) 10445–10454.

70. M. T. Ebrahimi, D. S. Balint, A. P. Sutton, D. Dini, A discrete crack dynamics model of toughening in brittle polycrystalline material by crack deflection. *Eng. Fract. Mech.* 214 (2019) 95–111.
71. F. Ma, C. Huang, J. Niu, L. Wang, H. Liu, X. Bai, Toughening mechanisms of Al_2O_3-matrix composites with SiC whiskers. *Ceram. Int.* 48 (12) (2022) 17556–17563.

6

Corrosion Resistance Improvement through Laser Cladding Techniques

Natarajan Jeyaprakash, Sundara Subramanian Karuppasamy, Ibrahim A. Alnaser, and Che-Hua Yang

6.1 Introduction

Along with human evolution, the use of pure metals and metallic alloys has evolved in order to meet the demands of each industrial revolution. During the Industrial Revolution era, every industrial sector has attained a lot of significant advancement, which paved the way for growth in each nation's gross domestic product (GDP) along with technological advancements [1–3]. Metallic alloys such as iron-based, nickel-based, titanium-based, and aluminum-based alloys have been listed as the predominantly used alloys in the aerospace, automobiles, nuclear, marine, and other industrial sectors. Each alloy has unique physical, mechanical, electrical, thermal, and electrochemical behaviors, which made them to get implemented while manufacturing the components in the above-mentioned sectors [4,5].

Iron-based alloys contain a significant amount of iron, along with other light and heavy refractory metals. These alloys are well known for their tensile properties, such as higher tensile and shear strength, good shock resistance, better hardness and wear resistance, and greater formability. These alloys contribute nearly 60% of the total weight of manufacturing components in almost all sectors [6,7]. Aluminum-based alloys are predominantly used in manufacturing applications where high strength-to-weight ratio and light weight are the key features. As said, these alloys are extremely light in weight with outstanding characteristics such as electrically active, excellent corrosion resistance, and better tensile strength with good resilience. Mostly, these alloys are used in manufacturing aerospace structures, space shuttles, food and beverage containers, and so on [8,9]. On the other hand, nickel-based alloys can withstand extremely high heat, nearly around 1,200°C, and offer resistance against stress corrosion cracks. Besides their high heat-withstanding capacity, these alloys can deliver efficient performance even near to their melting point along with greater hardness and toughness. These alloys are unavoidable in fabricating components of nuclear

DOI: 10.1201/9781032693576-6

reactors, pressurized water boilers, heat exchangers, boiler tubes, and so on [10,11]. Titanium and titanium-based alloys play a vital role in fabricating surgical instruments and biomedical implants. Due to their high degree of biocompatibility, these alloys are chemically inert to environments containing both acidic and alkaline solutions, such as body fluids, and blood, and resist corrosion to a greater extent [12,13]. Thus, the above-mentioned metallic alloys are unavoidable in fabricating the components that are employed in almost all the industrial sectors.

After long-term use, there is a reduction in the mechanical characteristics offered by these alloys which might decrease the service life of the metallic components made from these alloys. This reduction might be due to the prevailing environmental (working) conditions, such as high pressure, greater stress, radioactive, high temperature, environments containing chemicals, and dissolved oxygen [14]. The high pressure will induce a greater stress on the component's surface, which will lead to crack formation. The high temperature will alter the mechanical characteristics such as tensile strength, and modulus of elasticity which in turn lead to surface breathing cracks on the component's surface. This high pressure and high temperature along with dissolved oxygen and other salts will influence the surface properties of those components through chemical reactions, forming new by-products that severely render the mechanical characteristics offered by those components. In worst cases, these by-products will lead to catastrophic failures, followed by property loss and serve as an alarming danger to humankind [15,16].

One of the alarming and unavoidable issues that every industry faces nowadays is corrosion. Corrosion is defined as the chemical reaction that happens between the metal or metallic alloy's surface and the prevailing environments. This phenomenon is categorized as a serious issue in almost all industries and a recent survey reported that around 70% of the industrial accidents that occurred during the past four decades are related to corrosion. Moreover, 30%–40% of each industry's annual revenue is spent on repairing the issues correlated with corrosion. Hence, corrosion serves as a potential issue among the other issues faced by the industries [17,18].

Since corrosion is an unavoidable phenomenon, there are several ways to overcome corrosion. Temporarily, corrosion-resistant paints are often used which act as the potential barrier to resist corrosion. However, there are certain environments where the implication of these paints is unworthy. Also, corrosion is an electrochemical aspect that occurs on the surface, many researchers work on various methods to resist corrosion of components. One of the most used techniques to avoid the corrosion phenomenon is the surface treatment [19]. These techniques alter the surface properties, thereby uplifting the mechanical characteristics of the components. Also, it is reported that the durability (service life) of the components can be enhanced by treating the surfaces at the micron level [20].

There are several surface treatment techniques that could alter the surface properties of the parent materials. Some of these include deposition techniques, thermal spray, hardfacing, laser cladding, and so on. The deposition techniques, such as physical and chemical vapor deposition, focused on depositing the coating materials on the parent materials (substrates) via physical and chemical media. Both vapor deposition techniques require skilled labor and these techniques are not widely implemented because of their limitations such as (i) low (thin) coating thickness, (ii) high process cost, (iii) not being suitable to coat on complex shapes, (iv) release of hazardous gases, and (v) lack of precursors [21,22]. On the other hand, thermal spraying is a process where the substrate material is preheated to ensure a better deposition rate in order to obtain a fine coating finish. However, this process is expensive and is mostly suitable for a limited range of substrates [23,24]. Most of the surface treatment techniques repair surfaces by creating thin coatings on the substrates with high process costs. Also, it is hard to obtain coatings with surface defects, such as pores and cracks, which might act as nucleation sites and thus initiate corrosion [25].

Laser cladding is one of the most prominent surface treatment techniques, which employs a highly efficient laser beam to irradiate the substrate's surface and melt it, resulting in the formation of a molten metallic pool. The coating particles are deposited on this melt pool and solidified to form a coating on the substrate, with thickness in the range of microns to several millimeters [26]. This technique is reported to overcome the limitations of the other surface treatment techniques such as hardfacing, thermal spraying, and vapor deposition processes [27]. One of the main advantages of this technique is that it can be adapted for a wide range of substrates irrespective of material type, size and shape. By means of laser cladding, the substrate's surface can be repaired or treated in order to enhance the mechanical characteristics offered, such as service life, corrosion resistance, wear properties, and tensile behaviors. Moreover, this technique is quite advantageous over other techniques in terms of process cost, maintenance cost, labor cost, and ability to produce coatings with better surface finish and fewer or zero pores. Mostly, this technique is employed to elevate the surface properties of the substrates [28–30]. This chapter focuses on discussing the corrosion-resistant coatings fabricated via laser cladding. Moreover, it also details the cladding principle, corrosion-resistant powders, electrochemical reactions that take place on the clad surface, and the mechanisms behind passive film formation on the clad surfaces, along with relevant literature.

6.2 Corrosion and Its Types

As discussed earlier, corrosion is an electrochemical phenomenon that is unavoidable in most metals and metallic alloys. The corrosion can occur on the surface, subsurface, and inner of any metallic component. Based on the

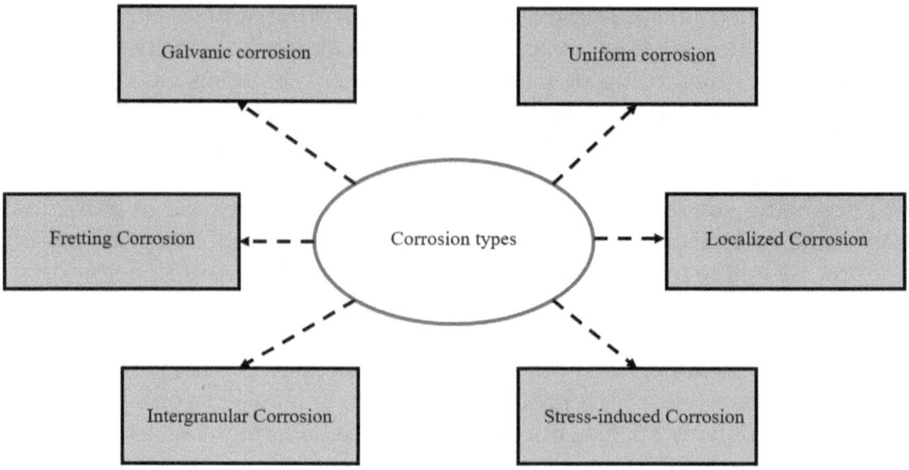

FIGURE 6.1
Types of corrosion.

prevailing environmental conditions and the nature of the material associated, corrosion can be classified into many categories. This section details the most important types of corrosion (Figure 6.1) that are usually associated with industries.

6.2.1 Uniform Corrosion

The most common type of corrosion that usually happens on the metallic surface is the uniform corrosion. As the name implies, the corrosion by-products are uniform throughout the sample's surface and this type is reported to spread over a larger surface area than the other corrosion types. In the presence of corrosive fluids (water, moisture, etc.), the metal's surface readily reacts with the corrosive media and thus forming uniform corrosion throughout the sample's surface. Moreover, the risk level of danger is quite low when compared to the other types since this type reacts with the sample's surface rather than the subsurface. Some examples of this kind include the formation of rust on any iron components, tarnishing of silver, and so on [31,32].

6.2.2 Galvanic (Bimetallic) Corrosion

In industries, different types of metallic alloys, such as iron, nickel, and aluminum, are used to manufacture components and are assembled to form assemblies. Thus, a mechanical assembly consists of parts made from various metallic alloys having different galvanic (electro) potentials. When exposed to corrosive media, some of the parts in the assembly might corrode in a faster manner than the other parts. This type of corrosion is termed as galvanic or bimetallic corrosion. The galvanic corrosion happens between the components that have different electropotential values. This difference in electropotentials caused some alloys to corrode rapidly while the other parts were left uncorroded [33,34].

6.2.3 Intergranular Corrosion

Most of the metallic components are fabricated to the respective shapes and sizes as per the demands. Traditional manufacturing methods such as casting and forging, and advanced manufacturing methods such as additive manufacturing, are employed in fabricating the metallic components. Both methods use heat energy to melt the metallic powders, and after solidification, the respective component is obtained. Intergranular corrosion happens due to the defects present during the production stage. During solidification, the impurities present in the alloy powders were segregated near the grain boundaries [35]. The segregated impurities act as the nucleation sites when exposed to corrosive media. Aggressive corrosion reaction happens in these sites and starts to spread throughout the microstructure. Also, this type of corrosion is seen in metals during the heat treatment process. Thus, the impurity regions act as the corrosion sites, while the other regions remain unattacked by the corrosive media [36].

6.2.4 Fretting Corrosion

There are some cases where the corrosion is experienced in rigidly coupled or riveted surfaces. Under cyclic loading (vibration), in the presence of corrosive media, the vibrations will create minute air gaps between the riveted or rigidly coupled surfaces. These air gaps react with the corrosive ions, thereby initiating corrosion at the coupled surfaces. This phenomenon is termed as fretting corrosion and is quite common in joints (riveted, fastened, and coupled joints) [37,38].

6.2.5 Stress-Induced Corrosion (Stress Corrosion Cracks)

One of the important factors that is reported to diminish the mechanical characteristics offered is the temperature. Most metals and metallic alloys are prone to severe damage with the effect of a rise in service temperatures. Usually, the tubes of boilers, heat exchangers, and pressurized water reactors transmit hot fluids and work in elevated temperatures with high pressure [39]. Prolonged exposure of these tubes in such environments induces stress on the surface of the tubes, which can lead to formation of cracks. These stress-induced cracks serve as the nucleation sites for stress corrosion cracking on the tubes [40].

6.2.6 Localized Corrosion

As the name implies, localized corrosionis corrosion attacks that occur in a specific area (localized areas), which in turn leads to catastrophic failure and fatal accidents. Localized corrosion is of three types, namely pitting corrosion, crevice corrosion, and filiform corrosion. Pitting corrosion is a serious issue faced by the marine industry. Pits are small pores that are formed on the surface of the metals. In due course, these pores will develop into leak holes, and if unnoticed, numerous leak holes cause sudden failure of the component [41]. Crevice corrosion usually happens in clogged regions in surfaces exposed to acidic environments [42]. On the other hand, filiform corrosion occurs between the interface of the painted layer and metals. Generally, to avoid corrosion, there are certain paints that are sprayed on the metallic surface, which can resist corrosion to some extent. However, in certain cases like galvanic corrosion, the metal surface reacts with the paint, initiating corrosion at the interface between the paint and metal surface, and starts to detach the paint from the surface, leading to filiform corrosion [43]. Thus, these are the most common corrosion types faced by the industries.

Some of the real-time issues are discussed as follows: the components in marine and offshore industries work in environments containing dissolved salts, dissolved oxygen, and moisture. The dissolved salts contain more aggressive chloride ions. In general, the chloride ion has the tendency to rigorously attack the metal surface and thus forming nucleation sites that promote corrosion [44].A pipeline serves as the transporting medium for transporting fluids such as water, oil, and chemical substances for long distances. Two or more pipes are welded together in order to transport without any leakage of the transporting substances. Galvanic and intergranular corrosion takes place in the welded region of the pipes which leads to failure and affects the transportation [45]. The fuel storage tanks of automobiles and space shuttles are reported to contain sediments which got settled on the inner surface. This is due to the sulfides and fluorides present in the fuel and these sediments could pave the way for inducing pitting corrosion and in due course, these pits are often developed into leak holes which will cause failures [46]. Thus, corrosion serves as an unavoidable phenomenon in almost all sectors.

6.3 Laser Cladding

Out of various surface treatment techniques, laser cladding holds good for producing highly efficient coatings on the substrate's surface. And, this technique can be adapted to a wide range of substrates than any other technique. The working principle behind the laser cladding process is described as follows. Here, a high-power (energy) laser beam is used which serves as the heat energy for the whole process. The sample (substrate) is rigidly clamped in the mounting table, and the whole process takes place in an inert gas (shielding gas) environment to avoid the influence of atmospheric gases. A high-energy laser beam is directed on the substrate's surface and this beam locally melts the surface to form a molten metallic pool [47,48]. The coating powders reach this molten metal pool and settle in the localized pool zone. After solidification, a fine layer of clad (coating) layer is formed on the substrate's surface. This clad layer is reported to act as the shield for the substrate, thereby improving the service life of the substrate [49]. Figure 6.2 represents a schematic illustration of the laser cladding process.

The major advantages of laser cladding over other surface treatment techniques are as follows [50]:

i. Substrate/coating materials – laser cladding can be well adapted to a wide range of substrate and coating materials, irrespective of their size and geometries.
ii. Durability – the clad (coating) layer produced via laser cladding has an excellent service life than the coatings prepared by other surface treatment techniques, such as thermal spraying or hardfacing.

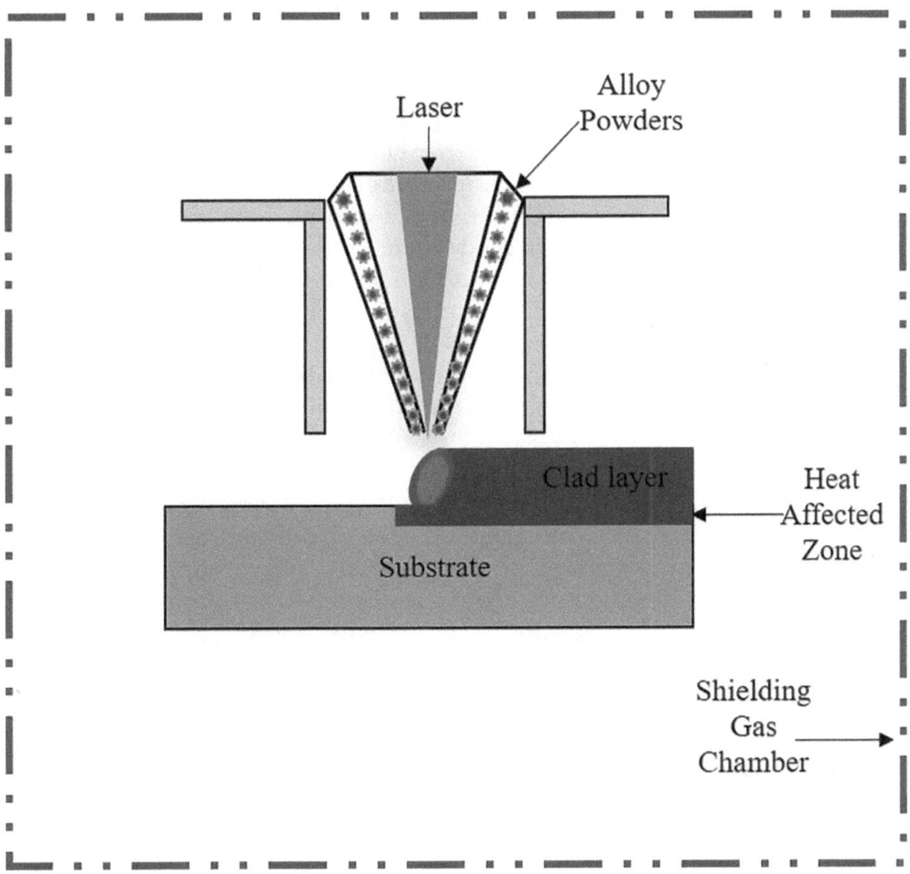

FIGURE 6.2
Laser cladding process.

iii. Distortion rate – the clad layer is reported to exhibit a low distortion rate and minimal heat-affected zone.
iv. Defect-free – since the coating powders are deposited on the molten melt pool, the obtained clad layer is reported to be free from voids or pores.
v. Microstructure and cooling rate – the cooling rate is said to be high which accounts for faster solidification and with this cooling rate, the clad layer exhibits finer microstructure, which is responsible for elevating the mechanical characteristics offered.

The deformation rate of the coatings produced by laser cladding is reported to be very low, and strong metallurgical bonding is exhibited between the substrate and the clad layer. With these advantages, laser cladding serves as the ideal technique for repairing or rebuilding worn-out surfaces with excellent mechanical characteristics.

Based on the feedstock, the laser cladding is broadly classified into two categories, namely wire and powder feedstock-based laser cladding. In the wire feedstock system, the coating particles represented in a wire form were used to coat on the substrate's surface [51]. On the other hand, the powder feedstock-based laser cladding is further classified into two types, namely preplaced powder and coaxial powder laser cladding systems. As the name implies, in preplaced powder cladding, the coating powder particles were mixed with certain solutions and made them into a paste. This powder paste is applied on the substrate's surface and when the laser beam hits this paste, the powder particles are deposited as a layer on the surface, while the solution used to make the paste is evaporated without any effects on the deposited (formed) clad layer [52]. In coaxial powder-based laser cladding, the powder particles have separate pathways which are aligned at some angle. In this system, the laser beam is directed on the substrate's surface forming the melt pool, while the powder particles reach this pool via coaxial powder pathways and solidify to form the clad layer [53]. Moreover, for obtaining a highly efficient clad layer on substrates, the cladding process parameters (Figure 6.3), such as laser power, powder feed rate, laser scanning speed, carrier gas flow rate, and percentage of overlap needed to be optimized [54]. This optimization process depends on the type of substrate and coating materials used during the laser cladding process.

6.4 Corrosion-Resistant Powders

To improve the mechanical characteristics offered by the substrates, various metallic alloy powders were deposited on different substrates. Each alloy powder is reported to improve specific mechanical behaviors, such as wear

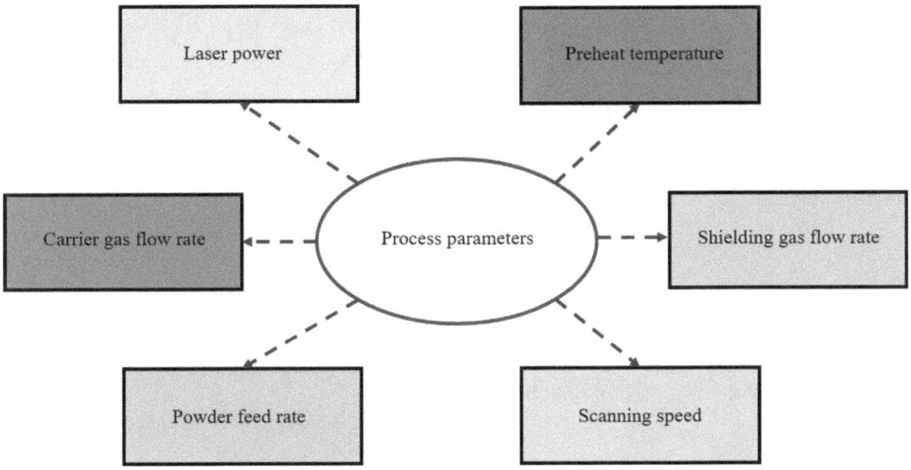

FIGURE 6.3
Laser cladding process parameters.

resistance, hardness, fatigue behaviors, tensile properties, and anticorrosion behaviors. Hence, care should be taken while choosing the particular alloy powders for enhancing particular mechanical behavior. There are certain alloy powders that could be well adapted to improve the corrosion-resistant behaviors of the substrate. Some of them are discussed as follows (Figure 6.4). The schematic illustration of the corrosion test setup is shown in Figure 6.5.

6.4.1 Nickel-Based Superalloys

Nickel-based superalloys belong to the class of superalloys which are used in fabricating components that work in high-temperature environments. These are considered as structural alloys that are obtained as the combined effect of solid solution strengthening and precipitation hardening. These alloys contain a higher percentage of nickel along with other refractory metals such as chromium, molybdenum, tungsten, and magnesium. These alloys are extensively used in manufacturing the components of nuclear reactors, heat exchangers, boilers, and power plants. Out of the various nickel-based superalloys, Inconel and Colmonoy series were extensively used to resist corrosion to a greater extent [55,56].

Inconel alloy belongs to the class of Ni-based superalloys which is reported to have an adequate combination of Ni, Fe, Cr, Mo, and Nb. This alloy was initially developed for the petrochemical industries but its various characteristics made this alloy as a suitable manufacturing material for aggressive applications. This precipitation-hardened alloy can perform outstandingly even near to its melting points. During cladding, these alloys are reported to produce finer microstructure with lave phases that are distributed within the gamma nickel matrix. These lave phases are said to proliferate the mechanical

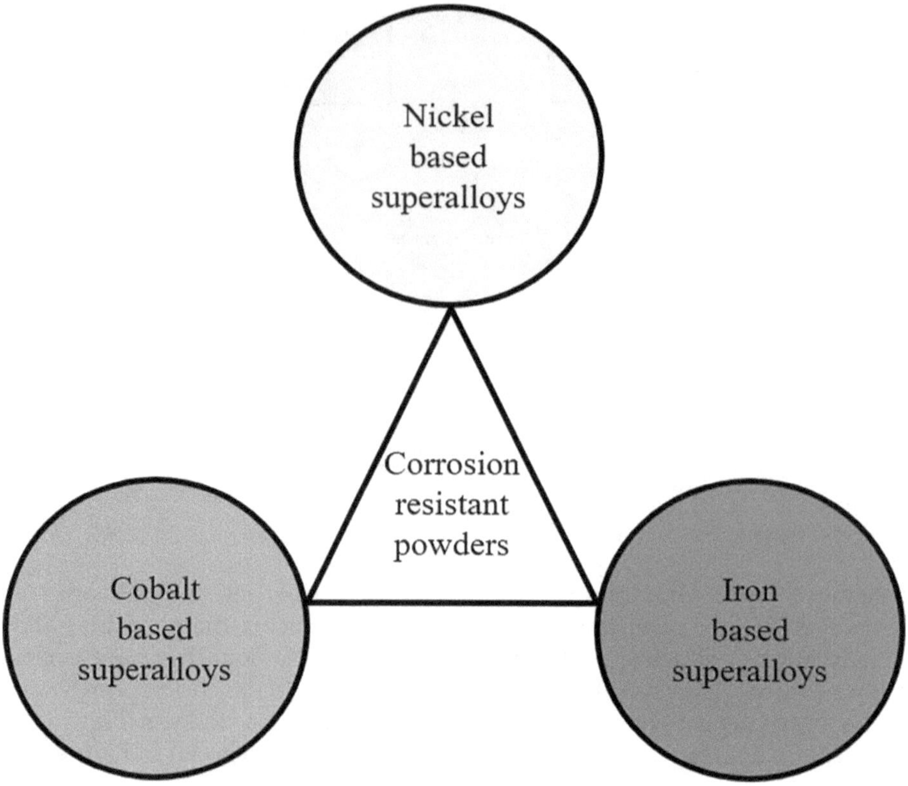

FIGURE 6.4
Corrosion-resistant powders used in laser cladding.

characteristics of this hardened alloy [57,58]. On the other hand, Colmonoy alloy, also belongs to the category of nickel-based superalloys, contains a relatively higher percentage of chromium and boron. During cladding, the Cr –B combination results in the formation of floret-like precipitates in the gamma nickel matrix. The grain boundaries of these precipitates were decorated with dark regions, which are termed as secondary precipitates. These secondary precipitates are said to be rich in CrB, which in turn improves the mechanical behaviors offered by this alloy [59,60]. Both the Inconel and Colmonoy series are reported to produce highly compact passive films when exposed to corrosive media, thereby providing greater resistance against corrosion. Some of the relevant literature is discussed as follows.

Jelvani et al. [61] evaluated the microstructure of laser-cladded Inconel 718 alloy at various process parameters and established the influence of powder feeding rate, laser power, and scanning speed on the microstructure formation during laser cladding. At various laser powers, the evolution of columnar dendrites to equiaxed dendrites was observed. The equiaxed dendrites were reported when the laser power is over 300 W. The cooling rate tends

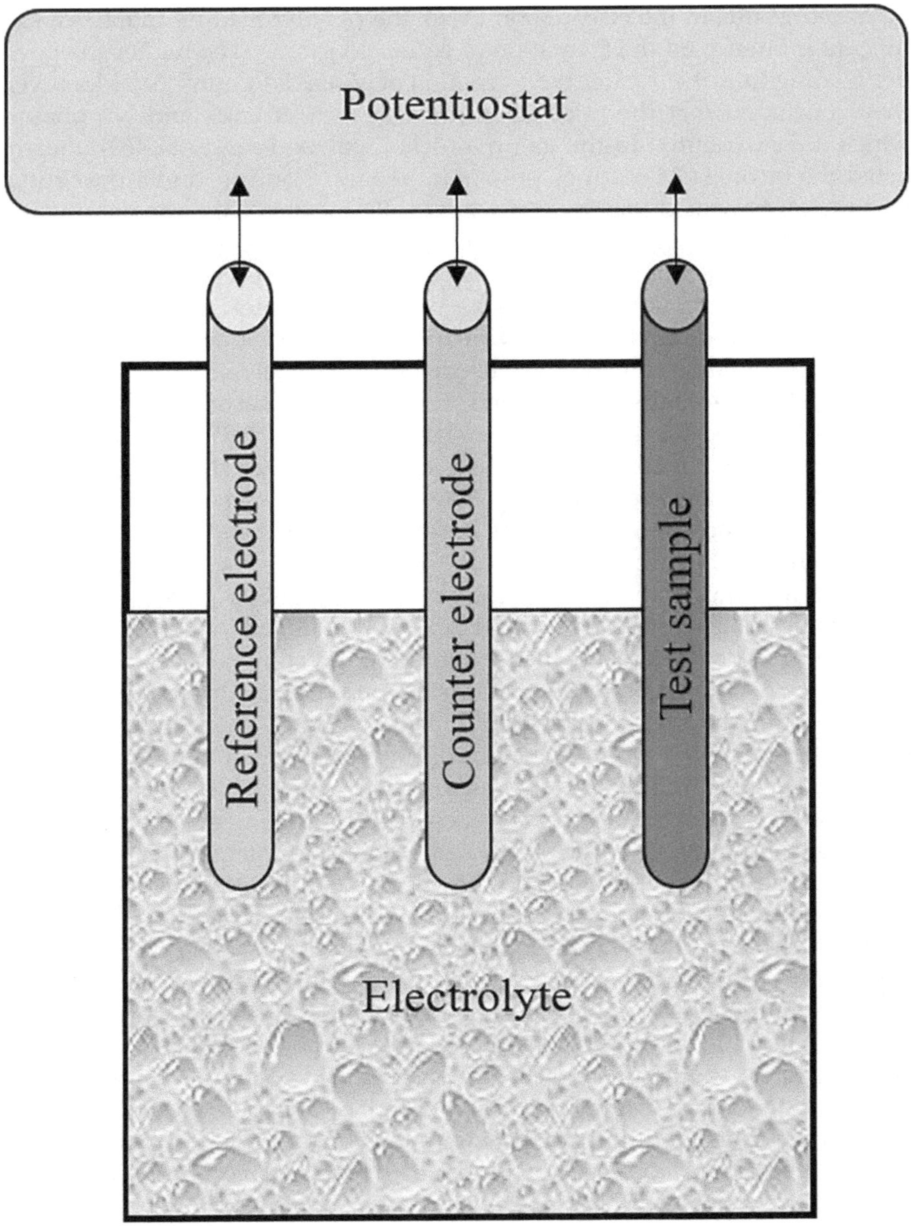

FIGURE 6.5
Corrosion test (three-electrode setup).

to be lower at higher laser power, which tends to the formation of equiaxed dendrites. The scanning rate is directly proportional to the equiaxed grain formation. A higher scanning rate is said to increase the tendency to form the equiaxed grains in the clad region. Also, the powder feeding rate has a significant influence on the formation of equiaxed grains. The higher the powder feeding rate, the greater the formation of equiaxed grains [62]. Moreover, they concluded that the microstructure was rich in laves and Nb phases, which are distributed in the gamma nickel matrix. Feng et al. [63] interrogated the Inconel 625 coatings prepared via laser cladding and arc welding, in terms of hardness and wear resistance. By means of the coaxial powder feed system, the Inconel 625 powder particles were deposited on the molten metal pool of martensitic steel under optimal process parameters. On comparing the microstructure, the laser-cladded Inconel 625 microstructure consists of columnar dendrites with niobium carbides as their secondary precipitates, whereas the microstructure of the Inconel 625 coating obtained by arc welding contains coarse dendrites with fewer secondary precipitates (NbC). Moreover, the Energy Dispersive Spectroscopy (EDS) area maps of the laser-cladded Inconel 625 coating exhibited better distribution of Ni, Fe, Cr, Mo, and Nb elements [64]. The hardness studies revealed that the laser-cladded Inconel 625 layer has higher hardness compared to the hardness offered by the arc-welded Inconel 625 layer. This increase in hardness is due to the columnar dendrites formation along with a greater amount of Nb carbides. The arc-welded Inconel 625 layer exhibited lower hardness due to the coarse dendrites and inadequate NbC precipitation. Abrasive wear is reported in the Inconel 625 coatings prepared by both arc welding and laser cladding, but the laser-cladded Inconel 625 layer has a lower wear rate due to the precipitation of large amounts of NbC's and columnar dendrites. Fesharaki et al. [65] compared the microstructure of the Inconel 625 coatings obtained on Inconel 738 through laser cladding and Tungsten inert gas (TIG) welding techniques. With the optimized process parameters, the Inconel 625 alloy particles were deposited, and the microstructure formed via laser cladding contains fine precipitation of coaxial and columnar dendrites, while the TIG-welded Inconel 625 layer exhibits coarse dendrites. Also, the clad layer experiences a higher cooling rate when compared to the TIG-welded layer and this high cooling rate leads to finer dendrites at the clad layer. Abioye et al. [66] investigated the Inconel 625 clad layer obtained on stainless steel 304 substrate by both wire and coaxial powder feed systems. Microstructure, dilution phenomenon, and hardness offered by the clad layers were analyzed. They concluded that both the wire feed and coaxial powder feed systems can be effectively used for producing the clad layer. Both clad layers were free from pores and voids. However, regarding the production of a continuous, single track of clad layer on the 304 substrates with a better surface finish, they suggested that the wire feed system holds good regarding the surface quality of the clad layers when compared to the powder feed system. On the other hand, a higher dilution rate was observed in the clad layers produced by the powder feed system. Moreover, in terms of mechanical characteristics such as hardness, the clad layer formed via the coaxial powder feed system has higher hardness compared to the hardness offered by the wire feed-clad layers. This high hardness might be due to the higher dilution rate obtained while feeding the Inconel 625 alloy via a coaxial powder feed system. Since laser cladding is a process of fabricating coatings on the substrate via irradiating the substrate's surface with a high-power laser beam, the heat flow decides the coating's behaviors, such as microstructure, hardness, wear resistance, and corrosion resistance. Wang et al. [67] established the relationship between heat flow and solidification via numerical modeling of laser-cladded Inconel 718 alloy. With the aid of this model, the thermal behavior exhibited by the molten melt pool was analyzed. Also, the relevant heat flow, which helps in attaining the laves phase, was studied along with the latent heat of heat change.

Other than the Inconel series, the Colmonoy series has been used to enhance the mechanical characteristics of the substrates. Zhang et al. [68] investigated the characteristics Colmonoy 6 clad layer obtained on the stainless steel 316L substrate. With the optimal process parameters, Colmonoy 6 powder particles were deposited on the 316L molten melt pool, and further studies have been carried out on the Colmonoy 6 clad layer. They also established the relationship between the various process parameters such as laser power, scanning speed, and powder feeding rate on the microstructural evolution in the clad layer. The Colmonoy 6 clad regions exhibited floret-like primary precipitates in the gamma nickel matrix [69]. In the case of the Inconel clad layer, the NbC contributed more in enhancing the hardness and wear resistance characteristics but here, the grain boundaries of the floret-like primary precipitates are decorated with dark (blocky) secondary precipitates, which are reported to be rich in chromium borides. Furthermore, they concluded that the clad bead's height, width, and dilution rate, increase with increasing the laser power. Also, the excellent wear resistance is provided by the formed Colmonoy 6 clad layer due to the segregated secondary precipitates. Ming et al. [70] evaluated the effect of powder feed rate and scanning speed on the microstructural evolution and mechanical properties exhibited by the two different Colmonoy alloy powders, namely Colmonoy 6 and Colmonoy 88. The major outcomes of their work are as follows. On inspecting the clad layers obtained from the same process parameters, the Colmonoy 6 clad layers exhibited nickel-rich phases such as NiB, NiSi, and NiC and chromium-rich phases like CrB and CrC. On the other hand, the Colmonoy 88 clad layer lacks in boride content. This layer contains a combination of nickel and nickel carbides. Moreover, the hardness property offered by these two Colmonoy alloy layers depends on the segregation of secondary precipitates such as borides of Cr and Ni. On comparing the hardness offered, the Colmonoy 6 layer exhibited better hardness than the Colmonoy 88 layer due to the adequate composition of CrB precipitates in its grain boundaries. Also, these precipitates are responsible for improving the wear resistance of the Colmonoy 6 clad layer. Kaul et al. [71] also reported that the Colmonoy 6 clad layer shows excellent mechanical characteristics and thus makes it as the most suitable alloy for fabricating coatings.

6.4.2 Cobalt-Based Superalloys

Cobalt-based superalloys are predominantly used in environments where hardness and corrosion resistance are the major key factors. These alloys contain significant percentages of Co and Cr, alloying with other elements. When coated on the substrate, this alloy is reported to exhibit excellent mechanical behaviors, and could enhance the service life of the components to a greater extent. One of the most used cobalt-based superalloys is the Stellite series. Alimardani et al. [72] evaluated the effect of process parameters while coating Stellite 1 alloy on the SAE 4340 steel via laser cladding. With the optimized process parameters, the Stellite 1 alloy was coated and through numerical simulations, the experimental results were verified. The Stellite 1 clad layer was rich in the Co–Cr matrix with the segregation of secondary precipitates. These secondary precipitates could proliferate the mechanical characteristics offered by this alloy. Ganesh et al. [73] interrogated the fracture kinetics of the Stellite 21 alloy cladded on the stainless steel 316L substrate. By means of various mechanical tests, the fracture kinetics of the Stellite 21 alloy was evaluated. The major outcomes of their results are as follows. Stellite 21 clad region exhibits fine dendrites with interdendritic precipitates which are distributed in the Co–Cr matrix. Moreover, these interdendritic precipitates are decorated as the grain boundaries of the primary dendritic phase. From the tensile test, it is reported that the Stellite 1 clad layer has maximum resistance against fracture when compared with the stainless steel substrate due to the formed secondary precipitates. D'Oliveira et al. [74] investigated the Stellite 6 clad layers obtained through consecutive depositions. When compared to single-layer deposition, multilayer (consecutive layer) deposition is reported to increase the residual stress, thereby decreasing the thermal gradients. Moreover, the hardness offered by the single-layer deposition is quite uniform throughout the layer, whereas the multilayer deposition is said to have non-uniform hardness. The single-layer deposition resulted in finer dendrites, whereas the multilayer deposition resulted in coarser dendrites which might be responsible for the non-uniform hardness as mentioned. Wu et al. [75] analyzed the microstructure evolution that happened while depositing the Stellite 6 alloy particles on mild steel. They reported that the formed microstructure consists of primary eutectic dendrites with the segregation of metal carbides. Furthermore, these metal carbides are reported to improve mechanical characteristics such as hardness, wear resistance, and high-temperature oxidation resistance [76].

6.4.3 Iron-Based Superalloys

Iron-based superalloys, such as stainless steel, hold good to resist corrosion. Stainless steel contains a higher percentage of iron and chromium, along with other alloying elements. When exposed to the corrosive media, a dual-layer oxide film is formed on the stainless steel's surface, thereby resisting corrosion to a greater extent. Alvarez et al. [77] evaluated the clad layer obtained by depositing 316L stainless steel particles on the carbon steel substrate at various process parameters, such as laser power, powder feeding rate, and scanning speed. Better metallurgical bonding between the substrate and the clad layer was noticed when the powder feed rate was in the range of 25–31 g/min. Moreover, this range combined with higher laser power results in finer microstructure of the clad layer, thereby enhancing the mechanical characteristics offered. Also, two different patterns of deposition were analyzed namely zig–zag and parallel patterns. Among these two patterns, regarding the thickness of the clad layer, the zig–zag pattern holds well since the height of the clad layer is reported to be higher than the parallel pattern. On the other hand, better homogeneity in clad layers was exhibited by the parallel pattern of deposition. Ding et al. [78] optimized the laser cladding process parameters for coating 316L stainless steel on U71Mn rail material and investigated the wear behaviors of the 316L clad layer. The laser power was varied in the range of 1,500–3,000 W, with the scanning speed in the range of 10–20 mm/s at a constant powder feed rate (15 g/min). The major outcomes of this research are as follows. The height and width of the clad layer tend to increase by higher laser power. Moreover, at higher laser power, the evolution of columnar dendrites to equiaxed cellular dendrites takes place in the clad region. The combination of higher laser power and higher scanning speed results in the formation of fine grains at the clad region, thereby providing excellent hardness when compared to the clad layers obtained at lower laser power and lower scanning speed. Also, these fine grains are responsible for providing better wear resistance than the U71Mn substrate. Furthermore, the mechanical behaviors of the stainless steel clad layer could be improved by doping molybdenum particles while performing cladding. Zhang et al. [79] evaluated the microstructure and the influence of Mo particles combined with stainless steel 420 particles via laser cladding on the mild steel A36 substrate. At higher laser power, the clad region contains adequate distribution of the Mo particles as grain boundaries of eutectic dendrites. Furthermore, this doping of Mo with 420 particles improved the segregation of precipitates at the clad layer and offered excellent hardness and wear resistance, compared with the mild steel substrate. Wang et al. [80] also confirmed that the microstructure of the stainless steel clad layer was rich in the hardened precipitates of Mo, Cr, and Fe. These secondary precipitates provide superior mechanical behaviors that will enhance the durability of the substrate.

6.4.4 Other Alloy Powders

Other than the nickel-, cobalt-, and iron-based superalloys, there are certain metallic alloys that exhibit excellent anticorrosion behaviors and can be utilized to clad on different substrates for improving the mechanical characteristics of the parent materials (substrates). Aluminum alloys are metallic alloys that are preferred in industrial environments where a high strength-to-weight ratio, along with excellent corrosion resistance, are the major key factors [81,82]. These alloys contain a significant amount of aluminum, alloying with other metallic elements, such as chromium, magnesium, and zinc, and offer outstanding mechanical performance. When exposed to corrosive media, this alloy is reported to form a highly stable aluminum oxide passive film on its surface, thereby resisting corrosion to a greater extent [83]. Grohol et al. [84] and Cheng et al. [85] investigated the effect of various cladding parameters, such as laser power, scanning speed, and powder feed rate, for fabricating aluminum alloy clad layers on substrates. They reported that the clad layer thickness is entirely dependent on the laser power and scanning speed. Higher laser power and scanning speed provide sufficient heating, thereby increasing the deposition rate of aluminum alloy powders, which in turn increases the clad layer's width and thickness on the substrate material. Moreover, this higher laser power and scanning speed aids in improving the precipitation rate of intermetallic precipitates, which are distributed at the grain boundaries in the α-Al matrix. These intermetallic precipitates are reported to elevate the hardness and wear resistance characteristics offered by the aluminum clad layer [23]. Song et al. [86] established a numerical model for examining the crack resistance offered by the aluminum alloy clad layer on the aircraft skin. They analyzed the temperature and stress field distributions on the clad layer by varying the laser power and scanning speed in the numerical model and concluded that residual stress increases with an increase in the laser power, which also results in obtaining a better temperature distribution in the molten melt pool. The crack resistance was improved at higher laser power and scanning speed due to the increase in the precipitation rate of secondary precipitates via better temperature distribution. Furthermore, the corrosion resistance offered by the aluminum alloy clad layer could be improved with the aid of composites. Feng et al. [87] analyzed the effect on corrosion resistance by mixing aluminum powder with Inconel 625 powder to obtain a combined Al and Inconel 625 clad layer. The corrosion test was carried in the NaCl environment and their results showed that the passive film formed on the Al+Inconel clad layer was rich in the oxides of aluminum and chromium. Moreover, the higher percentage of aluminum results in the formation of Al–Ni intermetallic precipitates, which aids in improving the hardness behaviors [88]. Riquelme et al. [89] investigated the combined effect of Al/SiC composites on the anticorrosion characteristics offered by the composite clad layer at various time intervals. By varying the SiC percentage, the precipitation rate of Al–Si secondary precipitates, such as aluminum carbides were improved. The corrosion test revealed that the passive film formed on the clad layer was rich in higher percentages of aluminum oxides and hydroxides, thereby resisting corrosion to a greater extent.

The biocompatibility and the biocorrosion resistance of the medical implants could be improved by cladding them with titanium alloys. Nabhani et al. [90] examined the corrosion characteristics offered by the Ti-6Al-4V clad layer in different corrosive media. Their study reported that the biocompatibility and the corrosion resistance offered by the Ti-6Al-4V clad layer were quite better than the substrate, due to the formation of compact oxide film on the clad layer. This passive film was rich in the oxides of titanium and aluminum, thereby proliferating the corrosion-resistant characteristics [91]. Obadele et al. [92] investigated the effect of tribocorrosion characteristics exhibited by the pure titanium clad layer obtained on the biomedical implants. They concluded that the polarization resistance offered by the pure Ti-clad layer was higher than the Ti-6Al-4V substrate and offers significant resistance to corrosion when compared with the substrate. Also, the pure Ti-clad layer resulted in improved wear resistance than the substrate layer due to the precipitation of secondary phases in its microstructure [93]. Furthermore, on a low scale, the magnesium and copper alloys were used as the cladding materials, with respect to their field of application [94–97].

6.5 Passivation

Passivation is the process in which an oxide film is formed on the surface of the metals or metal alloys, thereby protecting the metal against corrosion. This oxide layer is reported to act as a potential barrier against corrosion. The importance of laser cladding lies in its ability to enhance the corrosion resistance, thereby improving the service life of the components. After cladding, when exposed to corrosive media, the cladded elements react with the atmospheric oxygen or oxygen from the corrosive media, and thus forming an oxide layer (Figure 6.6) on the surface. Moreover, this passive film is a result of the electrochemical reaction that happens on the clad surface. The electrochemical reactions that take place behind the passive film formation in each superalloy are discussed as follows.

i. The clad layer obtained from the nickel-based superalloy contains nickel and chromium as their major constituent. Hence, their passive film is reported to be rich in oxides of nickel and chromium. Equations (6.1)–(6.7) represent the chemical reaction behind the passive film formation in nickel-based superalloys [98,99].

$$Ni \rightarrow Ni^{2+} + 2e^- \tag{6.1}$$

$$Cr \rightarrow Cr^{3+} + 3e^- \tag{6.2}$$

$$Ni^{2+} + 2H_2O \leftrightarrow Ni(OH)_2 + 2H^+ \tag{6.3}$$

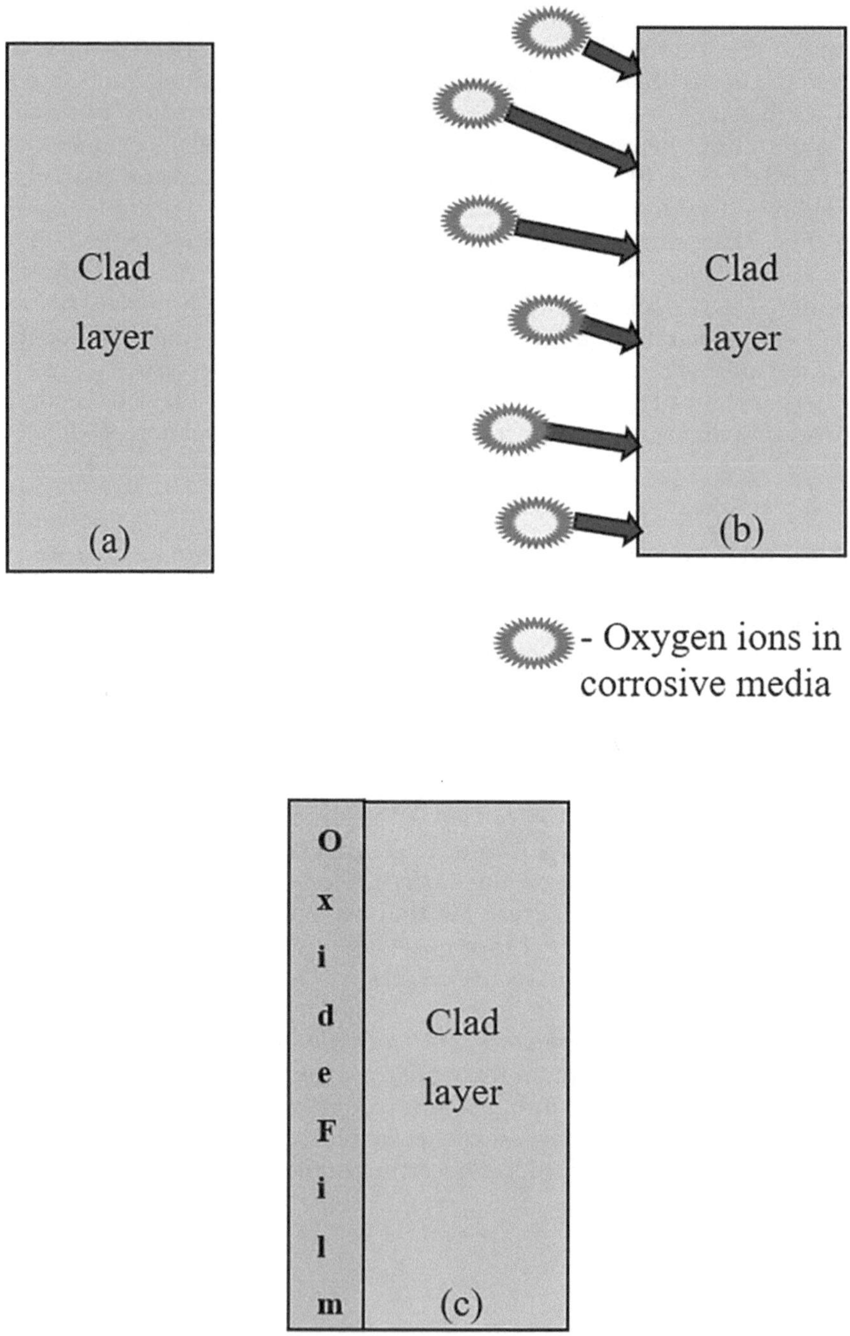

FIGURE 6.6
Passive film formation on the clad layer: (a) clad layer, (b) oxygen ions present in the corrosive media reacting with the clad layer, and (c) passive film formation on the clad layer.

$$Cr^{3+} + 3H_2O \leftrightarrow Cr(OH)_3 + 3H^+ \tag{6.4}$$

$$Cr(OH)_3 + Cr \rightarrow Cr_2O_3 + 3H^+ + 3e^- \tag{6.5}$$

$$Ni(OH)_{2(aq)} \leftrightarrow NiO_{(s)} + H_2O \tag{6.6}$$

$$2Cr^{3+} + 7H_2O \rightarrow Cr_2O_7{}^{2-} + 14H^+ + 6e^- \tag{6.7}$$

ii. The cobalt-based superalloy's clad layer was rich in Co and Cr phases. Hence, the passive film contains oxides of Co and Cr. Equations (6.8)–(6.14) represent the chemical reaction behind the passive film formation in cobalt-based superalloys [100,101].

$$3Co^+ + 6e^- + 2O_2 \rightarrow Co_3O_4 \tag{6.8}$$

$$4CoO \rightarrow Co_3O_4 + Co^+ + 2e^- \tag{6.9}$$

$$3CoO + 1/2O_2 \rightarrow Co_3O_4 \tag{6.10}$$

$$Cr \rightarrow Cr^{3+} + 3e^- \tag{6.11}$$

$$Cr^{3+} + 3H_2O \leftrightarrow Cr(OH)_3 + 3H^+ \tag{6.12}$$

$$Cr(OH)_3 + Cr \rightarrow Cr_2O_3 + 3H^+ + 3e^- \tag{6.13}$$

$$2Cr^{3+} + 7H_2O \rightarrow Cr_2O_7{}^{2-} + 14H^+ + 6e^- \tag{6.14}$$

iii. When exposed to corrosive media, the iron-based superalloys are reported to form a dual oxide film on it. The outer layer is the thin ferric oxide film, whereas the inner layer is the thick chromium oxide film. Equations (6.15)–(6.23) represent the chemical reaction behind the passive film formation in iron-based superalloys [102,103].

$$2Fe \rightarrow 2Fe^{2+} + 2e^- \quad (6.15)$$

$$Cr \rightarrow Cr^{3+} + 3e^- \quad (6.16)$$

$$\frac{1}{2}O_2 + H_2O + 2e^- \rightarrow 2OH^- \quad (6.17)$$

$$Cr^{3+} + 3H_2O \leftrightarrow Cr(OH)_3 + 3H^+ \quad (6.18)$$

$$2Fe + \frac{1}{2}O_2 + H_2O \rightarrow Fe(OH)_2 \rightarrow Fe_2O_3.xH_2O \quad (6.19)$$

$$Cr(OH)_3 + Cr \rightarrow Cr_2O_3 + 3H^+ + 3e^- \quad (6.20)$$

$$2Fe(OH)_2 + \frac{1}{2}O_2 + H_2O \rightarrow 2Fe(OH)_3 \rightarrow Fe_2O_3.H_2O \quad (6.21)$$

$$2Cr^{3+} + 7H_2O \rightarrow Cr_2O_7^{2-} + 14H^+ + 6e^- \quad (6.22)$$

$$Fe_2O_3 + 2Cr \rightarrow Cr_2O_3 + 2Fe \quad (6.23)$$

$$3Fe_3O_4 + 8Cr \rightarrow 4Cr_2O_3 + 9Fe \quad (6.24)$$

$$3FeO + 2Cr \rightarrow Cr_2O_3 + 3Fe \quad (6.25)$$

6.6 Conclusion

Laser cladding is one of the surface treatment techniques for repairing the worn-out surfaces. It involves a high-power laser beam that irradiates the sample's surface, thereby forming a molten melt pool. The coating materials are deposited on this pool and after solidification, a fine clad layer is obtained on the substrate's surface which will enhance the service life of the components. In most cases, laser cladding is meant to produce corrosion-resistant coating on the substrates. This chapter addressed the corrosion-resistant coatings prepared by laser cladding techniques. It details the type of corrosion, the cladding principle, the suitable alloy powders that can be used to coat different substrates, and the mechanisms behind the passive film formation on the clad layers with relevant literature.

Conflicts of Interest

The authors declare that there is no conflicts of interest.

Acknowledgment

The authors would like to acknowledge the editors of this book for providing an opportunity to write this section.

References

1. Petrillo A, De Felice F, Cioffi R, Zomparelli F. Fourth industrial revolution: Current practices, challenges, and opportunities. *Digital Transformation in Smart Manufacturing*. 2018;1:1–20.
2. Alaloul WS, Liew MS, Zawawi NA, Kennedy IB. Industrial revolution 4.0 in the construction industry: Challenges and opportunities for stakeholders. *Ain Shams Engineering Journal*. 2020;11(1):225–30.
3. Karuppasamy SS, Jeyaprakash N, Yang CH. 13 The sensors in future vehicles. In Jeyaprakash Natarajan, Mahendra Babu Kantipudi, Che-Hua Yang, Yaojung Shiao (Eds), *The Future of Road Transportation: Electrification and Automation*, 2023 (Vol. 12).
4. Faulkner JS. The modern of theory of alloys. *Progress in Materials Science*. 1982;27:1–187.
5. Karuppasamy SS, Jeyaprakash N, Yang CH. 7 Battery technology for hybrid and battery vehicles. In Jeyaprakash Natarajan, Mahendra Babu Kantipudi, Che-Hua Yang, Yaojung Shiao (Eds), *The Future of Road Transportation: Electrification and Automation*, 2023 (Vol. 136).

6. Tian X, Zhao J, Qin W, Gong F, Wang Y, Pan H. Performance of ceramic tools in high-speed cutting iron-based superalloys. *Machining Science and Technology*. 2017;21(2):279–90.
7. Mishra SB, Chandra K, Prakash S. Dry sliding wear behaviour of nickel-, iron-and cobalt-based superalloys. *Tribology-Materials, Surfaces & Interfaces*. 2013;7(3):122–8.
8. Murugesan P, Satheeshkumar V, Jeyaprakash N, Yang CH, Karuppasamy SS. Effect of α-Al and Si precipitates on microstructural evaluation and corrosion behavior of laser powder bed fusion printed AlSi10Mg plates in seawater environment. *Metals and Materials International*. 2023;29:2515–32.
9. Trevisan F, Calignano F, Lorusso M, Pakkanen J, Aversa A, Ambrosio EP, Lombardi M, Fino P, Manfredi D. On the selective laser melting (SLM) of the AlSi10Mg alloy: Process, microstructure, and mechanical properties. *Materials*. 2017;10(1):76.
10. Jeyaprakash N, Karuppasamy SS, Yang CH. 1 Application of wear-resistant laser claddings. In Hitesh Vasudev, Chander Prakash (Eds), *Handbook of Laser-Based Sustainable Surface Modification and Manufacturing Techniques*, 2023 (pp. 1–26). CRC Press.
11. Choudhury IA, El-Baradie MA. Machinability of nickel-base super alloys: A general review. *Journal of Materials Processing Technology*. 1998;77(1–3):278–84.
12. Liu S, Shin YC. Additive manufacturing of Ti6Al4V alloy: A review. *Materials & Design*. 2019;164:107552.
13. Jeyaprakash N, Yang CH, Karuppasamy SS, Rajendran DK. Correlation of microstructural with corrosion behaviour of Ti-6Al-4V specimens developed through selective laser melting technique. *Proceedings of the Institution of Mechanical Engineers, Part E: Journal of Process Mechanical Engineering*. 2022;236(5):2240–51.
14. Lancaster JF. Failures of boilers and pressure vessels: Their causes and prevention. *International Journal of Pressure Vessels and Piping*. 1973;1(2):155–70.
15. Renowicz D, Cieśla M. Crack initiation in steel parts working in boilers and steam pipelines. *Journal of Achievements in Materials and Manufacturing Engineering*. 2007;21(2):49–52.
16. Karuppasamy SS, Jeyaprakash N, Yang CH. 3 Alternative propulsion. In *The Future of Road Transportation: Electrification and Automation*, 2023 (Vol. 47, pp. 1–322) CRC Press.
17. Xia D, Song S, Wang J, Shi J, Bi H, Gao Z. Determination of corrosion types from electrochemical noise by phase space reconstruction theory. *Electrochemistry Communications*. 2012;15(1):88–92.
18. Karuppasamy SS, Jeyaprakash N, Yang CH. 27 Application of corrosion-resistant laser claddings. In Hitesh Vasudev, Chander Prakash (Eds), *Handbook of Laser-Based Sustainable Surface Modification and Manufacturing Techniques*, 2023 (Vol. 51, p. 27) CRC Press.
19. Holmberg K, Matthews A. *Coatings Tribology: Properties, Mechanisms, Techniques and Applications in Surface Engineering*, 2009. Elsevier.
20. Ramezani M, Mohd Ripin Z, Pasang T, Jiang CP. Surface engineering of metals: Techniques, characterizations and applications. *Metals*. 2023;13(7):1299.
21. Habig KH. Chemical vapor deposition and physical vapor deposition coatings: Properties, tribological behavior, and applications. *Journal of Vacuum Science & Technology A: Vacuum, Surfaces, and Films*. 1986;4(6):2832–43.

22. Schalk N, Tkadletz M, Mitterer C. Hard coatings for cutting applications: Physical vs. chemical vapor deposition and future challenges for the coatings community. *Surface and Coatings Technology*. 2022;429:127949.
23. Jeyaprakash N, Yang CH, Karuppasamy SS, Radhika N. Microstructural evolution and anti-corrosion characteristics of as-cast and LPBF printed Inconel plates in seawater environment. *Surface Review and Letters*. 2023;31:2450032.
24. Gérard B. Application of thermal spraying in the automobile industry. *Surface and Coatings Technology*. 2006;201(5):2028–31.
25. Burstein GT, Pistorius PC, Mattin SP. The nucleation and growth of corrosion pits on stainless steel. *Corrosion Science*. 1993;35(1–4):57–62.
26. Jeyaprakash N, Yang CH, Sivasankaran S. Laser cladding process of Cobalt and Nickel based hard-micron-layers on 316L-stainless-steel-substrate. *Materials and Manufacturing Processes*. 2020;35(2):142–51.
27. Zhu L, Xue P, Lan Q, Meng G, Ren Y, Yang Z, Xu P, Liu Z. Recent research and development status of laser cladding: A review. *Optics & Laser Technology*. 2021;138:106915.
28. Liu Y, Ding Y, Yang L, Sun R, Zhang T, Yang X. Research and progress of laser cladding on engineering alloys: A review. *Journal of Manufacturing Processes*. 2021;66:341–63.
29. Jeyaprakash N, Yang CH, Ramkumar KR. Microstructure and wear resistance of laser cladded Inconel 625 and Colmonoy 6 depositions on Inconel 625 substrate. *Applied Physics A*. 2020;126:455.
30. Santo L. Laser cladding of metals: A review. *International Journal of Surface Science and Engineering*. 2008;2(5):327–36.
31. Zhao Y, Karimi AR, Wong HS, Hu B, Buenfeld NR, Jin W. Comparison of uniform and non-uniform corrosion induced damage in reinforced concrete based on a Gaussian description of the corrosion layer. *Corrosion Science*. 2011;53(9):2803–14.
32. Jirarungsatian C, Prateepasen A. Pitting and uniform corrosion source recognition using acoustic emission parameters. *Corrosion Science*. 2010;52(1):187–97.
33. Jeyaprakash N, Yang CH, Duraiselvam M, Prabu G. Microstructure and tribological evolution during laser alloying WC-12% Co and Cr_3C_2– 25% NiCr powders on nodular iron surface. *Results in Physics*. 2019;12:1610–20.
34. Tavakkolizadeh M, Saadatmanesh H. Galvanic corrosion of carbon and steel in aggressive environments. *Journal of Composites for Construction*. 2001;5(3):200–10.
35. Davenport AJ, Yuan Y, Ambat R, Connolly BJ, Strangwood M, Afseth A, Scamans GM. Intergranular corrosion and stress corrosion cracking of sensitised AA5182. *Materials Science Forum*. 2006;519:641–6.
36. Jeyaprakash N, Yang CH, Tseng SP. Characterization and tribological evaluation of NiCrMoNb and NiCrBSiC laser cladding on near-α titanium alloy. *The International Journal of Advanced Manufacturing Technology*. 2020;106:2347–61.
37. Waterhouse RB. Fretting corrosion. *Proceedings of the Institution of Mechanical Engineers*. 1955;169(1):1157–72.
38. Swaminathan V, Gilbert JL. Fretting corrosion of CoCrMo and Ti6Al4V interfaces. *Biomaterials*. 2012;33(22):5487–503.
39. Jeyaprakash N, Duraiselvam M, Raju R. Modelling of Cr_3C_2–25% NiCr laser alloyed cast iron in high temperature sliding wear condition using response surface methodology. *Archives of Metallurgy and Materials*. 2018;63:1303–15.
40. Sieradzki K, Newman RC. Stress-corrosion cracking. *Journal of Physics and Chemistry of Solids*. 1987;48(11):1101–13.

41. Burstein GT, Liu C, Souto RM, Vines SP. Origins of pitting corrosion. *Corrosion Engineering, Science and Technology*. 2004;39(1):25–30.
42. Betts AJ, Boulton LH. Crevice corrosion: Review of mechanisms, modelling, and mitigation. *British Corrosion Journal*. 1993;28(4):279–96.
43. Weissenrieder J, Leygraf C. In situ studies of filiform corrosion of iron. *Journal of the Electrochemical Society*. 2004;151(3):B165.
44. Jeyaprakash N, Duraiselvam M, Aditya SV. Numerical modeling of WC-12% Co laser alloyed cast iron in high temperature sliding wear condition using response surface methodology. *Surface Review and Letters*. 2018;25(07):1950009.
45. Vanaei HR, Eslami A, Egbewande A. A review on pipeline corrosion, in-line inspection (ILI), and corrosion growth rate models. *International Journal of Pressure Vessels and Piping*. 2017;149:43–54.
46. Jeyaprakash N, Yang CH, Tseng SP. Wear Tribo-performances of laser cladding Colmonoy-6 and Stellite-6 Micron layers on stainless steel 304 using Yb: YAG disk laser. *Metals and Materials International*. 2021;27:1540–53.
47. Weng F, Chen C, Yu H. Research status of laser cladding on titanium and its alloys: A review. *Materials & Design*. 2014;58:412–25.
48. Wang K, Liu W, Hong Y, Sohan HS, Tong Y, Hu Y, Zhang M, Zhang J, Xiang D, Fu H, Ju J. An overview of technological parameter optimization in the case of laser cladding. *Coatings*. 2023;13(3):496.
49. Singh S, Goyal DK, Kumar P, Bansal A. Laser cladding technique for erosive wear applications: A review. *Materials Research Express*. 2020;7(1):012007.
50. Prabu G, Duraiselvam M, Jeyaprakash N, Yang CH. Microstructural evolution and wear behavior of AlCoCrCuFeNi high entropy alloy on Ti–6Al–4V through laser surface alloying. *Metals and Materials International*. 2021;27:2328–40.
51. Li J, Li HN, Liao Z, Axinte D. Analytical modelling of full single-track profile in wire-fed laser cladding. *Journal of Materials Processing Technology*. 2021;290:116978.
52. Das AK. Recent trends in laser cladding and alloying on magnesium alloys: A review. *Materials Today: Proceedings*. 2022;51:723–7.
53. Paul S, Gupta I, Singh RK. Characterization and modeling of microscale preplaced powder cladding via fiber laser. *Journal of Manufacturing Science and Engineering*. 2015;137(3):031019.
54. Jeyaprakash N, Yang CH. Comparative study of NiCrFeMoNb/FeCrMoVC laser cladding process on nickel-based superalloy. *Materials and Manufacturing Processes*. 2020;35(12):1383–91.
55. Darolia R. Development of strong, oxidation and corrosion resistant nickel-based superalloys: Critical review of challenges, progress and prospects. *International Materials Reviews*. 2019;64(6):355–80.
56. Jeyaprakash N, Yang CH, Karuppasamy SS, Dhineshkumar SR. Evaluation of microstructure, nanoindentation and corrosion behavior of laser cladded Stellite-6 alloy on Inconel-625 substrate. *Materials Today Communications*. 2022;31:103370.
57. Hosseini E, Popovich VA. A review of mechanical properties of additively manufactured Inconel 718. *Additive Manufacturing*. 2019;30:100877.
58. Dudzinski D, Devillez A, Moufki A, Larrouquere D, Zerrouki V, Vigneau J. A review of developments towards dry and high speed machining of Inconel 718 alloy. *International Journal of Machine Tools and Manufacture*. 2004;44(4):439–56.

59. Jeyaprakash N, Yang CH, Karuppasamy SS. Laser cladding of Colmonoy 6 particles on Inconel 625 substrate: Microstructure and corrosion resistance. *Surface Review and Letters*. 2022;29(08):2250102.
60. Balaguru S, Gupta M. Hardfacing studies of Ni alloys: A critical review. *Journal of Materials Research and Technology*. 2021;10:1210–42.
61. Jelvani S, Razavi RS, Barekat M, Dehnavi MR, Erfanmanesh M. Evaluation of solidification and microstructure in laser cladding Inconel 718 superalloy. *Optics & Laser Technology*. 2019;120:105761.
62. Jeyaprakash N, Yang CH, Susila P, Karuppasamy SS. Laser cladding of NiCrMoFeNbTa particles on Inconel 625 alloy: Microstructure and corrosion resistance. *Transactions of the Indian Institute of Metals*. 2023;76(2):599–612.
63. Feng K, Chen Y, Deng P, Li Y, Zhao H, Lu F, Li R, Huang J, Li Z. Improved high-temperature hardness and wear resistance of Inconel 625 coatings fabricated by laser cladding. *Journal of Materials Processing Technology*. 2017;243:82–91.
64. Jeyaprakash N, Yang CH, Prabu G, Clinktan R. Microstructure and tribological behaviour of Inconel-625 superalloy produced by selective laser melting. *Metals and Materials International*. 2022;28(12):2997–3015.
65. Fesharaki MN, Shoja-Razavi R, Mansouri HA, Jamali H. Microstructure investigation of Inconel 625 coating obtained by laser cladding and TIG cladding methods. *Surface and Coatings Technology*. 2018;353:25–31.
66. Abioye TE, Farayibi PK, Clare AT. A comparative study of Inconel 625 laser cladding by wire and powder feedstock. *Materials and Manufacturing Processes*. 2017;32(14):1653–9.
67. Wang C, Zhou J, Zhang T, Meng X, Li P, Huang S. Numerical simulation and solidification characteristics for laser cladding of Inconel 718. *Optics & Laser Technology*. 2022;149:107843.
68. Zhang H, Shi Y, Kutsuna M, Xu GJ. Laser cladding of Colmonoy 6 powder on AISI316L austenitic stainless steel. *Nuclear Engineering and Design*. 2010;240(10):2691–6.
69. Murugesan P, Satheeshkumar V, Jeyaprakash N, Prabu G, Yang CH. Nano-level mechanical and tribological behavior of additively manufactured AlSi10Mg plates. *Archives of Civil and Mechanical Engineering*. 2023;23(1):62.
70. Ming Q, Lim LC, Chen ZD. Laser cladding of nickel-based hardfacing alloys. *Surface and Coatings Technology*. 1998;106(2–3):174–82.
71. Kaul R, Ganesh P, Albert SK, Jaiswal A, Lalla NP, Gupta A, Paul CP, Nath AK. Laser cladding of austenitic stainless steel with hardfacing alloy nickel base. *Surface Engineering*. 2003;19(4):269–73.
72. Alimardani M, Fallah V, Khajepour A, Toyserkani E. The effect of localized dynamic surface preheating in laser cladding of Stellite 1. *Surface and Coatings Technology*. 2010;204(23):3911–9.
73. Ganesh P, Moitra A, Tiwari P, Sathyanarayanan S, Kumar H, Rai SK, Kaul R, Paul CP, Prasad RC, Kukreja LM. Fracture behavior of laser-clad joint of Stellite 21 on AISI 316L stainless steel. *Materials Science and Engineering: A*. 2010;527(16–17):3748–56.
74. D'Oliveira AS, da Silva PS, Vilar RM. Microstructural features of consecutive layers of Stellite 6 deposited by laser cladding. *Surface and Coatings Technology*. 2002;153(2–3):203–9.
75. Wu Z, Qian M, Brandt M, Matthews N. Ultra-high-speed laser cladding of Stellite® 6 alloy on mild steel. *Jom*. 2020;72:4632–8.

76. Kao CM, Jeyaprakash N, Yang CH. Modelling and non-destructive measurement of dispersion behaviour of stainless steel and alumina ceramic porous plates. *Ceramics International*. 2021;47(10):14233–43.
77. Alvarez P, Montealegre MÁ, Pulido-Jiménez JF, Arrizubieta JI. Analysis of the process parameter influence in laser cladding of 316L stainless steel. *Journal of Manufacturing and Materials Processing*. 2018;2(3):55.
78. Ding H, Yang T, Wang W, Zhu Y, Lin Q, Guo J, Xiao Q, Gan L, Liu Q. Optimization and wear behaviors of 316L stainless steel laser cladding on rail material. *Wear*. 2023;523:204830.
79. Zhang Z, Farahmand P, Kovacevic R. Laser cladding of 420 stainless steel with molybdenum on mild steel A36 by a high power direct diode laser. *Materials & Design*. 2016;109:686–99.
80. Wang K, Chang B, Chen J, Fu H, Lin Y, Lei Y. Effect of molybdenum on the microstructures and properties of stainless steel coatings by laser cladding. *Applied Sciences*. 2017;7(10):1065.
81. Karuppasamy SS, Yang CH. Adapting the time-domain synthetic aperture focusing technique (T-SAFT) to laser ultrasonics for imaging the subsurface defects. *Sensors*. 2023;23(19):8036.
82. Ni C, Shi Y, Liu J, Huang G. Characterization of Al0. 5FeCu0. 7NiCoCr high-entropy alloy coating on aluminum alloy by laser cladding. *Optics & Laser Technology*. 2018;105:257–63.
83. Karuppasamy SS, Jeyaprakash N, Yang CH. The sensors in future vehicles. In Jeyaprakash Natarajan, Mahendra Babu Kantipudi, Che-Hua Yang, Yaojung Shiao (Eds), *The Future of Road Transportation*, 2023 (pp.261–283).CRC Press.
84. Grohol CM, Shin YC, Frank A. Laser cladding of aluminum alloy 6061 via off-axis powder injection. *Surface and Coatings Technology*. 2021;415:127099.
85. Cheng Q, Guo N, Fu Y, Wang G, Yu M, He J. Investigation on in-situ laser cladding 5356 aluminum alloy coating on 5052 aluminum alloy substrate in water environment. *Journal of Materials Research and Technology*. 2021;15:4343–52.
86. Song M, Wu L, Liu J, Hu Y. Effects of laser cladding on crack resistance improvement for aluminum alloy used in aircraft skin. *Optics & Laser Technology*. 2021;133:106531.
87. Feng X, Wang H, Liu X, Wang C, Cui H, Song Q, Huang K, Li N, Jiang X. Effect of Al content on wear and corrosion resistance of Ni-based alloy coatings by laser cladding. *Surface and Coatings Technology*. 2021;412:126976.
88. Karuppasamy SS, Jeyaprakash N, Yang CH. Battery technology for hybrid and battery vehicles. In Jeyaprakash Natarajan, Mahendra Babu Kantipudi, Che-Hua Yang, Yaojung Shiao (Eds), *The Future of Road Transportation*, 2023 (pp.136–158).CRC Press.
89. Riquelme A, Rodrigo P, Escalera-Rodríguez MD, Rams J. Corrosion resistance of Al/SiC laser cladding coatings on AA6082. *Coatings*. 2020;10(7):673.
90. Nabhani M, Razavi RS, Barekat M. Corrosion study of laser cladded Ti-6Al-4V alloy in different corrosive environments. *Engineering Failure Analysis*. 2019;97:234–41.
91. Diao Y, Zhang K. Microstructure and corrosion resistance of TC2 Ti alloy by laser cladding with Ti/TiC/TiB_2 powders. *Applied Surface Science*. 2015;352:163–8.
92. Obadele BA, Andrews A, Olubambi PA, Mathew MT, Pityana S. Tribocorrosion behaviour of laser cladded biomedical grade titanium alloy. *Materials and Corrosion*. 2015;66(10):1133–9.

93. Jing Z, Cao Q, Jun H. Corrosion, wear and biocompatibility of hydroxyapatite bio-functionally graded coating on titanium alloy surface prepared by laser cladding. *Ceramics International*. 2021;47(17):24641–51.
94. Volovitch P, Masse JE, Fabre A, Barrallier L, Saikaly W. Microstructure and corrosion resistance of magnesium alloy ZE41 with laser surface cladding by Al–Si powder. *Surface and Coatings Technology*. 2008;202(20):4901–14.
95. Karuppasamy SS, Jeyaprakash N, Yang CH. Alternative propulsion systems. In Jeyaprakash Natarajan, Mahendra Babu Kantipudi, Che-Hua Yang, Yaojung Shiao (Eds), *The Future of Road Transportation*, 2023 (pp. 47–68). CRC Press.
96. Wu Q, Zhang J. Corrosion behavior of laser-clad Cu-9Ni-6Sn coating. *Surface and Coatings Technology*. 2018;349:1117–29.
97. Jin L, Jiang K, Ren H, Liu J, Lu X, Liu W, Wu D, Liu H, Liu B, Wang F, Gu Y. A review of laser cladding on copper and copper alloys. *International Journal of Electrochemical Science*. 2022;17(9):220920.
98. Karuppasamy SS, Jeyaprakash N, Yang CH. Microstructure, nanoindentation and corrosion behavior of colmonoy-5 deposition on SS410 substrate using laser cladding process. *Arabian Journal for Science and Engineering*. 2022;47:8751–67.
99. Fesharaki MN, Shoja-Razavi R, Mansouri HA, Jamali H. Evaluation of the hot corrosion behavior of Inconel 625 coatings on the Inconel 738 substrate by laser and TIG cladding techniques. *Optics & Laser Technology*. 2019;111:744–53.
100. Jeyaprakash N, Yang CH, Karuppasamy SS, Duraiselvam M. Stellite 6 cladding on AISI Type 316L stainless steel: microstructure, nanohardness and corrosion resistance. *Transactions of the Indian Institute of Metals*. 2023;76(2):491–503.
101. Behazin M, Biesinger MC, Noël JJ, Wren JC. Comparative study of film formation on high-purity Co and Stellite-6: Probing the roles of a chromium oxide layer and gamma-radiation. *Corrosion Science*. 2012;63:40–50.
102. Ma L, Wiame F, Maurice V, Marcus P. Origin of nanoscale heterogeneity in the surface oxide film protecting stainless steel against corrosion. *Npj Materials Degradation*. 2019;3(1):29.
103. Ohmi T, Nakagawa Y, Nakamura M, Ohki A, Koyama T. Formation of chromium oxide on 316L austenitic stainless steel. *Journal of Vacuum Science & Technology A: Vacuum, Surfaces, and Films*. 1996;14(4):2505–10.

7

High-Temperature Oxidation of Laser Cladding Material

T. Sathies, N. Jayanth, and B. Mallikarjuna

7.1 Introduction

Engineering alloys find several applications owing to their corrosion resistance, low density, oxidation resistance and high specific strength. However, the applications of several engineering alloys are limited because the alloys lose their properties at higher temperatures and exhibit low hardness, less wear resistance and limited high-temperature oxidation resistance. Hence, the components are subjected to several surface modification techniques and some of the commonly used surface modification methods are thin film coating (vapour deposition (CVD/PVD)), laser surface modification, welding, thermal spraying, diffusion process (chromizing, nitriding, carbonitriding, carburizing, etc.), surface hardening and plating [1]. Laser-based surface modification is one of the most preferred methods because the dilution and heat-affected zone (HAZ) generated during the process is controllable. Laser cladding (LC) is a surface modification technique in which the heat source from the high-power laser is applied to melt and deposit a cladding layer on the base material and a strong metallurgical bonding is expected between the cladding and base material. Strong physicochemical compatibility between the clad layer and base material may lead to crack-free coating with better properties. Moreover, the choice of cladding material depends on the intended application. The cladding materials can be broadly classified as metal and alloys, ceramics, metal matrix composites and rare-earth (RE) oxides. Some of the commonly used metals and alloys are single metals or alloys, amorphous alloys, single crystal alloys and high-entropy alloys (HEAs)[2]. The LC sample generally consists of four zones and they are base material (substrate), interfacial zone (IZ), cladding zone (CZ), and HAZ. Based on the type of coating material feed arrangement, LC can be classified as a pre-placed powder system, coaxial powder feeding, wire feeding and off-axis powder feeding method. In addition, the application

 DOI: 10.1201/9781032693576-7

capability of the LC technique is expanded through ultrasonic vibration-assisted LC, electromagnetic-assisted LC and underwater LC. The application range of the LC method is expanded through high-speed LC technology [3,4]. With ultra-high-speed LC, the surface rate and speed of the cladding line could reach up to 500 cm^2/min and 200 m/min. It is possible to obtain crack-free coatings of thickness about 10–250 μm through the high-speed LC technique. Yuan et al. coated Ni45 alloy on steel base material through high-speed LC and traditional low-speed LC processes and analysed the coating hardness, cross-sectional microstructure, cladding efficiency, corrosion and wear resistance. With high-speed LC, a much thinner clad layer with a cladding efficiency and cladding speed of 156.79 cm^2/min and 76.86 m/min was achieved. Denser and finer grains were formed on high-speed LC samples because of the faster cooling rate. Moreover, the high-speed laser-cladded samples exhibited better corrosion and wear resistance in comparison to the traditional laser-cladded samples [5]. Laser cladding improves the functionality of the components by augmenting the oxidation, wear and corrosion resistance. For instance, Inconel 625 alloy laser cladded with NiCrMoFeNbTa [6] and 316L stainless steel cladded with Stellite 6 [7] exhibited better corrosion resistance. The passive film formed by the LC protects the substrate against corrosion and improves its corrosion resistance. Another major application of the LC technique is to reduce the wear rate of the component. The elements significantly influencing the oxidation rate of metallic components are Al (aluminium), Ni (nickel), Cr (chromium), Fe (iron), Si (silicon), Co (cobalt) and Mo (molybdenum). Laser-cladded coatings such as SS420 [8], CoCrCuFeNiTix [9], FeCoNiCrAl0.5Ti0.5 [10], Colmonoy-6 and Stellite-6 [11], In625/(Nbx+SiC0.5) composite [12], TiC/micro TiB_2/C-276 composite [13], NiCrMoNb and NiCrBSiC [14] made on Ti–6Al–4V, 45 steel substrates, AISI 1045 steel, stainless steel 304, Inconel 625, 316L stainless steel, near-α titanium alloy demonstrated remarkable wear resistance and improved the life of the components without affecting their functionality. Clad properties such as geometry, microstructure, hardness, cracks, pores, residual stress, surface roughness, dilution, wear resistance and oxidation resistance at high temperatures are influenced by the laser parameters, powder properties, powder feeding system and ambient conditions. Several components involved in energy, aerospace, petroleum and chemical industries are subjected to extreme temperatures and corrosive environments. As a result, the functional components get oxidized and fail because of diffusion, volatilization, decomposition and oxide scale growth. To overcome the said failures, LC was successfully employed by industries and researchers to develop high-temperature oxidation-resistant coatings on numerous alloys.

7.2 High-Temperature Oxidation of Metal

Upon exposure to an oxidizing atmosphere, pure metallic objects form oxide scales in four stages. In the first stage, oxygen gets absorbed on the metal surface and the second stage involves the development of oxygen nucleation. The growth of the nuclei along the lateral direction happens in the third stage, and a denser oxide scale is generated in the fourth stage. When the nuclei lateral growth was completed, the oxide film separated the substrate from the gaseous environment [15]. The high-temperature oxidation behaviour of the component is influenced by the characteristics of the oxide layer formed on the component. Several oxides developed on the metallic substrate are stable at low temperatures but pores and voids are generated at high temperatures and the oxides scale gets loosened. As a result, oxygen diffuses inwards and metallic elements diffuse outwards, leading to thicker and unstable oxide scale formation. In the case of alloys, the alloying element also has control over the oxidation rate. The rate of oxidation is governed by the nature of the reaction product and the oxidation rate gets reduced when the transport of oxidizing reactants is restricted. The materials considered for the high-temperature oxidation resistance are the ones which form oxides with slowest growth rates (usually α-Al_2O_3 or SiO_2). For instance, low-alloy steels form M_3O_4 (M=Fe, Cr) surface layers and it is stable up to 500°C. Titanium dioxide (TiO_2) layers are developed on titanium alloys and can withstand 600°C. Ferritic stainless steels, which generate chromium oxide (Cr_2O_3) surface layers, are employed in applications involving 650°C. Austenitic Fe–Ni–Cr alloys and Ni–Cr alloys which develop Cr_2O_3 surface layers are used to about 850°C and 950°C, respectively. MCrAlY (M=Fe, Co or Ni) coatings, Austenitic Ni-Cr-Al alloys and aluminide, which generate aluminium oxide (Al_2O_3) surface layers, are employed in applications where the temperature reaches 1,100°C. Ceramics or refractory metals are used in applications requiring 1,100°C or above [16]. The oxide growth rate can be described by principal laws namely (i) linear law, (ii) logarithmic law and (iii) parabolic law. Linear law represents reaction in which the reaction rate is independent of time and the reaction rate is influenced by diffusion through the gas phase or surface-reaction step. In the case of parabolic law, the oxidation rate is inversely proportional to $\sqrt{t}$. The parabolic law is obeyed when the penetration of the oxidizing elements through the oxide scale is the oxidation rate–determining process. The logarithmic law is applicable for low-temperature processes and it is obeyed when a thin layer of oxide is formed. It is possible to measure the oxidation rate by measuring oxygen or metal consumption, or by computing oxide accumulation over time [17,18]. The above factors must be considered while developing a coating material for high-temperature applications. Further, compatibility between the base material and coating must be ensured so that strong metallurgical bonding is established between the base material and coating. In the case of LC, the process parameters used and the powder morphology also affect the oxidation resistance of the coatings made by the LC process. The forthcoming section discusses the advancement in oxidation resistance ability of LC made on ferrous alloys, titanium alloys and nickel-based alloys.

7.3 High-Temperature Oxidation Behaviour of LC Made on Ferrous Alloys

The microstructure of the coating made by LC is usually enhanced by the inclusion of several alloying elements. RE oxides are widely used as reinforcement for refining the microstructure and enhancing the oxidation, corrosion and wear resistance of clad coating. Commonly used RE oxides in LC are lanthanum oxide (La_2O_3), cerium oxide (Ce_2O_3) and yttrium oxide (Y_2O_3). Zhang et al. deposited iron-based composite coating containing multiple ceramics ($TiC–TiB_2$-based reinforced phase) and different weight percentages of Y_2O_3 on 5CrNiMo die steel and investigated the oxidation behaviour of the said coating at high temperature (600°C). The addition of Y_2O_3 facilitated reinforcement refinement and coating purification and helped in the generation of compact oxide film. Further, the oxidation mass of the coating was reduced because of the inclusion of RE oxide (2 wt.%) [19]. TP347H is a typical boiler superheater steel and it usually fails in the application site owing to oxidation. Hence, Kong et al. prepared Hastelloy C22 powder with different Si concentrations (0–3 wt.%) and deposited it on a TP347H commercial hot-rolled tube to enhance its oxidation resistance. The oxidation process of the coating with and without silicon is shown in Figure 7.1. The Si addition (3 wt%) helped in increasing the oxidation resistance of the C22 clad. The

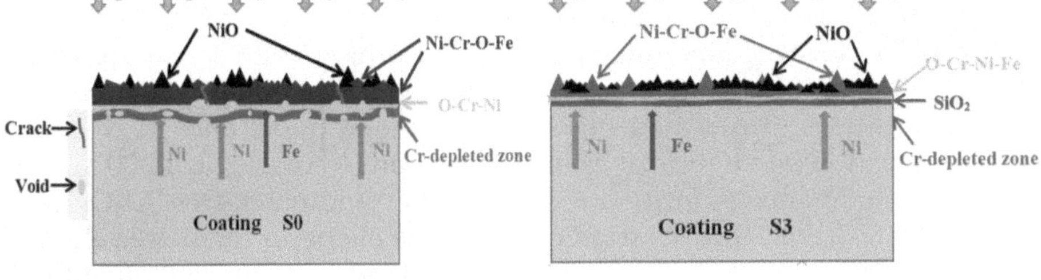

FIGURE 7.1
Effect of silicon addition on oxidation resistance of C22 coating [20].

oxide layer consists of silicon dioxide (SiO_2), nickel oxide (NiO), Cr_2O_3 and numerous mixed oxides. SiO_2 layer restricted the oxygen diffusion and compressed the Cr-depletion region width. As a result, the formation of voids and cavities on the Cr-depletion zone of the coating was reduced. In addition, the silicon addition reduced the grain size in the outer oxide layer, made the inner oxide layer compact and denser and enhanced the coating's oxidation resistance. The coating without Si had an oxidation weight gain of 3.9478 mg/cm^2 and the coating with 3 wt% Si had the lowest weight gain of 0.4421 mg/cm^2 at 700°C [20]. Nie et al. coated NiCoCrAlY on 304 stainless steel via LC and investigated the oxide layer formation at 950°C for a different duration. Cr_2O_3 film and Al_2O_3 existed on the coating surface and internal oxidation zone and served as a barrier against high-temperature oxidation. When the oxidation duration increased, the protective effect of the Cr_2O_3 reduced and the thickness of the Al_2O_3 layer (internal oxidation zone) increased which helped in controlling the subsequent oxidation [21].

Cr5Mo steel with better thermal strength is a preferred material for the construction of hydrocracking furnace tubes. However, Cr5Mo tubes are susceptible to damage at high temperatures due to fatigue. Hence, Xu et al. used an LC technique to deposit IN718 coating on the Cr5Mo substrate and investigated the influence of LC parameters on the high-temperature oxidation resistance of the coating. The eutectic quantity precipitation in the CZ was affected by the laser power and scanning speed. At higher laser scanning speed, the precipitation of eutectic elements increased and martensite was formed in the HAZ. Moreover, the mass gain was also higher at a higher scanning speed and it played a major role in the oxidation resistance of the coating. The oxide formation was influenced by the Ni3Nb phase distribution. Material shedding happened because of cracks and corrosion [22]. TiC–TiB_2 composite ceramics usually show excellent chemical stability, high hardness and wear resistance at higher temperatures. Zhang et al. adopted LC for the in situ fabrication of Fe–Ti–B–Cr–Mo–C composite coatings on 5CrNiMo die steel substrates with different molybdenum contents. The developed coating contains flower-like (Ti, Mo) C-reinforced particles and block-like or cuboidal Mo_2B and TiB_2 particles. The coating containing 9 wt% $FeMo_{70}$ demonstrated improved high-temperature oxidation resistance, lower oxidation rate and better wear resistance. The coatings had lesser oxidation mass gain and it was about 1/4 to 1/6 of that of 5CrNiMo die steel substrate [23]. Jin et al. prepared aluminium-bronze coatings with different percentages of Fe and Ni elements on 316 stainless steel substrates and revealed the influence made by Fe and Ni elements on the oxidation behaviour of aluminium-bronze coating at 1,000°C. The coating with Fe and Ni formed a denser, finer and crack-free Al_2O_3 layer (Figure 7.2). Further, the inclusion of Fi and Ne elements facilitated the formation of κ phase and Fe–Cr phase and it helped in grain refinement and grain boundary density enhancement. The aluminium-bronze coatings containing Ni and Fe elements possessed lesser oxidation rate constant

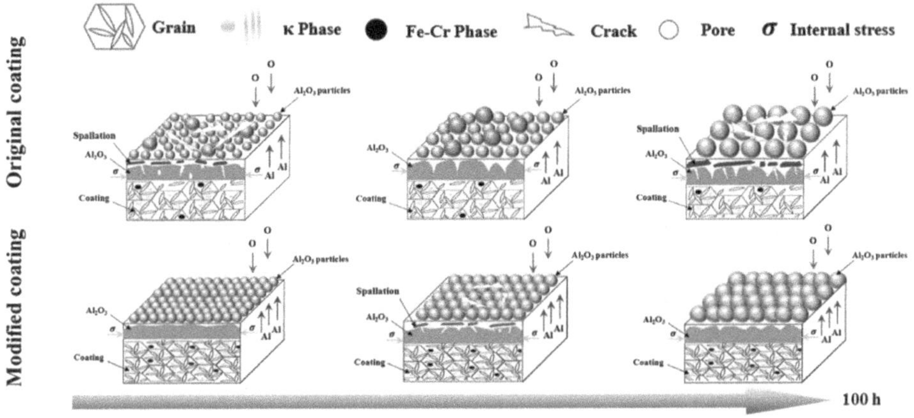

FIGURE 7.2
Oxidation mechanism of aluminium-bronze coating with and without Fe and Ni elements [24].

(kp) and oxidation scale thickness than the coating without Ni and Fe elements. Superior recovery ability and internal stress resistance were displayed by modified coating [24].

Zhang et al. made WC/Co–Cr alloy coating on an H13 steel plate and analysed the oxidation behaviour of the LC coating at 600°C. The coating mainly consists of ε-Co, CW3, Cr23C6, Cr7C3 and Co6W6C. During the oxidation test, ε-Co converted to α-Co, whereas carbide phases remained unchanged and it enabled the coating to be thermally stable and high-temperature oxidation resistant. After the oxidation test, the coated substrate showed a minimum mass gain rate and increased microhardness in comparison to the uncoated base material [25]. Similarly, Sun et al. deposited CrMnFeCoNi/xTiC HEA composite coatings on AISI H13 steel and studied the oxidation and wear behaviour of the proposed coatings at 600°C. The inclusion of TiC leads to short-circuit diffusion and the coating undergoes an oxidation process. An intermediate with high wear resistance was generated during the high-temperature wear test and it protected the sample from brittle cracking. The samples subjected to the oxidation process contain Cr_2O_3 and manganese oxide (Mn_2O_3) layers and traceable amounts of TiO_2. The composite protective film layers generated during the test helped the sample to exhibit a lower wear rate and maximum friction coefficient [26]. The properties of HEAs coatings produced through the LC technique can be improved through direct addition or in situ formation of ceramic phases. Yu et al. prepared $AlCoCrFeNiMo_x(TiC)_{2-x}$ with different mole ratios (x=0, 0.5, 1, 1.5 and 2) on Q235 steel substrate in which different weight percentage of Mo was added and TiC particles were in situ formed. The coating exhibited the BCC phase and Mo addition helped in the development of an oxide layer with a protective effect. The coating's microhardness was enhanced because of precipitation and a solid solution-strengthening mechanism. The phase boundaries containing vacancies and dislocations lead to short-circuit diffusion of various elements. As a result, alloying element's diffusion activation energy was lowered and alloying elements moved outwards along the diffusion channel and formed oxide by interacting with oxygen. As the x value increases in a coating, the phase boundary and TiC quantity decrease, and oxidation resistance increases. However, the excessive inclusion of Mo led to the generation of excessive BCC/σ and BCC/NiCrCoMo phase boundaries and enhanced the oxidation of the coating. Among the combinations considered, the AlCoCrFeNiMo(TiC) (x=1) demonstrated higher corrosion resistance and oxidation resistance at higher temperatures. The outwards diffusion of metal cations during oxidation was limited by the Cr_2O_3 and Al_2O_3 oxide films generated on the coating surface (Figure 7.3). The corrosion resistance of the coating with appropriate Mo (x=1) increases because of the presence of Cr_2O_3 in the protective passivation film [27]. Cai et al. synthesized $FeCoCrNiAl_x$-based HEA coating on Cr12MoV die steel and revealed the effect of aluminium addition on the oxidation behaviour of the said coating. The aluminium addition facilitated the generation of BCC solid solution in coating and a change in solidification pattern from columnar to equiaxed crystals was observed. The samples subjected to high-temperature oxidation tests contain bottom oxidation film (Cr_2O_3, $(Co, Ni)(Al, Cr, Fe)_2O_4$) and surface oxidation film (Al_2O_3 and $(Al, Cr)_2O_3$). The bottom oxidation film protected the coating from oxidation at high temperatures [28]. Zhang et al. involved traditional and extreme high-speed LC techniques for depositing TiC/IN625 coatings on steel substrates (AISI 1045). The optimum cladding speed was identified as 49.8 m/min. The TiC/IN625 coatings mainly consist of dendrites and no signification transition layer was observed between the layers of coating. The oxidation mass gain observed from the high-speed laser-cladded sample was lesser in comparison to the conventional laser-cladded sample and it was attributed to the presence of smaller grains, smaller dendrites, lower surface roughness and large volume of low-angle boundary. In the outer oxide layer, Cr and Nb oxides dominate, whereas Ni and Fe oxides dominate in the inner oxide layer [29].

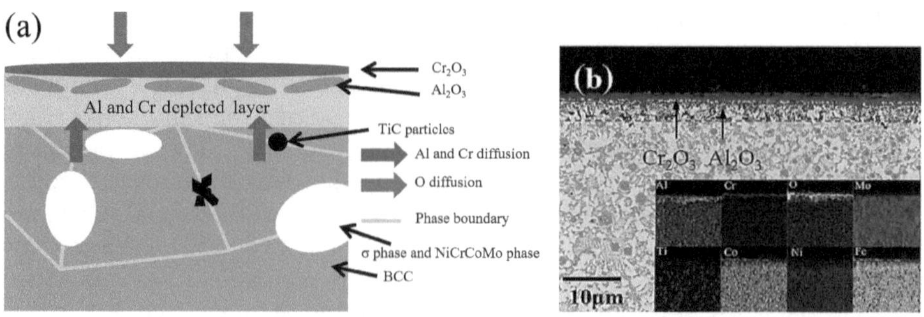

FIGURE 7.3
(a) Oxidation mechanism and (b) cross-sectional images and EDS elemental mappings of AlCoCrFeNiMox coating and steel substrate subjected to oxidation at 800°C for 50 hours [27].

The oxidation behaviour of TiC-reinforced CrMnFeCoNi coating made on AISI H13 was reported by Xiong et al. The high-temperature experiments were conducted at 600°C, 750°C and 900°C and the mechanism behind oxidation was explored. TiO_2, Cr_2O_3 and manganese oxide (Mn_2O_3) were the major oxides found in the coatings when oxidized at 750°C and 600°C. At higher temperatures (900°C), a significant quantity of Fe present in the base material moved into the coating and changed the oxidation characteristics of the coatings. The oxide layer of specimens treated at 900°C consists of iron oxide (Fe_3O_4). A 'short-circuit diffusion' channel was formed in TiC-reinforced CrMnFeCoNi composite coating and it lowered the coating's oxidation resistance. To avoid variation in coating characteristics, it is important to consider the chemical composition of the base material while choosing the elements of HEA coating [30].

HEA usually forms simple solid solutions and exhibits slow diffusion, cocktail effects and lattice distortion. These characteristics enable HEA to possess excellent hardness, corrosion and wear resistance and high-temperature properties. Metallic elements namely copper (Cu), Fe, Mo, Co, Ti, Cr, zirconium (Zr), manganese (Mn), nickel (Ni), niobium (Nb), yttrium (Y), tantalum (Ta) and Al and a few non-metallic elements like boron (B), nitrogen (N), Si, and carbon (C) are involved in HEA development. The Si component of the HEA may contribute to the development of a protective SiO_2 layer and hinder further oxidation. Hence, Liu et al. developed a novel $FeCrMnVSi_x$ HEA system (x represents molar ratio, $x=0$ to 2) and coated it on 1Cr13 steel substrate via LC technique. The HEA coating contains a BCC solid solution structure and a Si-rich dendritic structure. As the wt% of Si addition increased, the lattice constant of the coating reduced and lattice distortion was enhanced. The oxidation test conducted by the authors showed that HEA coating without Si generated Cr_2O_3 and Mn_2VO_4 spinel along with many loose and porous oxides. As the percentage of Si addition increased, SiO_2 was generated and the oxide layer became dense protecting the coating from further damage. The mass gain of HEA coating with Si2.0 was very low in comparison to substrate and HEA coating without Si [31]. Similarly, Ma et al. developed a $CoCr_2FeNb_{0.5}Ni$ HEA coating system with and without Si content and investigated the role played by Si during high-temperature oxidation tests. The HEA coating with Si had an FCC and rich Laves phases and the lattice constant of the Laves phase reduced upon the inclusion of silicon. The main oxidation product of the coating was Cr_2O_3. During continuous oxidation, Cr_2O_3 decomposed into Cr_2O_3 and Cr and Kirkendall pores were formed because of the outwards diffusion of chromium. When the percentage of chromium decreases in the Laves phase, the oxidation of silicon begins and the (Cr, Si)Ox amorphous oxide scale develops on the coating. This amorphous oxide scale prevented the oxygen from diffusing inwards, thereby increasing the oxidation resistance of the coating [32]. Laser-cladded and laser-alloyed samples subjected to wear usually form a protective oxidative film and reduce the wear rate.

The nodular cast iron laser alloyed with Cr_3C_2-25% NiCr underwent oxidation wear during the high-temperature wear test and possessed improved wear resistance than the unalloyed substrate [33]. Similarly, nodular cast iron laser alloyed with WC-12% Co also exhibited a minimum wear rate at high-temperature conditions. The oxide formation at the interface region helped to reduce the material removal rate during the wear test [34]. From the works reviewed, it is learned that LC is an effective strategy for promoting the oxidation resistance of ferrous alloys at high-temperature conditions. Oxides with better diffusion resistance are produced on the coating surface and it helps in promoting the wear, oxidation and corrosion resistance of the base material.

7.4 High-Temperature Oxidation Behaviour of LC Made on Titanium Alloys

Titanium alloys find application in the aerospace field owing to their superior resistance to oxidation and creep, high specific strength and thermal stability. However, titanium alloys lose their properties at high-temperature conditions and fail to meet stringent service conditions. Hence, several surface modification techniques especially LC are used to apply thermal barrier coatings on the titanium alloy surface. The schematics of oxygen diffusion zones and oxide scales that can form on titanium-base alloys are shown in Figure 7.4 [35]. Li and Yamaguchi developed a $NiSi_{0.5}CrCoMoNb_{0.75}$ HEA coating system which can be coated on the TC4 alloy (Ti-6Al-4V) to increase the oxidation resistance at higher temperatures. The coating mainly

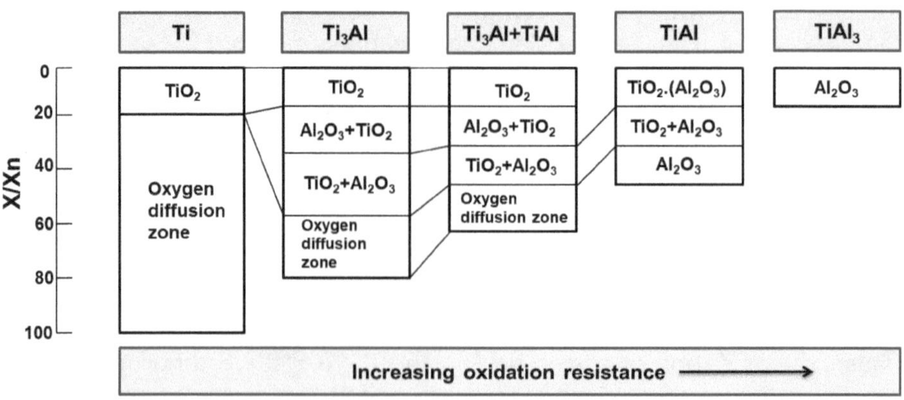

FIGURE 7.4
Schematic representation of oxygen diffusion zones and oxide scales formed on various titanium alloys [35].

contains three different structures and they are the lumpy σ phase (rhombohedral structure), HCP structure and netlike μ phase (FCC structure). The coating subjected to oxidation test consists of $CoCr_2O_4$, $Ni_3(VO_4)_2$, Ni_3TiO_5, $Nb_{0.6}Cr_{0.4}O_2$ and $NiCrO_3$. The developed oxidation layer protected the coating from further oxidation and the thickness of the oxide scale generated on the cladding was significantly lesser than the substrate without coating [36]. Li et al. coated Ti–Al–Si–SiCx with different percentages of SiC content ($x=1$, 2, 3, 4) on a TC4 substrate and investigated its high-temperature tribological properties and oxidation behaviour at 800°C. The Ti–Al–Si–SiCx deposition made through the LC process consists of Ti_5Si_3/Ti_3Al eutectic structure and TiC. During oxidation, Al_2O_3 and TiO_2 oxide layers were formed and the Ti_5Si_3 phase demonstrated superior resistance to oxidation. Among the combinations considered by the authors, the Ti–Al–Si–SiC_3 system demonstrated high oxidation and wear resistance [37].

Zhang et al. tested the role of Ti addition on the wear resistance and high-temperature oxidation resistance of CoCrMoNbTix cladding made on Ti alloy. The cladded layer consists of the intermetallics and BCC. The coating subjected to oxidation test had three different layers, the outer layer consisted of Co oxide, the intermediate layer contained chromium, niobium and titanium elements and the inner layer had Cr. The microhardness and wear resistance of the laser-cladded material surged when the addition of Ti element enhanced, while the oxidation mass gain decreased [38]. Li et al. discussed the benefits of adding Y_2O_3 to the high-temperature oxidation resistance, microstructure, hot corrosion resistance and wear resistance of CoCrFeNiTiNb HEA coating made on the TC4 substrate. Y_2O_3 addition resulted in grain refinement and the microhardness increased and the wear rate decreased because of the fine-grained strengthening mechanism. Various oxides present on the developed coating were TiO_2, Cr_2O_3, Al_2O_3, $NiCr_2O_4$ and $Al_{0.2}Nb_{0.2}Ti_{0.6}O_2$. The oxide layers of minimum thickness were generated on the coating (9.6 μm) and it was significantly thinner than the oxide layer developed on pure substrate (141.3 μm). The CoCrFeNiTiNb coating resistance against oxidation and hot corrosion increased because of Y_2O_3 addition [39]. Similarly, LaB_6 inclusion improved the oxidation resistance of (TiC+TiBx)/Ti coating made on titanium. Even though the (TiC+TiBx)/Ti coating system had better oxidation resistance than the substrate, cracks were generated between the titanium matrix and coarse rod-like TiB_2 present in the coating during the oxidation test and it decreased the resistance of (TiC+TiBx)/Ti coating against oxidation. By adding Y_2O_3 (2 wt.%) to TiC+TiBx/Ti, grain refinement occurs and La_2O_3 was in situ formed during the oxidation test. In turn, the oxide scale becomes denser and more protective and the oxidation mass gain of the cladded material with Y_2O_3 decreases drastically [40]. The Ti–Cu–NiCoCrAlTaY coating system was introduced by Li-Yan et al. to increase the flame retardancy and oxidation resistance of Ti alloy. Double-layered composite oxide scales (Al_2O_3 and Cr_2O_3) developed on the cladded material during the oxidation test and these oxides resist further oxygen diffusion thereby coating's oxidation resistance increased. Ti–Cu eutectic transition zone was generated above the base material and it helped in improving the burn resistance of the base material [41]. Barekat et al. formed Co–Cr–Mo laser clad on γ-TiAl material and studied its oxidation behaviour at 900°C for 10 and 480 hours. The coating contains γ-Co (FCC), $Cr_{23}C_6$ and ε-Co (HCP) structures. The specific mass gain during oxidation followed parabolic law and at a shorter oxidation test (10 hours), the Cr_2O_3 oxidation layer developed on the coating surface. As the oxidation duration increased to 480 hours, porosity and microchannels were generated on the Cr_2O_3 oxidation layer and Co reached the surface by diffusion through the defects and formed a protective $CoCr_2O_4$ film. The resistance of the coating to high-temperature oxidation was promoted by the $CoCr_2O_4$ layer [42].

Lv et al. laser-cladded TiNi/Ti_2Ni coatings on TC4 alloy by mixing NiCrBSi alloy powder and tantalum carbide (TaC) powder in different proportions. Coatings with 40 wt.% of TaC showed better high-temperature wear and oxidation resistance. The oxidation film developed on the coating surface contained TiO_2, Al_2O_3, NiO, SiO_2, Cr_2O_3, TaC and Ta_2O_5 and these oxides acted as a barrier against coating damage [43]. TC4 alloy coated with AlTiVMoNb had excellent oxidation resistance when oxidized at 800°C for 120 hours, and the coating's oxidation weight gain was 10.58% of the substrate. The major oxides formed on the coating were TiO_2, Al_2O_3, $Al_{0.2}Nb_{0.2}Ti_{0.6}O_2$, $Nb_{10.7}V_{2.38}O_{32.7}$ and $AlMoVO_7$ [44]. Likewise, $CoCrFeNiMo_{0.2}$ HEA coating made on titanium substrate displayed improved oxidation resistance than the TC4 alloy because of the generation of a Cr-rich oxide scale. Grain refinement hardening and solid solution strengthening enabled the laser-cladded coating with superior corrosion and wear resistance [45]. Ti_3Al intermetallic composite usually shows low density, high hardness and thermal stability at high temperatures. Liu et al. in situ fabricated TiN/Ti_3Al coatings on TC4 titanium alloy by using pure Ti and AlN as the coating material. The developed coating (molar ratio of 4Ti:1AlN) possessed excellent high-temperature oxidation resistance and it was because of the formation of TiO_2, Al_2O_3 and TiN during the oxidation test. The development of continuous and dense hybrid Al_2O_3/TiO_2 oxides within the coating decreased the growth rate of porous titanium dioxide and it helped in promoting oxidation resistance [46]. Similarly, FENG et al. in situ formed Ti_3Al and TiB on a titanium substrate by depositing Ti and AlB_2 powder mixed in the weight ratio of 4:1. When compared to the TiO_2 layer formed on the titanium-base material, the $TiO_2+Al_2O_3$ layer formed on the coating was denser and well bonded and it effectively hindered oxygen diffusion. The wear rate of the coating was also decreased by the protective effect of the $TiO_2+Al_2O_3$ layer. The titanium substrate without coating had a TiO_2 oxide layer with several cracks and pores whereas the laser-cladded (Ti_3Al+TiB)/Ti composites generated a protective $TiO_2+Al_2O_3$ layer during the oxidation test and it helped in promoting the oxidation resistance of the titanium-base material [47]. The gradient Ni–Ti cladding made on the TA2 substrate also contained $NiTi_2$, NiTi and Ni_3Ti intermetallic compounds. The oxidation resistance and microhardness of the cladded layer were enhanced because of the presence of dense intermetallic compounds [48].

Jiang et al. prepared FeCoNiCrAl /AlSi12 coating on TC4 substrate and examined the effect of LC on mechanical characteristics and oxidation resistance of the coatings. Highly dense homogenous coating with white particle phase, dendrite (DR) and interdendrite (IR) phases was achieved through LC. Figure 7.5 schematically represents the mechanism behind the oxidation of the FeCoNiCrAl /AlSi12 coating. The main phases observed on the oxidized coating were FCC solid solution phase, BCC solid solution phase, Al_2O_3, $NiAl_2O_4$, Fe_3O_4/Fe_2O_3, SiO_2, Cr_2O_3 and TiO_2. The sluggish diffusion behaviour of coating and generation of Al_2O_3 film during oxidation stopped the penetration of oxygen and improved the oxidation resistance of LC coating [49]. To enhance the oxidation resistance at high temperatures and reduce the wear rate of the titanium alloy, Yu et al. adopted LC for depositing NiCrBSiFe coatings on the TC4 alloy substrate. The coating contained γ-(Ni, Cr, Fe) matrix phase in which TiC, TiB_2 and CrB particulates got reinforced. During the oxidation test, thin and dense oxide layers (NiO, Cr_2O_3, TiO_2 and FeO) formed above the coating surface and it prevented oxygen diffusion [50].

Sui et al. unveiled the influence made by TiN inclusion on the oxidation resistance of Ti–Al–Cr cladding on TC21 titanium alloy plate. The coating with TiN mainly consists of $(Al_{0.89}Cr_{0.11})_3Ti$, TiN, Ti_3Al and Cr_2Ti phases and its strength improved because of the refinement of grains and dispersion strengthening mechanism. The schematic representation of the oxidation mechanism of

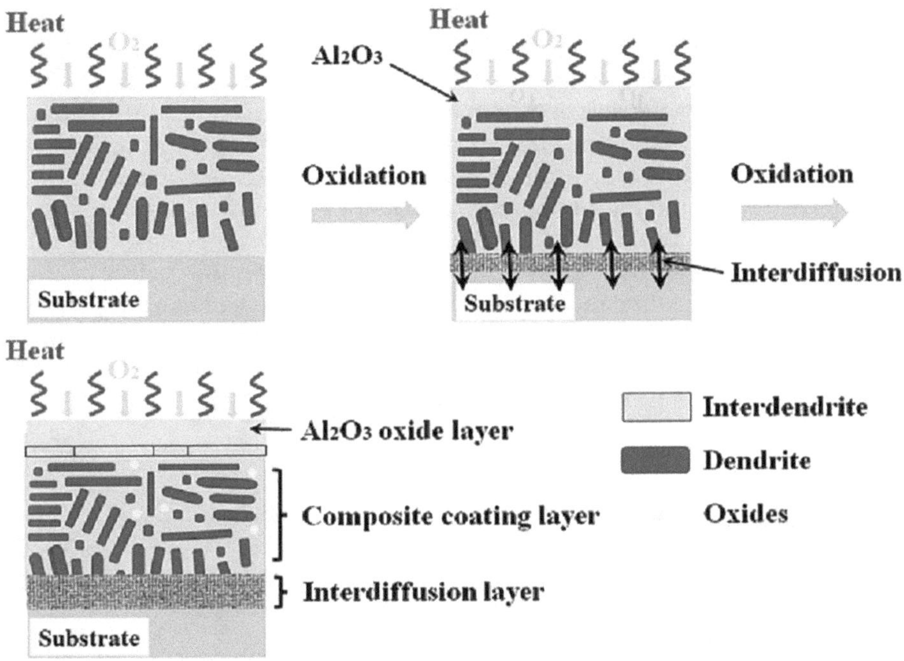

FIGURE 7.5
Oxidation mechanism exhibited by FeCoNiCrAl-AlSi12 composite coating [49].

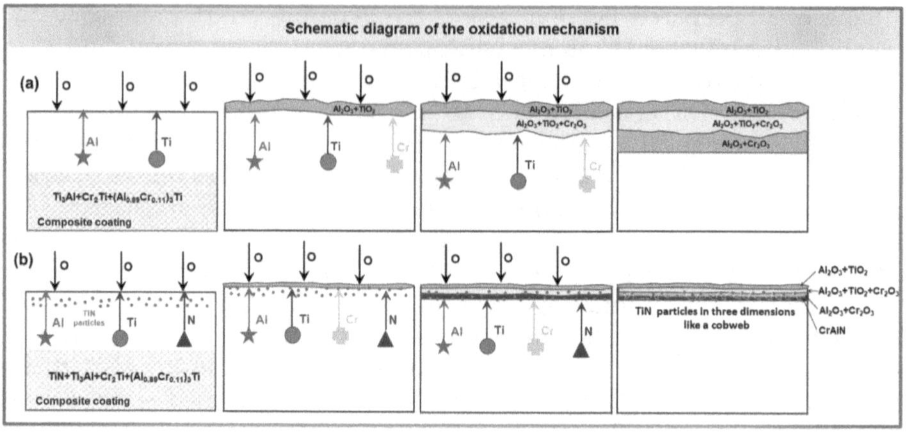

FIGURE 7.6
Difference in oxidation mechanism of Ti–Al–Cr coatings (a) without TiN and (b) with TiN addition [51].

coating with and without TiN is shown in Figure 7.6. Upon oxidation at 850°C, the coating with TiN had an oxidation mass gain of 0.56 mg/cm^2, and it was 40.1% lesser in comparison to the coating without TiN. The 4 wt.% TiN addition prolonged and diminished the oxygen diffusion during the oxidation test by forming protective CrAlN, Al_2O_3 and Cr_2O_3 layers [51]. The effect of AlNbTaZrx HEA coating with 5 different mole ratios (x=0.2 to 1) on the high-temperature oxidation and wear resistance of TC4 alloy was reported by Li et al. The coating comprised two solid solutions namely, α' with an HCP structure and β with a BCC structure and the β got transformed into α' when the proportion of Zr (x value) in the coating increased. The coating with x=0.8 demonstrated better resistance to wear and high-temperature oxidation and it was attributed to the presence of Al_2O_3, TiO_2, ZrO_2, Ta_2O_5 and Nb_2O_5 [52]. During the oxidation of titanium alloys, loose and porous TiO_2+Al_2O_3 develop on the surface of alloys and it affects its oxidation resistance. As a result, the usage of titanium alloys is limited. However, LC helps in overcoming these limitations. The laser-cladded coating limits the development of non-protective TiO_2 by forming dense and protective oxides such as Cr_2O_3, Al_2O_3, SiO_2 and La_2O_3.

7.5 High-Temperature Oxidation Behaviour of LC Made on Nickel Alloys

Nickel is usually found in the aerospace and energy sector owing to its ability to maintain strength at high-temperature conditions. Several post-processing techniques are used to develop coatings on nickel-based alloy to improve its oxidation and corrosion resistance at higher temperatures. The present section discusses the high-temperature oxidation resistance of laser-cladded coating made on nickel-based superalloys. MCrAlYX (where M indicates Co or/and Ni and X represents hafnium (Hf), Si or platinum (Pt)) coatings usually exhibit hot corrosion and high-temperature oxidation resistance and are employed as thermal barrier coatings (TBCs) for industrial gas turbines. The protective capability of MCrAlYX coatings is due to the development of continuous and dense α-Al_2O_3 scale over the coating surface upon exposure to high-temperature oxidation conditions. The thermally grown oxides (TGO) shield the base material by resisting the diffusion of oxygen. The TGO growth is affected by the microstructure, chemical composition and surface roughness of the coating. To mitigate the defects developed during the LC, the coating can be subjected to intense energy flux like a laser beam or electron beam. The high-current pulsed electron beam (HCPEB) system produces an electron beam with an energy of 20–30 keV and 60 mm spot diameter for a pulse duration of 1–5 μs. The benefits of HCPEB modification on the oxidation resistance of NiCoCrAlYSi claddings deposited through LC were reported by Cai et al. The NiCoCrAlYSi coatings made on nickel-based superalloy GH4169 substrate contain γ/γ' phases and dendritic segregation and pores were observed on the coating. Upon HCPEB treatment, original defects were removed and a dense and uniform remelted layer was generated. The oxidation rate of the irradiated coating was lowered because a continuous and uniform Al_2O_3 scale was formed and it protected the coating. The influence made by HCPEB on the oxidation behaviour of laser-cladded NiCoCrAlYSi coatings is illustrated in Figure 7.7 [53]. Xu et al. involved a high-speed LC technique to deposit CuAlNiCrFe HEA bond coat on Inconel718 alloy and compared the oxidation behaviour of laser-cladded CuAlNiCrFe coating with thermal sprayed coating. The laser-cladded coating shortened the initial oxidation stage and avoided the generation of spinel structures by forming dense α-Al_2O_3 TGO. The developed bond coat had a block-like structure and it enabled the supply of aluminium elements to the TGO layer during the isothermal oxidation process. So, the laser-cladded bond coat had excellent diffusion and oxidation resistance [54]. When the aluminium content in the CuAlxNiCrFe HEA bond coat increased, their morphology changed from columnar to equiaxed. A 4 μm thick Al_2O_3 TGO was developed on the bond coat during oxidation at 1,100°C and $CuAl_{1.5}NiCrFe$ had the lowest diffusion coefficient of $1.43\times10^{-16}m^2/s$ [55].

Wang et al. deposited refractory $Al_{0.75}NbNi_{0.1}Mo_{0.1}Cr_{0.5}Si_{0.1}Wx$ (X=1 to 2) HEA coating on nickel-based GH4099 superalloy and investigated its oxidation resistance. Upon high-temperature oxidation, the coating with x=1.5 had a minimum specific mass gain of 8.5 mg/cm^2. The protective oxide film developed on coating mainly contained SiO_2, Al_2O_3, $NiWO_4$ and $NiNb_2O_6$ and it limited the inwards penetration of O_2. However, the formation of Al_2O_3 and SiO_2 was reduced when W elements were added excessively [56]. For high-temperature protective coatings, the Mo–Si–B alloys are preferred due to their high melting points (above 2,000°C) and ability to maintain strength at high temperatures. When Mo–Si–B alloy was heated to high temperatures,

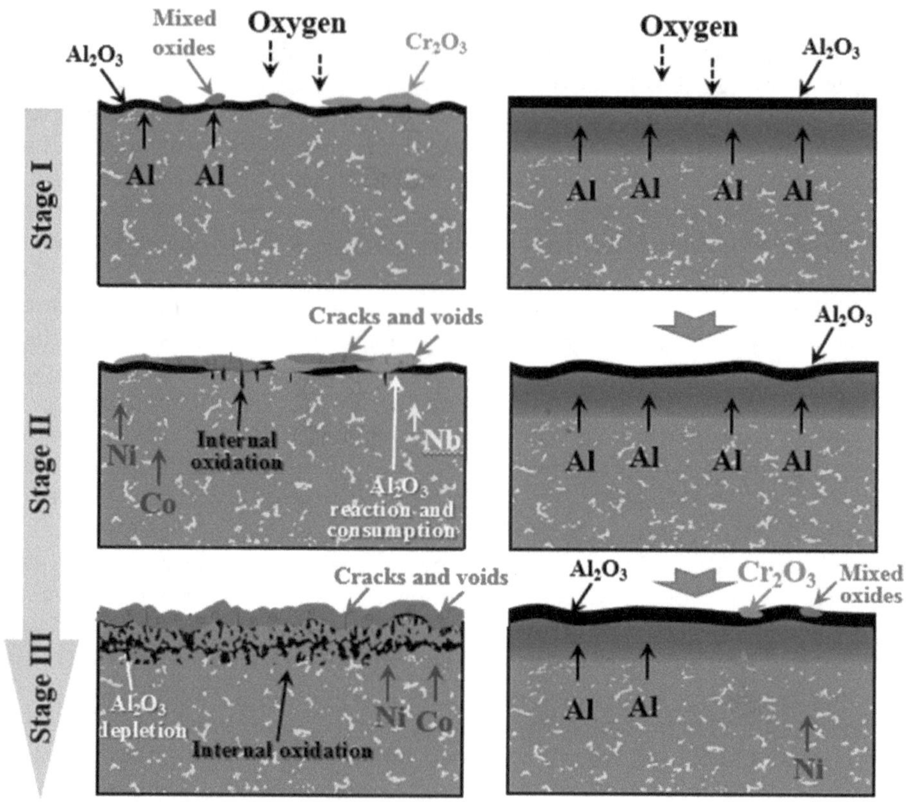

FIGURE 7.7
Oxidation mechanism observed on NiCoCrAlYSi coatings with and without HCPEB treatment [53].

borosilicate glass (SiO_2-B_2O_3) formed and it hindered the inwards diffusion of oxygen. Ding et al. coated Mo-12Si-8.5B-Y_2O_3 on GH3600 alloy (nickel alloy) and analysed its oxidation behaviour. The effect of Y_2O_3 inclusion on the properties of the coating was also examined. The coating with 1 wt.% Y_2O_3 possessed a mixture of α-Mo, Mo_3Si and Mo_5SiB_2 (T2) phases and had higher mirohardness and oxidation resistance. The SiO_2–B_2O_3 oxide film developed on the coating surface at high temperature and it resisted the oxygen diffusion. In addition, $Y_5Mo_2O_{12}$ was produced by the reaction between Y_2O_3 and MoO_3, which provided high-temperature oxidation resistance to coatings [57]. Deng et al. in situ synthesized CoNiCrAlY-xAl_2O_3 coating with different fractions of Al_2O_3 on K438 nickel-based superalloy and evaluated the coating's resistance to oxidation at higher temperatures. The in situ formed Al_2O_3 (15 wt%) reduced the average grain size and the oxidation resistance of the coating was enhanced because (Co, Ni)(Al, Cr)$_2O_4$ produced during high-temperature oxidization protected the coating [58]. Wang et al. studied the effect of Al_2O_3p nanoparticles addition on the oxidation resistance of low Al

NiCoCrAlY coatings made on Ni-based superalloy GH4033. After 100 hours of oxidation at 1,050°C, the coating's (1 wt.% Al_2O_3p) oxidation weight gain was 1.72 mg/cm^2 and it was 50% lesser than the weight gain of coating without Al_2O_3p nanoparticle. The Al_2O_3p addition reduced the consumption of Cr during long-term isothermal oxidation and inhibited the volatilization of Cr_2O_3 oxide scales thereby enhancing its self-repairing ability. Further, nano-Al_2O_3p got enriched at the grain boundary of coating and blocked microchannels; thereby, the inwards oxygen diffusion and Cr consumption during oxidation were limited [59]. Yong et al. coated CoCrAlYSi with various HfH_2 contents on Ni-based superalloy GH586 and analysed coating properties after subjecting it to an oxidation test. The HfH_2 inclusion helped in grain refinement and micro-cracks elimination and postponed the failure of the coating by delaying the formation of CoO and $CoCr_2O_4$ [60]. Self-fusing alloy powders (cobalt-based alloy, nickel and iron-based alloy) usually exhibit excellent deoxidation and slagging performance and are widely used as the binding phase. Zhang et al. used laser-cladded Stellite3–Ti_3SiC_2–Cu coatings on Inconel 718 to increase its wear and oxidation resistance at severe working conditions. Ti_3SiC_2 addition facilitated the generation of dense oxide film and enhanced the oxidation resistance of the coatings. On the other hand, Cu promoted the continuous generation of oxide film. The oxidation process had two stages namely rapid and stable oxidation stage. A stable oxidation stage was reached after 20 hours, and the oxide layer formed on the surface inhibited the reaction between oxygen and interior, resulting in a significant reduction of the curve rate. The oxide layer developed on the cladded layer consists of $NiCr_2O_4$ and $FeCr_2O_4$ spinel, Cr_2O_3, TiO_2 and SiO_2 [61].

For enhancing the tribological properties of Inconel 718 superalloy at high-temperature conditions, He et al. created Co/TiN composite coatings on Inconel 718 substrate. The coating contained γ-Co, intermetallics $FeNi_3$ and TiN and (Fe, Ni) solid solution. The composite coatings with 6% TiN had higher microhardness and better wear resistance. However, the coating's oxidation resistance was almost the same as that of the base material and the oxide layer of the coating mainly consisted of CoOx, TiO_2 and Fe_2O_3 [62]. The impact of $La_2Zr_2O_7$ coating on the oxidation resistance of GH4169 superalloy was reported by Huang et al. Since the $La_2Zr_2O_7$ ceramic layer was stable at 1,100°C, oxygen entered through the pores and cracks of coating and formed oxides such as Nb_2O_5, Cr_2O_3, Fe_2O_3 and $NiCr_2O_4$ by reacting with the base material at high temperature. However, the thinner oxide layer was formed on the coating [63]. Jeyaprakash et al. laser-cladded FeCrMoVC and NiCrFeMoNb on nickel-based superalloy IN 625 and investigated the wear resistance of the coating. The tribo oxidation layer formed on the FeCrMoVC cladding enabled it to exhibit a lower wear rate than the base material and NiCrFeMoNb-coated substrate. The tribo oxidation layer protected the coating against unexpected damage and enhanced the coating's wear resistance [64]. Table 7.1 summarizes the oxidation weight gain on various samples subjected to LC. It can be inferred from Table 7.1 that laser-cladded coating effectively reduces oxidation and increases the high-temperature oxidation resistance of the base material.

TABLE 7.1

Oxidation Weight Gain on Laser-Cladded Samples

Substrate	Coating	Oxidation Temperature and Duration	Oxidation Weight Gain for Substrate	Oxidation Weight Gain for Coating	Major Oxides Formed	Ref
5CrNiMo steel	60%$FeTi_{30}$-8.33B_4C 12$FeCr_{67}$-9$FeMo_{70}$-10.67Fe	600°C, 72 hours	~26	~3.5	TiO_2, Cr_2O_3	[23]
5CrNiMo die steel	60$FeTi_{30}$-8.33B_4C-12$FeCr_{67}$-9 $FeMo_{70}$-8.67Fe-2Y_2O_3	600°C, 55 hours	3.58	1.7	TiO_2, Cr_2O_3	[19]
TP347H commercial hot-rolled tube	C22 powder and Si	700°C, 168 hours	0.4956	0.169	NiO, Fe_2O_3, SiO_2 and Cr_2O_3	[20]
304 stainless steel	NiCoCrAlY	950°C, 80 hours	3.92	1.265	Cr_2O_3 and Al_2O_3	[21]
Q235 steel	AlCoCrFeNiMTiC	800°C, 50 hours	~136	~8	Al_2O_3, Cr_2O_3, $CoMoO_4$ and Co_2MoO_8	[27]
H13 steel	WC/Co–Cr alloy coating	600°C, 40 hours	1.5	0.5	γ-Co, Co_6W_6C, Cr_7C_3	[25]
1Cr13 steel	$FeCrMnVSi_2$	900°C, 100 hours	1.67	0.42	SiO_2, Cr_2O_3 and Mn_2VO_4	[31]
Ti-6Al-4V	Ti–Al–Si–SiC_3	800°C, 200 hours	5.532	0.191	TiO_2, Al_2O_3 and SiO_2	[37]
Ti-6Al-4V	CoCrMoNbTi	800°C, 100 hours	21.82	6.05	$CrNbO_4$, $CoMoO_4$, TiO_2	[38]

(Continued)

TABLE 7.1 (*Continued*)

Oxidation Weight Gain on Laser-Cladded Samples

Substrate	Coating	Oxidation Temperature and Duration	Oxidation Weight Gain for Substrate	Oxidation Weight Gain for Coating	Major Oxides Formed	Ref
Ti-6Al-4V	CoCrFeNiTiNb/ Y_2O_3	800°C, 50 hours	9.6	1.3	Al_2O_3, Cr_2O_3, $Al_{0.2}Nb_{0.2}Ti_{0.6}O_2$, and $NiCr_2O_4$	[39]
Ti-6Al-4V	(TiC+TiBx)/Ti/LaB_6	600°C, 50 hours	4.9	0.87	La_2O_3	[40]
Ti-6Al-4V	Ti–Cu–NiCoCrAlTaY	800°C, 50 hours	15.6	5.93	Cr_2O_3, Al_2O_3.	[41]
Ti6Al4V	TiNi/Ti_2Ni/ TaC	1,000°C, 50 hours	83.27	26.10	TiO_2, Ta_2O_5	[43]
Ti–6Al–4V	AlTiVNbMo	800°C, 120 hours	46.7	4.94	$Al_{0.2}Nb_{0.2}Ti_{0.6}O_2$, $AlMoVO_7$, and $Nb_{10.7}V_{2.38}O_{32.7}$	[44]
Ti-6Al-4 V	$CoCrFeNiMo_{0.2}$	600°C, 48 hours	5.95	3.6	Cr_2O_3	[45]
Ti-6Al-4 V	FeCoNiCrAl /AlSi12	850°C, 100 hours	~72	~2	Al_2O_3, SiO_2, $NiAl_2O_4$, Cr_2O_3 and Fe_3O_4/Fe_2O_3	[49]
TC21 Ti alloy	Ti–Al–Cr	850° 70 hours	3.39	0.56	Al_2O_3 and Cr_2O_3	[51]
Ti-6Al-4 V	AlNbTaZr	1,000°C, 50 hours	10.78	2.60	Ta_2O_3, Nb_2O_5, ZrO_2, Al_2O_3	[52]
IN718	Stellite3–10% Ti_3SiC_2	800°C, 50 hours	26.56	14.14	SiO_2, Cr_2O_3	[61]
Inconel718	Co/TiN composite coatings	800°C, 50 hours	56.55	66.16	CoOx, Fe_2O_3, TiO_2	[62]

7.6 Conclusion

High-temperature alloys based on Ni, Fe, Ti or Co usually form moderately stable oxide layers at high-temperature condition and their usage at high temperatures is limited. LC is an effective way to develop wear, corrosion and high-temperature oxidation-resistant coating on base material. Numerous HEAs and composites were successfully deposited by LC to increase the oxidation resistance of base material. The coating made through the LC technique undergoes selective oxidation at high temperatures and forms a protective oxide layer. Elements such as Al, Cr and Si are included as part of LC material and the laser-cladded coating is successful in improving the high-temperature oxidation resistance of the base component by forming protective oxides such as Cr_2O_3, Al_2O_3 and SiO_2. These oxides enhance the life of the coating by restricting the diffusion of various oxidants at high temperatures.

References

1. Unune, D.R., Brown, G.R. and Reilly, G.C., 2022. Thermal based surface modification techniques for enhancing the corrosion and wear resistance of metallic implants: A review. *Vacuum*, 203, 111298. https://doi.org/10.1016/j.vacuum.2022.111298.
2. Zhu, L., Xue, P., Lan, Q., Meng, G., Ren, Y., Yang, Z., Xu, P. and Liu, Z., 2021. Recent research and development status of laser cladding: A review. *Optics & Laser Technology*, 138, 106915. https://doi.org/10.1016/j.optlastec.2021.106915.
3. Liu, Y., Ding, Y., Yang, L., Sun, R., Zhang, T. and Yang, X., 2021. Research and progress of laser cladding on engineering alloys: A review. *Journal of Manufacturing Processes*, 66, 341–363. https://doi.org/10.1016/j.jmapro.2021.03.061.
4. Das, A.K., 2022. A review on coating with high entropy alloy developed by laser energy based surfacing process. *Materials Today: Proceedings*, 52, 1551–1557. https://doi.org/10.1016/j.matpr.2021.11.234.
5. Yuan, W., Li, R., Chen, Z., Gu, J. and Tian, Y., 2021. A comparative study on microstructure and properties of traditional laser cladding and high-speed laser cladding of Ni45 alloy coatings. *Surface and Coatings Technology*, 405, 126582.
6. Jeyaprakash, N., Yang, C.H., Susila, P. and Karuppasamy, S.S., 2023. Laser cladding of NiCrMoFeNbTa particles on Inconel 625 alloy: Microstructure and corrosion resistance. *Transactions of the Indian Institute of Metals*, 76(2), 599–612. https://doi.org/10.1007/s12666-022-02701-7.
7. Jeyaprakash, N., Yang, C.H., Karuppasamy, S.S. and Duraiselvam, M., 2023. Stellite 6 cladding on AISI Type 316L stainless steel: Microstructure, nanohardness and corrosion resistance. *Transactions of the Indian Institute of Metals*, 76(2), 491–503. https://doi.org/10.1007/s12666-022-02731-1.

8. Jeyaprakash, N. and Yang, C.H., 2020. Microstructure and wear behaviour of SS420 micron layers on Ti–6Al–4V substrate using laser cladding process. *Transactions of the Indian Institute of Metals*, 73, 1527–1533. https://doi.org/10.1007/s12666-020-01927-7.
9. Huang, Y., Hu, Y., Zhang, M., Mao, C., Tong, Y., Zhang, J., Li, K. and Wang, K., 2022. On the enhanced wear resistance of laser-clad CoCrCuFeNiTix high-entropy alloy coatings at elevated temperature. *Tribology International*, 174, 107767. https://doi.org/10.1016/j.triboint.2022.107767.
10. Gao, S., Cao, J., Qiu, Z. and Yan, X., 2022. A novel wear-resistant FeCoNiCrAl0. 5Ti0. 5 coating fabricated by laser cladding technology. *Materials Letters*, 321, 132393. https://doi.org/10.1016/j.matlet.2022.132393.
11. Jeyaprakash, N., Yang, C.H. and Tseng, S.P., 2021. Wear Tribo-performances of laser cladding Colmonoy-6 and Stellite-6 micron layers on stainless steel 304 using Yb: YAG disk laser. *Metals and Materials International*, 27, 1540–1553. https://doi.org/10.1007/s12540-019-00526-6.
12. Wang, H., Wu, M., Miao, X., Jin, X., Cui, C., Ma, C. and Wang, Q., 2023. Effect of Nb on the microstructure and wear resistance of In625/(Nbx+ SiC0. 5) composite coatings by laser cladding. *Ceramics International*, 49, 38420–38431. https://doi.org/10.1016/j.ceramint.2023.09.247.
13. Liu, H., Wang, X., Yu, Y., Yang, W. and Gao, H., 2023. Microstructure and wear resistance of laser cladding nano TiC/micro TiB2/C-276 composite coatings on 316L stainless steel. *Materials Research Express*, 10(3), 036512. https://doi.org/10.1088/2053-1591/acc447.
14. Jeyaprakash, N., Yang, C.H. and Tseng, S.P., 2020. Characterization and tribological evaluation of NiCrMoNb and NiCrBSiC laser cladding on near-α titanium alloy. *The International Journal of Advanced Manufacturing Technology*, 106, 2347–2361. https://doi.org/10.1007/s00170-019-04755-2.
15. Dai, J., Zhu, J., Chen, C. and Weng, F., 2016. High temperature oxidation behavior and research status of modifications on improving high temperature oxidation resistance of titanium alloys and titanium aluminides: A review. *Journal of Alloys and Compounds*, 685, 784–798. https://doi.org/10.1016/j.jallcom.2016.06.212.
16. Birks, N., Meier, G.H. and Pettit, F.S., 2006. *Introduction to the High Temperature Oxidation of Metals*. Cambridge University Press.
17. Samal, S., 2016. High-temperature oxidation of metals. *High Temperature Corrosion*, 12, 11–17.
18. Taylor, C.D. and Tossey, B.M., 2021. High temperature oxidation of corrosion resistant alloys from machine learning. *npj Materials Degradation*, 5(1), 38.
19. Zhang, M., Wang, X. H., Qu, K. L. and Liu, S. S., 2019. Effect of rare earth oxide on microstructure and high temperature oxidation properties of laser cladding coatings on 5CrNiMo die steel substrate. *Optics & Laser Technology*, 119, 105597. https://doi.org/10.1016/j.optlastec.2019.105597.
20. Kong, Y., Liu, Z., Wang, X., Li, J. and Liu, Q., 2023. Evolution in microstructure and oxidation behaviors of the laser cladding C22 coatings with the Si addition concentrations. *Journal of Alloys and Compounds*, 932, 167536. https://doi.org/10.1016/j.jallcom.2022.167536.
21. Nie, J., Li, Y., Chen, B., Yang, Y., Zhang, H., Liu, S., Wei, S., Cai, J. and Guan, Q., 2021. Evolution of oxide layer during high-temperature oxidation of NiCoCrAlY coating via laser cladding on 304 stainless steel. *Materials Letters*, 286, 129233. https://doi.org/10.1016/j.matlet.2020.129233.

22. Xu, Z., Wang, F., Peng, S., Liu, W. and Guo, J. (2023). Effects of process parameters on microstructure and high-temperature oxidation resistance of laser-clad IN718 coating on Cr5Mo steel. *Coatings*, 13(1), 197.
23. Zhang, M., Liu, S. S., Luo, S. X. and Qu, K. L., 2018. Effect of molybdenum on the high-temperature properties of TiC-TiB2 reinforced Fe-based composite laser cladding coatings. *Journal of Alloys and Compounds*, 742, 383–390. https://doi.org/10.1016/j.jallcom.2018.01.275.
24. Jin, F., Yin, T. Y., Zhang, S., Wu, C. L., Zhang, C. H. and Chen, J., 2023. Effect of Ni and Fe elements on microstructure and high temperature oxidation behavior of laser cladding aluminum bronze coating. *Materials Chemistry and Physics*, 297, 127383. https://doi.org/10.1016/j.matchemphys.2023.127383.
25. Zhang, X., Jie, X., Zhang, L., Luo, S. and Zheng, Q., 2016. Improving the high-temperature oxidation resistance of H13 steel by laser cladding with a WC/Co-Cr alloy coating. *Anti-Corrosion Methods and Materials*, 63(3), 171–176. https://doi.org/10.1108/ACMM-11-2015-1606.
26. Sun, D., Cai, Y., Zhu, L., Gao, F., Shan, M., Manladan, S.M., Geng, K., Han, J. and Jiang, Z., 2022. High-temperature oxidation and wear properties of TiC-reinforced CrMnFeCoNi high entropy alloy composite coatings produced by laser cladding. *Surface and Coatings Technology*, 438, 128407. https://doi.org/10.1016/j.surfcoat.2022.128407.
27. Yu, K., Zhao, W., Li, Z., Guo, N., Xiao, G. and Zhang, H., 2023. High-temperature oxidation behavior and corrosion resistance of in-situ TiC and Mo reinforced AlCoCrFeNi-based high entropy alloy coatings by laser cladding. *Ceramics International*, 49(6), 10151–10164. https://doi.org/10.1016/j.ceramint.2022.11.198.
28. Cai, Y., Zhu, L., Cui, Y., Geng, K., Manladan, S.M. and Luo, Z., 2019. High-temperature oxidation behavior of FeCoCrNiAlx high-entropy alloy coatings. *Materials Research Express*, 6(12), 126552. https://doi.org/10.1088/2053-1591/ab562d.
29. Zhang, X., Sun, Y., Yu, G., Chen, C., Ren, X. and Chen, L., 2023. Microstructural evolution and high-temperature oxidation of TiC/IN625 coatings fabricated by multi-layer extreme high-speed laser cladding. *Optics & Laser Technology*, 158, 108838. https://doi.org/10.1016/j.optlastec.2022.108838.
30. Xiong, J., Sun, D., Dong, N., Cai, Y., Wang, D. and Luo, Z., 2022. Oxidation mechanism of TiC reinforced CrMnFeCoNi composite coatings by laser cladding at 600° C/750° C/900° C. *Materials Letters*, 324, 132710. https://doi.org/10.1016/j.matlet.2022.132710.
31. Liu, Y., Yao, Z., Zhang, P., Xu, Z., Lin, S., He, M., Lu, S. and Wu, X., 2022. Tailoring high-temperature oxidation resistance of FeCrMnVSix high entropy alloy coatings via building Si-rich dendrite microstructure. *Applied Surface Science*, 606, 154862. https://doi.org/10.1016/j.apsusc.2022.154862.
32. Ma, M., Han, A., Zhang, Z., Lian, Y., Zhao, C. and Zhang, J., 2021. The role of Si on microstructure and high-temperature oxidation of CoCr2FeNb0.5Ni high-entropy alloy coating. *Corrosion Science*, 185, 109417. https://doi.org/10.1016/j.corsci.2021.109417.
33. Jeyaprakash, N., Duraiselvam, M. and Raju, R., 2018. Modelling of Cr3C2–25% NiCr laser alloyed cast iron in high temperature sliding wear condition using response surface methodology. *Archives of Metallurgy and Materials*, 63(3), 1303–1315. https://doi.org/10.24425/123805.

34. Jeyaprakash, N., Duraiselvam, M. and Aditya, S.V., 2018. Numerical modeling of WC-12% Co laser alloyed cast iron in high temperature sliding wear condition using response surface methodology. *Surface Review and Letters*, 25(07), 1950009. https://doi.org/10.1142/S0218625X19500094.
35. Leyens, C. and Peters, M. eds., 2006. *Titanium and Titanium Alloys: Fundamentals and Applications*. Wiley-VCH.
36. Li, S. and Yamaguchi, T., 2022. High-temperature oxidation behavior of laser-cladded refractory NiSi0. 5CrCoMoNb0. 75 high-entropy coating. *Journal of Materials Research and Technology*, 17, 1616–1627. https://doi.org/10.1016/j.jmrt.2022.01.115.
37. Li, J., Cui, X., Guan, Y., Jin, G., Zheng, W., Su, W., Wan, S. and Shi, Z., 2023. Analysis of oxidation behavior of laser cladding SiC-Ti based composite strengthening coating. *Materials Characterization*, 204, 113210. https://doi.org/10.1016/j.matchar.2023.113210.
38. Zhang, S., Qi, W. and Zhang, R., 2023. Effect of Ti content on wear and high-temperature oxidation resistances of laser-clad CoCrMoNbTix high-entropy alloy coatings. *Materials Letters*, 345, 134490. https://doi.org/10.1016/j.matlet.2023.134490.
39. Li, Z., Zhao, W., Yu, K., Guo, N., Xiao, G., Wang, Z. and Zhang, H., 2023. Effect of Y2O3 on microstructure and properties of CoCrFeNiTiNb high entropy alloy coating on Ti–6Al–4V surface by laser cladding. *Journal of Rare Earths*, 42, 586–599. https://doi.org/10.1016/j.jre.2023.02.015.
40. Yin, X., Liang, J., Gao, Y., Lin, Z., Chen, S., Liu, C., Tian, K., Zhang, H. and Tang, G., 2021. Effects of LaB6 on the high-temperature oxidation behavior of TiC+ TiBx reinforced titanium matrix composite coatings fabricated by laser cladding. *Surface and Coatings Technology*, 421, 127445. https://doi.org/10.1016/j.surfcoat.2021.127445.
41. Li-Yan, L., Yu, Z., Yun-Jie, J., Yan, L., Hong-Fang, T., Yu-Jun, C. and Cheng-Xin, L., 2020. High speed laser cladded Ti-Cu-NiCoCrAlTaY burn resistant coating and its oxidation behavior. *Surface and Coatings Technology*, 392, 125697. https://doi.org/10.1016/j.surfcoat.2020.125697.
42. Barekat, M., Shoja Razavi, R. and Ghasemi, A., 2016. High temperature oxidation behavior of laser clad Co–Cr–Mo coating on γ-TiAl substrate. *Journal of Laser Applications*, 28(4), 042005. https://doi.org/10.2351/1.4958971.
43. Lv, Y. H., Li, J., Tao, Y. F. and Hu, L. F., 2017. High-temperature wear and oxidation behaviors of TiNi/Ti2Ni matrix composite coatings with TaC addition prepared on Ti6Al4V by laser cladding. *Applied Surface Science*, 402, 478–494. https://doi.org/10.1016/j.apsusc.2017.01.118.
44. Chen, L., Wang, Y., Hao, X., Zhang, X. and Liu, H., 2021. Lightweight refractory high entropy alloy coating by laser cladding on Ti–6Al–4V surface. *Vacuum*, 183, 109823. https://doi.org/10.1016/j.vacuum.2020.109823.
45. Deng, C., Wang, C., Chai, L., Wang, T. and Luo, J., 2022. Mechanical and chemical properties of CoCrFeNiMo0. 2 high entropy alloy coating fabricated on Ti6Al4V by laser cladding. *Intermetallics*, 144, 107504. https://doi.org/10.1016/j.intermet.2022.107504.
46. Liu, H., Zhang, X., Jiang, Y. and Zhou, R., 2016. Microstructure and high temperature oxidation resistance of in-situ synthesized TiN/Ti3Al intermetallic composite coatings on Ti6Al4V alloy by laser cladding process. *Journal of Alloys and Compounds*, 670, 268–274. https://doi.org/10.1016/j.jallcom.2015.10.168.

47. Feng, Y., Feng, K., Yao, C., Li, Z. and Sun, J., 2019. High temperature oxidation and wear resistance of in situ synthesized (Ti 3 Al+ TiB)/Ti composites by laser cladding. *Metallurgical and Materials Transactions A*, 50, 3414–3428. https://doi.org/10.1007/s11661-019-05241-w.
48. Liu, F., Mao, Y., Lin, X., Zhou, B. and Qian, T., 2016. Microstructure and high temperature oxidation resistance of Ti-Ni gradient coating on TA2 titanium alloy fabricated by laser cladding. *Optics & Laser Technology*, 83, 140–147. https://doi.org/10.1016/j.optlastec.2016.04.005.
49. Jiang, J., Hou, W., Feng, X. and Shen, Y., 2023. Oxidation resistant FeCoNiCrAl high entropy alloy/AlSi12 composite coatings with excellent adhesion on Ti-6Al-4 V alloy substrate via mechanical alloying and subsequent laser cladding. *Surface and Coatings Technology*, 464, 129577. https://doi.org/10.1016/j.surfcoat.2023.129577.
50. Yu, P. C., Liu, X. B., Lu, X. L., Qiao, S. J., Zhai, Y. J., Zhu, G. X., Wang, Y. G. and Chen, Y., 2015. Tribology and high-temperature oxidation behaviors of NiCrBSiFe composite coatings on Ti6Al4V alloy by laser cladding. *RSC Advances*, 5(93), 76516–76525. https://doi.org/10.1039/C5RA14732H.
51. Sui, X., Lu, J., Wei, D., Zhang, L., Wang, R., Zhao, W. and Zhang, W., 2022. Unveiling the influence of TiN on the microstructure and high-temperature oxidation behavior of Ti-Al-Cr composite coating. *Corrosion Science*, 206, 110539. https://doi.org/10.1016/j.corsci.2022.110539.
52. Hao, P., Li, J., Zhang, Y., Li, X., Xia, M. M. and Yuan, B. G., 2021. Wear and high-temperature oxidation resistances of AlNbTaZrx high-entropy alloys coatings fabricated on Ti6Al4V by laser cladding. *Journal of Alloys and Compounds*, 862, 158405. https://doi.org/10.1016/j.jallcom.2020.158405.
53. Cai, J., Yao, Y., Gao, C., Lyu, P., Meng, X., Guan, Q., Li, Y. and Han, Z., 2021. Comparison of microstructure and oxidation behavior of NiCoCrAlYSi laser cladding coating before and after high-current pulsed electron beam modification. *Journal of Alloys and Compounds*, 881, 160651. https://doi.org/10.1016/j.jallcom.2021.160651.
54. Xu, Q. L., Zhang, Y., Liu, S. H., Li, C. J. and Li, C. X., 2020. High-temperature oxidation behavior of CuAlNiCrFe high-entropy alloy bond coats deposited using high-speed laser cladding process. *Surface and Coatings Technology*, 398, 126093. https://doi.org/10.1016/j.surfcoat.2020.126093.
55. Xu, Q. L., Liu, K. C., Wang, K. Y., Lou, L. Y., Zhang, Y., Li, C. J. and Li, C. X., 2021. TGO and Al diffusion behavior of CuAlxNiCrFe high-entropy alloys fabricated by high-speed laser cladding for TBC bond coats. *Corrosion Science*, 192, 109781. https://doi.org/10.1016/j.corsci.2021.109781.
56. Wang, Z., Zhang, X., Cui, X., Zhang, Q., Qi, M. and Jin, G., 2023. High-temperature oxidation behavior of laser-cladded refractory Al0. 75NbNi0. 1Mo0. 1Cr0. 5Si0. 1Wx high-entropy coatings. *Materials Chemistry and Physics*, 309, 128331. https://doi.org/10.1016/j.matchemphys.2023.128331.
57. Ding, Z., Fang, Z., Wang, G., Wu, C. and Yang, S., 2023. Microstructure and properties of Mo-Si-B-Y2O3 composite coating on nickel-based alloy by laser cladding. *Optics & Laser Technology*, 164, 109473. https://doi.org/10.1016/j.optlastec.2023.109473.
58. Deng, C., Yi, Y., Jiang, M., Hu, L. and Zhou, S., 2023. Microstructure and high-temperature resistance of Al_2O_3/CoNiCrAlY coatings by laser cladding. *Ceramics International*, 49, 32885–32895. https://doi.org/10.1016/j.ceramint.2023.07.262.

59. Wang, H., Zuo, D., Yan, J., Huang, M. and Li, X., 2010. Effects of nanometer Al 2 O 3 particles on oxidation behaviors of laser cladding low Al NiCoCrAlY coatings. *Oxidation of Metals*, 74, 49–60. https://doi.org/10.1007/s11085-010-9197-3.
60. Yong, Z., Xie, D., Liu, Y., Jiang, S., Chang, L., Shen, L. and Tian, Z., 2023. High temperature oxidation behavior of HfH2-CoCrAlYSi coating fabricated by laser cladding. *Journal of Materials Engineering and Performance*, 33, 3077–3086. https://doi.org/10.1007/s11665-023-08203-x.
61. Zhang, S. Y., Liu, X. B., Zhu, Y., Liu, Y. F., Meng, Y., Liang, J. and Zhang, S. H., 2022. Stellite3-Ti3SiC2-Cu composite coatings on IN718 by laser cladding towards improved wear and oxidation resistance. *Surface and Coatings Technology*, 446, 128766. https://doi.org/10.1016/j.surfcoat.2022.128766.
62. He, B. M., Liu, X. B., Zhang, F. Z., Zhang, S. Y., Liu, Z. Y. and Zhang, S. H., 2023. Tribological and oxidation behaviors of TiN reinforced Co matrix composite coatings on Inconel718 alloy by laser cladding. *Tribology International*, 188, 108781. https://doi.org/10.1016/j.triboint.2023.108781.
63. Huang, K., Li, W., Pan, K., Lin, X. and Wang, A., 2020. High temperature oxidation and thermal shock properties of La2Zr2O7 thermal barrier coatings deposited on nickel-based superalloy by laser-cladding. *Coatings*, 10(4), 370. https://doi.org/10.3390/coatings10040370.
64. Jeyaprakash, N. and Yang, C.H., 2020. Comparative study of NiCrFeMoNb/FeCrMoVC laser cladding process on nickel-based superalloy. *Materials and Manufacturing Processes*, 35(12), 1383–1391. https://doi.org/10.1080/10426914.2020.1779933.

8

Influence of Laser Cladding on Mechanical Properties

Govindarajan Prabu, Ibrahim A. Alnaser, Natarajan Jeyaprakash, and Che-Hua Yang

8.1 Introduction

The study of material behavior was important for adapting it in a particular application. The mechanism involved in the material behaviors was mainly deepened on the microstructures, presence of phases, chemical composition, distribution of elements, etc. These factors were the building blocks of material properties. Based on the above-depending factors, the material properties were built. Most of the fabrication process and the applications depended on the mechanical properties of the material. The flowability, suitability to form into particular shapes, strength, deformably, etc. depended on their mechanical property. Hence, the analysis of the mechanical properties of the material was essential. Laser cladding was one of the surface modification techniques to fabricate the desired mechanical properties on the substrate surface. The metallurgical bonding was obtained between the cladding and the substrate material due to the rapid heating and cooling process. In addition, the laser cladding process was used to remanufacture the mechanical parts with low heat-affected region, dense coating, high precision, and excellence in mechanical properties. As a result, the surface mechanical properties of the material were improved with environmentally friendly technology. It confirmed that the different types of coating were easily implemented on the selected substrate. This type of laser-based coating was utilized in different fields, namely defense, aerospace, metallurgy, chemical processing, automobiles, and other industries [1–3].

Enormous research was conducted on the enhancement of the mechanical properties of the material through the laser cladding process. In addition, these types of research works were done in the atmosphere and at elevated temperatures [4]. Moreover, corrosion resistance, microstructures, abrasion resistance, microhardness, and phase analysis of laser cladding region were performed on different materials [5–7]. The drawback of this fabrication

DOI: 10.1201/9781032693576-8

process was the crack formation. Optimizing the process parameters was an important step to overcome this drawback [8]. The residual stress induced in the fabrication component was an unavoidable issue in the laser cladding process because the presence of the crack affected the continuity of the microstructure. As a result, mechanical properties such as tensile strength, hardness, and fatigue strength were affected. Cui et al. [9] investigated the influence of laser beam scanning velocity on the microstructural formation of nickel-tungsten carbide coating on the carbon steel substrate [10]. Shepeleva et al. [11] performed two different processes, namely the laser cladding process and plasma base cladding for the enhancement of microhardness of the material. Riveiro et al. [12] correlated the geometric characteristics of the laser cladding region with the process parameters of the laser cladding process. In this study, the aluminum material was deposited on the stainless-steel substrate. Chryssolouris [13] varied the feeding rate of Cu powder in the laser cladding region. In addition, the scanning speed, supply of inert gas, and process parameters of different working ranges were analyzed along with the supply of Cu alloy powder. This Cu alloy was deposited on the aluminum alloy and studied the formation of microstructure at different process parameters.

The laser cladding region possessed porosity and different dilution levels when one material was deposited on the other material [14]. For example, the laser cladding region of Inconel 625 had porosity, cracks, and various dilution levels when it was deposited on the Inconel 713 [15]. The chemical composition of Inconel 625 is shown in Table 8.1. The formation of microstructure on the laser cladding region of Inconel 625 alloy on the Inconel 713 was helpful in enhancing the mechanical properties such as microhardness when compared to the TIG welding process. This output was used to repair the Inconel 713 alloy when small cracks formed after a particular period of service time. This laser cladding process enhanced the properties of the material for utilizing it in different applications such as rails in high-speed trains because the rail had the strength to carry the weight of the train during the running stage and the stationary position. It provided safe working conditions. The friction and damage on the rail surface were the main issues during the carrying of heavy weight.

The contact fatigue of the rail was a key parameter that affected the surface and induced damage during the working conditions. This type of damage occurred at the contact point of the rail and wheel at the time of service. It required periodic maintenance to avoid the severe accident of high-speed trains [16]. The peening and case hardening were used to improve the surface

TABLE 8.1

Chemical Composition of Inconel 625

Elements	Ni	Cr	Mo	Fe	Nb	Others
Wt. (%)	Bal.	22.12	9.3	4.3	3.4	<0.3

property of the component as a traditional technique [17]. The safety and the development of the high-speed train transportation industry required the larger service life of the rail component [18]. Fatigue strength can be improved by the heat treatment process but the cost of the component was increased due to this additional process. The rail had a length of more than 10 m, which was impractical to perform the surface hardening process. Hence, coating on the rail surface through the laser cladding process was one of the viable solutions to enhance the fatigue strength and other mechanical properties of the rail.

8.2 Types of Mechanical Properties

The material response to the applied load was known as the mechanical properties of the material. This study was useful in determining the extent of material deformation during the applied normal load. Some of the important mechanical properties of the material were elasticity, plasticity, ductility, hardness, strength, toughness, and stiffness. The types of mechanical properties are listed in Figure 8.1. Different elements were used to enhance the mechanical properties of the material. The Cr content decreased the

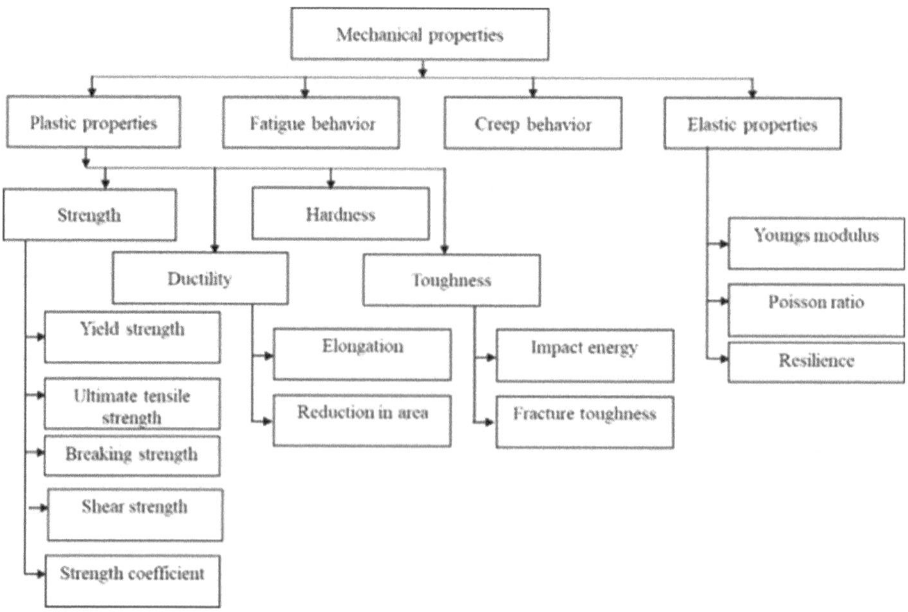

FIGURE 8.1
Types of mechanical properties of material.

melting point and improved the nucleation rate with the high viscosity of the molten liquid. It led to refining the grain structure of the laser cladding region and improved the mechanical properties [19,20]. In addition, the Cr content improved the plasticity and hardness of the material at elevated temperatures. However, the higher Cr content led to deposits around the grain boundary region and resulted in a decrease in the mechanical properties. In addition, the elemental crystalline also affected the mechanical properties of the material. The nanocrystalline Fe element improved the intrinsic ductility of the material. The ball-milled Fe element improved the hardness of the coated material when the grain size was smaller than that of the initial coarse-grained Fe element. The failure of nanocrystalline Fe-coated material was brittle in nature when it was subjected to a tensile test. At the same time, the shear band was present at the local region of the fractured surface due to the local ductility of the material. The compressive behavior of Fe-coated material led to behave as plastic material.

The mechanical properties of the material mainly deal with external force acting on it and its stability when force was applied on the material surface. The material deformation and the corresponding fracture that occurred during the applied normal force on the surface mainly depend on the structure of the material. The material responds to the applied force in the form of deformation or change in its shape in a macroscopic size. In some cases, the material was disintegrated into two or more parts due to the influence of a large value of force. The applied external force was higher than that of the interatomic forces to cause the material deformation. The changes that occurred in the material were measured in force per area, which was represented as stress, and the ratio of displacement to the initial size was called strain. In addition, the defect-free lattice surface withstood the higher amount of force and was difficult to cause the deformation of the material. However, the real material contained a lot of defects such as lattice dislocations, which resulted in the deformation of the material at low load conditions. These types of defects occurred in the machine components. Based on the laboratory test, the type of mechanical properties that the particular material withstands was found. The yield strength of the material was mostly used in the engineering application to calculate the strength during the tensile loading. The stress value before the plastic deformation of the material was called yield strength. The stress value was calculated in MPa. The material deformed permanently when the load exceeded the yield point.

The plastic property was considered as the first property of the material in the classification. The strength of the material was defined as the opposing force to the deformation of the material when the external force was applied. The material must have enough strength to withstand the different forces when it was used as the engineering component. The resistance against the permanent deformation of material was denoted as the hardness of the material [21]. It was measured by passing the indenter on the external surface of the material. The indenter has the diamond tip that was dropped from the particular predefined height on the material surface. The toughness was one of the material properties to absorb energy when the material was subjected to sudden loading or impacted with other materials. The toughness value was measured as the energy per cubic meter with the unit of J/m^3. The stress–strain curve was used to characterize the toughness of the engineering material. This toughness material possessed higher strength as well as superior ductility of the material. The brittle material had higher strength and lower ductility, which resulted in the low toughness. Moreover, the material had good ductility with low strength and also denoted as low tough material. Hence, the tough material had withstood both high stress and high strain.

8.3 Improvement of Hardness through Laser Cladding Process

The hardness of the material was generally measured across the cross-section of the laser cladding region. An indenter was used to measure the hardness, which had the value in terms of micron or nano level. The indenter is made of diamond, which is one of the highest hardening materials. The scanning speed of the laser beam and the powder supplied to the cladding region affected the distribution of the hardness of the material. The grain structures were significantly decreased due to rapid heating and cooling rates, which resulted in the increase of the microhardness of the material. Generally, the microhardness was increased at higher scanning velocity conditions than that of at lower scanning velocity during the higher feeding rate of power in the cladding region. Moreover, the ratio between the temperature gradient and the solidification rate was used to assess the grain size. This ratio increased with the increase of the laser scanning velocity, which resulted in the reduction of grain size. As a result, the microstructure transferred from coarse grain size to fine grains. The microhardness of laser cladding material differed from the higher value to the lower value due to the formation of the laser-clad layer, the intersection between the clad-substrate layer and the substrate layer. The hardness fluctuated slightly across the cross-section of the laser cladding region. The higher microhardness was observed at the clad layer and the lower microhardness was observed at the substrate region. The microhardness value transferred from the higher value to the lower value at the intersection region. The microhardness values were caruncculated based on an average of three or more hardness-tested values due to the slight fluctuation across the cross-section region. The different phases and compound formation in the clad region were the main reasons for the increase in hardness behavior. The nanohardness was measured similar to microhardness and the corresponding FESEM image of nanohardness indentation impression is shown in Figure 8.2.

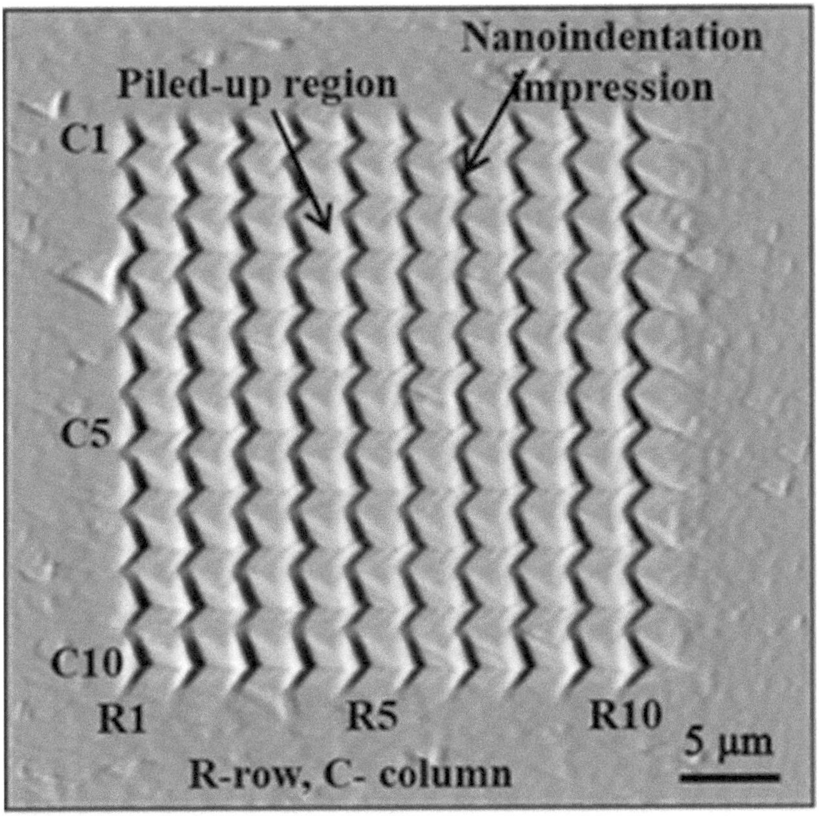

FIGURE 8.2
FESEM image of nanohardness indentation impression.

Most of the compound formation was deepened on the oxide growth during the cladding process. It led to inducing different forms of oxides such as Al_2O_3, TiO_2, ZrO_2, etc., which may result in an increase or decrease of hardness of clad region. The formation of these oxides was confirmed through the X-ray diffraction analysis. The temperature at the melt pool region was higher than that at the substrate region. The bottom layer remelted during the deposition of the new layer, which resulted in the modification of grain formation. It led to a significant effect on the microhardness of the clad region. Moreover, incorporation of hard material such as TiC in the clad region caused a drastic improvement in microhardness. This improvement of the hardness mainly depended on the hard phase rather than the pores, cracks, etc. The hardness improvement in the clad region was nearly 3–10 times higher than that in the substrate material. The schematic representation of the microhardness curve is shown in Figure 8.3. It contained three different regions, namely the clad region, heat-affected region, and unaffected

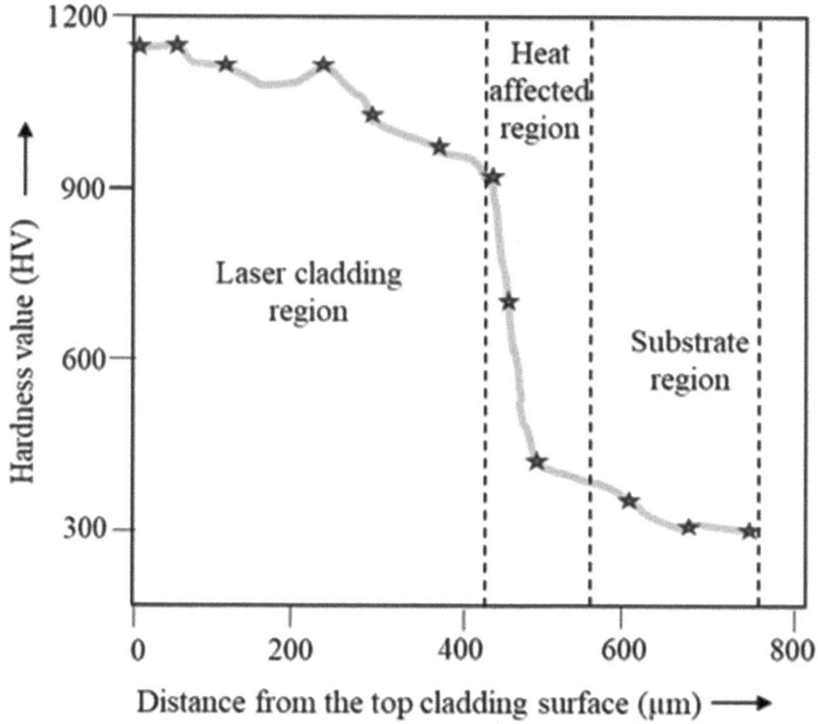

FIGURE 8.3
Schematic representation of the microhardness curve.

substrate region. The microhardness was maximum at area I, which was the clad region. The remaining two areas possessed less microhardness than area I. The increase in the hardness of the laser-clad region improved the resistance of heat dissipation by incorporating some specific elements such as TiC with B_4C, etc.

The reaction was induced between the elements during the cladding process, which led to enhancement of specific properties such as heat dissipation resistance. Moreover, the incorporation of lower hardness material to the higher hardness material, namely B_4C ceramics, led to increased hardness by compactness with other structures. The formation of graphite and the distribution of dense material improved the hardness of the cladding region [22,23]. The presence of unmelted particles in the cladding region reduced the microhardness due to improper bonding with the nearer region. In addition, the presence of alloying elements in the cladding region promoted the diffusion of specific elements. The temperature gradient decreased with the progress of the cladding process, which resulted in the decrease of elemental reaction in the clad region with the other element. The heat-affected region increased with the increase of the laser power. It resulted in the increase of the microhardness compared to the substrate region [24]. The grain size from the top of the cladding region to the substrate surface was in descending order due to the decrease in temperature. The enhancement of the hardness in the cladding region was dependent on the martensite and carbide hard phase formation. Moreover, the hard phase martensite formation may be affected in some of the circumferences when the austenite-like content was completely dissolved with the alloying elements in the clad region [25].

8.4 Tensile Behavior of Laser Cladding Material

Industrial development required higher flexibility of the component shape design with excellent performance [26]. The tensile strength was one of the important properties used to assess the yield value during the drawing and other ductile-related work. The laser cladding process was superior to their fabrication process to improve the tensile strength at the surface up to some micron thickness. Hence, this fabrication method was widely used in the industry to improve the performance of the component. A variety of parameters were available in the laser cladding process to obtain the desired tensile properties of the material [27]. The optimized parameter was important to enhance the tensile strength as the higher laser energy input led to an increase in the porosity of the cladding region. It resulted in the induction of cracks and provided the path for product failure due to loss of tensile strength. The thickness of the clad region, presence of alloying elements, formation of phases, and surface roughness of the cladding region affected the tensile strength. The tensile test was used to identify the strength of the laser-clad lap joint. The stain rate was varied to identify the appropriate tensile strength of the lap joint. This strain rate was selected in the range of 3–10 mm/min. The inhomogeneous distribution of microstructure in the cladding region resulted in poor metallurgical bonding between the overlapping regions. It resulted in less tensile strength than the homogeneously distributed cladding layer [28]. The full-lap joint was made through the laser cladding process, which led to minimizing the defect in the laser cladding overlapping region and improved the tensile strength.

The grain boundary formation in the clad region increased with the reduction of grain size [29]. This fine grain formation improved the tensile strength of the clad region. The grain boundaries are crossed into a small area when the clad material was subjected to the external tensile load. It led to improved strength due to small damage to the grain boundary when compared to large-sized grains. The uniform distribution of fine grains in the clad region improved the resistance against the deformation and provided the small plastic deformation with enhanced toughness. In addition, the formation of different directions of microstructure in the cladding region affected the tensile strength of the component. The metallurgical bonding mainly occurred in the overlap area of the clad layer. The stretching occurred in the perpendicular direction of the overlap region during the tensile test. It avoided the formation of defects at the clad region due to fine grain formation. Hence, the time of failure during the tensile test differed with respect to the orientation of grains in the cladding region. However, the failure mode was the same even with the different orientations of the grains. The direction of stretching during the tensile test should be perpendicular to the clad region. The dendrite structure was present in the overlap cladding region, which improved the tensile strength value. The clad region increased the time period between the maximum tensile stress and the fracture of the component. It confirmed that the laser cladding material avoided the immediate crack formation during the tensile test. It exhibited deformation for a particular period of time before failure, which led to protecting the machine before huge damage.

The strain value of 0.001 was used to eliminate the specimen clapping effect on the tensile property of the material. The fracture morphology was varied with respect to the deposition and thickness of the clad region. The elongation of the laser-clad layer during the tensile test was increased by incorporating the nanoparticles in the cladding region. The elongation of nanoparticle-incorporated clad region was nearly 10% higher than that of the pure cladding region. Hence, the elongation in the stress–strain curve was increased, which led to improve the ductility of the laser cladding region than that of the substrate. The toughness was used to predict the repairing quality when cracks or tears occurred at the surface of the specimen during the service time. The nanoparticles with the laser cladding process improved the toughness to a particular level. The refining of the microstructure was the main reason for the improvement of the ductility in the stress–strain curve and the toughness of the material. The vibration was also used to improve the tensile strength of the clad material. In addition, the elongation of the sample was increased under the condition of the vibration. The dimple pattern was generally present in the fractured surface of the laser cladding specimen during the tensile test. Moreover, the river pattern was also present in the laser-clad fracture surface [30].

8.5 Laser Cladding Coatings on Creep Strength

The creep behavior of material was considered as one of the important parameters to assess the material deformation with respect to time under a particular load. The load acted on the material may be the external load or self-weight of the component. This type of material behavior was evaluated when the material was subjected to high temperature during its service. Moreover, the involvement of certain stress for a long duration also induced the creep behavior on the material. Different processes were used to improve

TABLE 8.2

Chemical Composition of Inconel 713

Elements	Ni	Cr	Al	Mo	Nb	Ti	C	Others
Wt. (%)	Bal.	13.2	6.19	4.60	2.41	0.95	0.13	<0.3

the creep resistance of the material by changing the grain size and shape with different phases. In addition, the creep behavior proved the material performance details at the local region. Materials used in different applications were involved in creep behavior during the service time. Hence, the creep behavior study was used to confirm the material capability to fulfill the specific purpose. The GTD-111 nickel superalloy was used in the turbine blade in the gas turbine engine and Inconel 713 was cladded on the GTD-111 material using the laser cladding process [31]. The chemical composition of Inconel 713 is listed in Table 8.2.

The creep test was generally performed at the intersection region of cladding and substrate material. This creep resistance of laser cladding Inconel 713 behavior was improved over the casting, aging, and solution heat treatment processes. The main contribution to improving the creep resistance property was the restriction of dislocation movement at the Inconel 713 coating with the GTD-111 substrate interface. The presence of phases such as γ' phase in the laser-clad Inconel 713 region improved the creep resistance of the material. The probability of the formation of a crack in the heat-affected region in the laser-clad specimen was less compared to other processed specimens. The different stages of material behavior during the creep test is shown in Figure 8.4.

The formation of γ-γ' eutectic in the casting specimen induced the crack formation at the heat-affected region. The GTD-111 casting specimen was heat treated before the laser cladding process to avoid the formation of γ-γ' eutectic. The evaluation of the creep property was performed on the laser cladding material using a small punch creep tester. In this analysis, the cladding region, heat-affected region, and substrate were evaluated. The maximum creep resistance occurred at the cladding region. The formation of different microstructures in the cladding region affected the creep resistance. The equiaxed structure formation in the cladding region improved the creep resistance with γ' phase. Moreover, the presence of NbC carbide in the cladding region and the large volume of fine γ' phase increases the creep resistance of the material. Even the heat-affected region tried to deform during the creep test, but the cladding region restricted its movement. The creep test was usually carried out at different load conditions based on the application. The load of 200 N, 400 N, etc. at elevated temperatures such as 750°C was used to evaluate the creep test. In addition, this test was carried out in a uniaxial or biaxial direction. The resultant creep curve between the

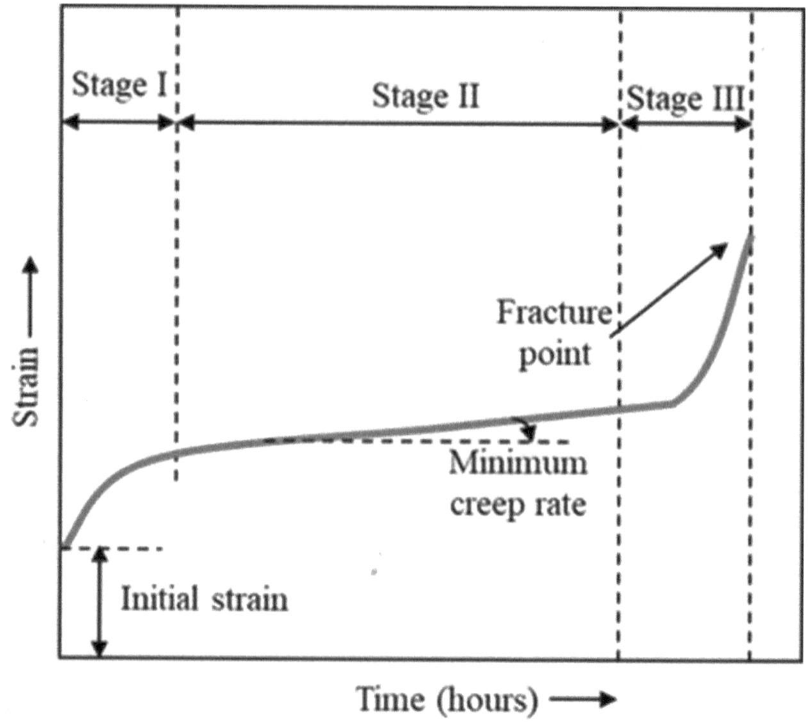

FIGURE 8.4
Different stages of material behavior during the creep test.

deformation of material against the duration of the test is shown in Figure 8.3. The first region of the creep curve consisted of a vertical line that was parallel to the deformation.

This deformation occurred during the initial stage of all testing samples due to the small thickness of the sample. Moreover, the adjustment between the punch ball with the test sample was performed manually to transfer the load to the sample. The point contact was made between the load ball and the laser cladding sample. The maximum stress occurred on the laser cladding sample due to the point contact. This stress value was usually higher than that of the yield stress of the material [32–34]. The high displacement happened at the time nearer to zero hour, which was considered as the first stage of the creep test. Most of the materials had similar shapes to the curves when they were subjected to the creep test [35–37]. The material that has the shortest creep resistance exhibited the highest displacement under the same loading condition. The elastic-to-plastic transformation occurred during the creep test. The uneven spreading of γ-γ' at the intersection region controlled the dislocation movement, which indirectly induced the creep resistance.

In addition, the formation of precipitation at different regions affected the different properties such as tribology and creep behavior of the laser cladding region [38–40]. The induction of stress on the specimen decreased with respect to the increase in test time. The contact between the load-applied ball and the test sample was increased, which reduced the stress. This was the second step of material deformation during the creep test. In this stage, the deformation was slowly increased with time. The dissolving of one phase into the other phases also reduced the dislocation movement during the creep test.

8.6 Enhancing Ductility of Material through Laser Cladding Process

The ductility of the material was used to assess the material deformation before its fracture and loss its service. The laser cladding process was the rapid heating and cooling process, which improved the hardness and wear resistance of the material. The ductility of the material was controverted to the hardness and wear resistance of the material. Generally, the laser cladding specimen has less ductility due to the hard phase of the clad region. The important factor affecting the ductility of the laser cladding material was discussed in this section. The laser cladding was performed on top and bottom, i.e., both sides to enhance the ductility along with strength. The heterogeneous structures with different hard phases formed in the clad region. The straightforward movement of the laser beam movement on the alloying element led to the production of the heterostructure in the clad region. It provided an excellent combination of ductility with the strength of the clad material. Nearly more than two or more materials were deposited on the substrate material one over the other. The thickness of the cladding layers was maintained in the range of 0.5 mm excluding the polishing operation. During this fabrication, the laser scanning direction during the deposition of the alloying material was changed up to 90°, which may be followed in a clockwise or counterclockwise direction. The grain growth with the previous deposition layer was controlled to columnar or epitaxial structure [41–43]. The thickness reached up to 6 mm when multilayers were deposited on the material. The ductility was significantly increased without the reduction of tensile strength [44]. The strain hardening occurred during the laser cladding process that contained four different stages. The presence of twins in the cladding region was higher compared to another manufacturing process. These twins supported the deformation of the cladding region and improved the ductility property of the material [45,46]. However, the grain boundaries in the twins effectively protected the material from the failure of the component by reducing the dislocation [47,48].

8.7 Toughness of Laser Cladding Engineering Material

The formation of structure in the laser cladding region was dense as well as fine due to the fast heating and cooling processes. Moreover, the bonding strength between the grains was superior to that of other processes [49,50]. The crystal formation in the cladding region deviated from that of the equilibrium process due to rapid solidification. The toughness of the laser cladding specimen was improved by the heat treatment process. In the heat treatment process, the normalizing and the annealing process were used to improve the toughness of the material. The toughness of the heat-treated laser cladding specimen was superior to that of the un-heat-treated specimen due to the refinement of grains. The toughness was generally analyzed using the Charpy impact test. The heat treatment temperature greater than 550°C and the time of 2hours were used to improve the toughness of the cladding region to that of before the heat treatment test. The enhancement of the toughness was around two times higher than that before the heat treatment process [51,52]. The secondary process of annealing heat treatment and normalizing processes improved the toughness. The microstructure of the cladding region was significantly refined during the two consecutive normalizing processes. The fine grains with M7C3 carbide precipitate at the grain boundary region improved the pinning of dislocation and restricted the dislocation slip. The impact resistance of the material depended on the morphology of the grains and their size. It resulted in the enhancement of plasticity and toughness by gaining a higher impact resistance in the laser cladding region [53]. Moreover, the residual stress from the laser cladding region was effectively relieved, which improved the impact resistance.

The failure surface of the specimen after the impact test was considered the dimples, which confirmed the ductile fracture. The crystal structures had grown at the heat dissipation direction, which resulted in the control of grain growth direction. In addition, the toughness of the material was improved through the addition of alloying elements such as La_2O_3 to the substrate surface through the laser cladding process. Moreover, the toughness value increased up to the particular percentage of alloying elements, and above the critical value, the toughness had decreased. The toughness of the Mo_2FeB_2-WC composite was increased through the addition of 0.3wt% of La_2O_3. The improvement of toughness was around 3.8% when compared to the substrate. The toughness was decreased when 0.5wt% of La_2O_3 was added to the Mo_2FeB_2-WC composite [54]. The reduction in the toughness of the cladding material was 5.6% compared to the substrate material. The cladding region formation also affected the toughness of the cladding material in addition to the microstructure. The alloying elements were hindered at the grain boundary region due to dislocation slip. It reduced the pores in the laser cladding region and formed a defect-free surface. The stress concentration in the clad region was decreased by the reduction of pores in the clad region. It led to create a barrier for the crack propagation in the clad region, which eliminated the stress concentration effect. It resulted in the reduction of deformation of the material and increased the fracture toughness [55,56]. However, the growth of grains increased with the addition of a large quantity of alloying elements. The presence of hard phases increased with the increase of alloying elements.

The presence of hard phases in the laser cladding region increased the brittleness and a corresponding reduction in the toughness occurred. These hard phases were dense in nature and growth at particular direction in the cladding region. It allowed the crack to easily propagate in the hard phases during the toughness test. Moreover, the grain boundary was unable to stop the crack propagated as it flowed through the crack source [57,58]. The alloying elements in the cladding region improved the heat transfer in convection form from the melt pool region. It relieved the gas from the melt pool region to the atmosphere, which led to a reduction in the pore formation. It resulted in the inhabitation of crack propagation in the clad region. Some of the alloying elements such as S, La, and so on, improved the slag removal and purified the melt pool region. It reduced the formation of porosity and improved the removal of impurities. The energy was observed during the propagation of the crack and traveled toward the lower energy direction. The inherent property of the pinning effect of alloying elements reduced the crack propagation and improved the toughness of the coating.

8.8 Influence of Laser Cladding on Fatigue Property of Engineering Material

The fatigue property of the material depended on the microstructural variation and distribution of residual stress. The fine grains were formed at the clad region with nanotwins structure, which had a maximum of 30nm thickness. Moreover, high-density dislocations occurred at the clad region, which resulted in the reduced deformation. The tensile residual stress was reduced by converting it into the compressive stress. The laser cladding process improved the fatigue of the substrate up to three times due to the addition of alloying elements [59,60]. The formation of specific structures such as Widmanstatten reduced the fatigue of material, which improved the fatigue strength by nearly 164% than that of the substrate surface. The length of the crack increased when the number of test cycles increased. The crack induction during the fatigue test is schematically represented in Figure 8.5. The crack formation in the ceramic cladding region was the main reason for the reduction of the fatigue property of the material.

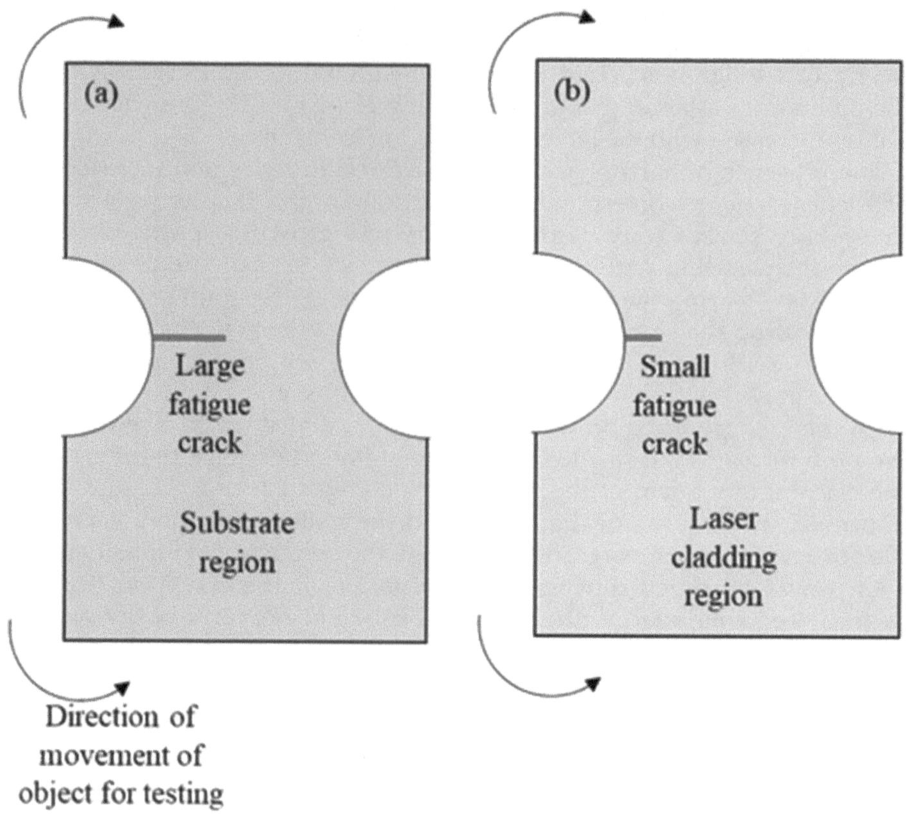

FIGURE 8.5
Schematic representation of crack induction in the material during the fatigue test.

The residual stress was induced during the rapid solidification process. The thermal-physical properties and other characteristics of the ceramic material, substrate, and binder material were quite varied. The thermal-physical properties included the melting point of the material, thermal expansion during the heat input, etc. The crack was induced in the material when the intensity of residual stress was higher than the toughness of the material. The residual stress formation in the coating region depended on the laser processing parameters. This value was high when the test was performed in the perpendicular direction to the laser scanning direction. The ceramic material such as tungsten carbide improved properties such as hardness, elastic modulus, and so on. Moreover, it increased the residual stress of the cladding region, which led to an increase in the tendency of crack formation. The FESEM image of fatigue strength failure of material is shown in Figure 8.6.

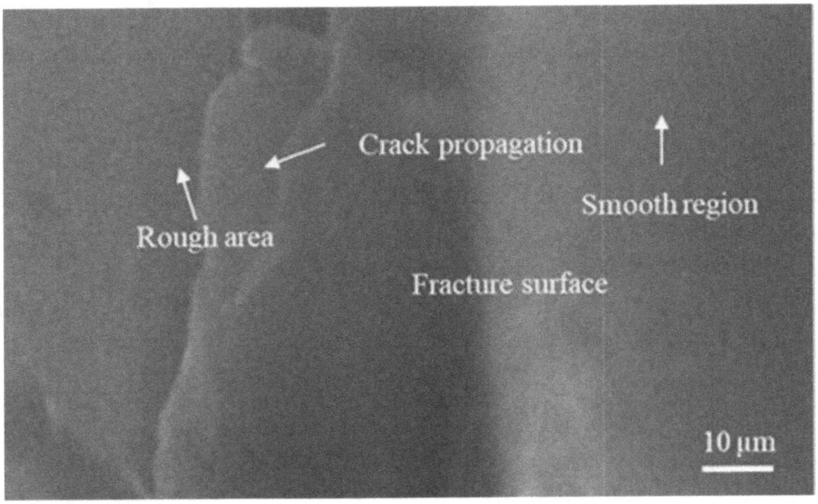

FIGURE 8.6
Fatigue failure of material.

It led to reduced fatigue life of the component during the working conditions. Hence, the optimal value of ceramic material was used to obtain the maximum fatigue strength of laser cladding materials.

8.9 Conclusion

This chapter discussed the impact of the laser cladding process on the mechanical behavior of the material. The classification of mechanical properties was discussed in a detailed manner. Different mechanical properties such as hardness, tensile strength, creep strength, ductility, toughness, and fatigue strength. This study gave an overview of the laser cladding process to enhance the properties of the material for utilizing it in different applications such as rails in high-speed trains. Different elements were discussed in this chapter that improved the mechanical properties of the material, such as Cr content decreased the melting point and improved the nucleation rate with high viscosity of the molten liquid. Different forms of oxides such as Al_2O_3, TiO_2, ZrO_2, and so on, were discussed in this chapter, which may be useful to increase the hardness of the clad region. The uniform distribution of fine grain in the clad region was explained in this chapter to improve the resistance against the deformation and provide the small plastic deformation with enhanced toughness. Hence, the essential requirement of the laser cladding was explained in detail in this chapter, which would be useful for industrial-related workers, researchers, and general learners.

Acknowledgment

We are grateful for the support from the Ministry of Science and Technology, Taiwan, for providing the financial support.

References

1. S. Dong, S. Yan, B. Xu, Y. Wang, W. Ren. Laser cladding remanufacturing technology of cast iron cylinder head and its quality evaluation. *Journal of Academy of Armored Force Engineering*, 27 (1) (2013). Corpus ID: 138794807
2. S. Dong, X. Zhang, B. Xu, Z. Wang, S. Yan. Laser cladding remanufacturing of 45 steel Camshaft WornCam. *Journal of Academy of Armored Force Engineering*, 25 (2) (2011) 85–87.
3. C. Navas, A. Conde, B.J. Fernandez, F. Zubri, J. de Damborenea. Laser coatings to improve wear resistance of mould steel. *Surface and Coatings Technology*, 194 (1) (2005) 136–142.
4. N. Jeyaprakash, C.-H. Yang, G. Prabu, R. Clinktan. Microstructure and tribological behaviour of Inconel-625 superalloy produced by selective laser melting. *Metals and Materials International*, 28 (2022) 2997–3015.
5. X. Wang, M. Zhang, Z. Zou, S. Qu. Microstructure and properties of laser clad TiCp/Ni-basedalloys composite coating. *Journal of Mechanical Engineering*, 39 (2) (2003) 37–41.
6. S. Zhou, X. Dai, H. Zheng. Characteristics on structure and properties of WC-Ni60A coatings by laser cladding and laser-induction hybrid cladding. *Chinese Journal of Mechanical Engineering*, 48 (7) (2012) 114–118.
7. A. Conde, R. Colaco, R. Vilar, J. de Damborenea. Corrosion behavior of steels after laser surface melting. *Materials and Design*, 21 (2000) 441–445.
8. G. Prabu, N. Jeyaprakash, C.-H. Yang, I.A. Alnaser. Investigation on deformation of nano-twins of LPBF produced Cu alloy through Triboindenter. *Tribology International*, 191 (2024) 109117.
9. G. Cui, B. Han, N. Cui, M. Yu. Effects of scanning speed on microstructure and properties of laser cladding Ni-based WC alloy coating. *China Surface Engineering*, 27 (4) (2014) 82–88.
10. N. Jeyaprakash, C.-H. Yang, M. Duraiselvam, G. Prabu, S.-P. Tseng, D. Raj Kumar. Investigation of high temperature wear performance on laser processed nodular iron using optimization technique. *Results in Physics*, 15 (2019) 102585.

11. L. Shepeleva, B. Medres, W.D. Kaplan. Laser cladding of turbine blades. *Surface and Coatings Technology*, 125 (2000) 45–48.
12. A. Riveiro, A. Mejias, F. Lusquinos, et al. Optimization of laser cladding for Al coating production. *Physics Procedia*, 41 (2013) 327–334.
13. G. Chryssolouris, S. Zannis, K. Tsirbas, C. Lalas. An experimental investigation of laser cladding. *CIRP Annals*, 51 (1) (2002) 145–148.
14. N. Jeyaprakash, C.-H. Yang, M. Duraiselvam, G. Prabu, Microstructure and tribological evolution during laser alloying WC-12% Co and Cr3C2– 25% NiCr powders on nodular iron surface. *Results in Physics*, 12 (2019) 1610–1620.
15. L. Sexton, S. Lavin, G. Byrne, A. Kennedy. Laser cladding of aerospace materials. *Journal of Materials Processing Technology*, 122 (2002) 63–68.
16. Q. Lai, R. Abrahams, W. Yan, et al. Influences of depositing materials, processing parameters and heating conditions on material characteristics of laser-cladded hypereutectoid rails. *Journal of Materials Processing Technology*, 263 (2019) 1–20.
17. N. Jeyaprakash, G. Prabu, C.-H. Yang. Formation of different phases and their influences on the mechanical and tribological properties of surface alloyed FeCoCrNiAl particles using PTA technique. Intermetallics, 142 (2022) 107457.
18. L. Biazon, B.P. Ferrer, A. Toro, T. Cousseau. Correlations between rail grease formulation and friction, wear and RCF of a wheel-rail tribological pair. *Tribology International*, 153 (2021) 106566.
19. G. Prabu, M. Duraiselvam, N. Jeyaprakash, C.-H. Yang. Microstructural evolution and wear behavior of AlCoCrCuFeNi high entropy alloy on Ti-6Al-4V through laser surface alloying. *Metals and Materials International*, 27 (2021) 2328–2340.
20. Z. Huda. Mechanical testing and properties of materials. In *Mechanical Behavior of Materials. Mechanical Engineering Series*, 2022. Springer, Cham. https://doi.org/10.1007/978-3-030-84927-6_3.
21. N. Jeyaprakash, G. Prabu, C.-H. Yang. Inclusion of nano silicon particle in SS316L through LPBF and its responses on corrosion behaviour. *Ceramics International*, 49 (16) (2023) 26450–26468.
22. M. Mashhadi, E. Taheri-nassaj, M. Mashhadi, V.M. Sglavo. Pressureless sintering of B4C–TiB2 composites with Al additions. *Ceramics International*, 37 (2011) 3229–3235.
23. C. Xu, Y. Cai, K. Flodström, Z. Li, S. Esmaeilzadeh, G. Zhang. Spark plasma sintering of B4C ceramics: The effects of milling medium and TiB2 addition. *International Journal of Refractory Metals and Hard Materials*, 30 (2012) 139–144.
24. G. Prabu, N. Jeyaprakash, C.-H. Yang, N. Radhika. Analysis on the transformation of melt wear to self-healing crack of wire arc additive manufactured Al 5356 alloy. *Tribology International*, 188 (2023) 108802.
25. R. Colaco, R. Vilar. Phase selection during laser surface melting of martensitic stainless tool steels. *Scripta Materialia*, 36 (2) (1997) 199–205.
26. E. Williams, N. Lavery. Laser processing of bulk metallic glass: A review. *Journal of Materials Processing Technology*, 247 (2017) 73–91.
27. K.Y. Nagulin, F.R. Iskhakov, A.I. Shpilev, A.K. Gilmutdinov. Optical diagnostics and optimization of the gas-powder flow in the nozzles for laser cladding. *Optics & Laser Technology*, 108 (2018) 310–320.

28. Y. Zhao, T. Yu, J. Sun, L. Chen, Y. Chen. Effect of laser cladding on forming microhardness and tensile strength of YCF101 alloy powder in the different full lap joint modes. *Journal of Alloys and Compounds*, 820 (2020) 150230.
29. N. Jeyaprakash, G. Prabu, C.-H. Yang, M. Prasad Banda, E. Mohan. Structural and nanohardness of laser melted CM247LC nickel-based superalloy. *Transactions of the Indian Institute of Metals*, 76 (2) (2022) 287–295.
30. Z. Yao, J. Chen, H. Qian, Y. Nie, Q. Zhang, J. Yao. Microstructure and tensile property of laser cladding assisted with multidimensional high-frequency vibration. *Materials*, 15 (2022) 4295.
31. M. Taheri. Development of a novel method for measuring the interfacial creep strength of laser cladding coatings. *Results in Optics*, 7 (2022) 100226.
32. Y. Cao, N. Farouk, M. Taheri, A.V. Yumashev, S.F.K. Bozorg, O.O. Ojo, Evolution of solidification and microstructure in laser-clad IN625 superalloy powder on GTD-111 superalloy. *Surface and Coatings Technology*, 412 (2021) 127010.
33. Y. Chew, J.H.L. Pang, G. Bi, B. Song. Thermo-mechanical model for simulating laser cladding induced residual stresses with single and multiple clad beads. *Journal of Materials Processing Technology*, 224 (2015) 89–101.
34. Y. Zhao, T. Yu, J. Sun, L. Chen, Y. Chen, Effect of laser cladding on forming microhardness and tensile strength of YCF101 alloy powder in the different full lap joint modes, *Journal of Alloys and Compounds*, 820 (2020), 150230.
35. G. Prabu, N. Jeyaprakash, C.-H. Yang, S. Sivasankaran. Synergetic effect of piled-up material around the nanoindentation cavity on additively manufactured Cu-Cr-Zr alloy. *Material Today Communication*, 38 (2024) 107691.
36. J.H. Kim, U. Ro, H. Lee, S.J. Kang, B.H. Lee, M.K. Kim. A direct assessment of creep life based on small punch creep test. *Theoretical and Applied Fracture Mechanics*, 104 (2019) 102346.
37. S. Yang, X. Ling, Y. Zheng, R. Ma. Creep life analysis by an energy model of small punch creep test. *Materials & Design*, 91 (2016) 98–103.
38. S.A. Sajjadi, S. Nategh, R.I. Guthrie. Study of microstructure and mechanical properties of high performance Ni-base superalloy GTD-111. *Materials Science and Engineering A*, 325 (1–2) (2002) 484–489.
39. N. Jeyaprakash, C.-H. Yang, G. Prabu, N. Radhika. Mechanism correlating microstructure and wear behaviour of Ti-6Al-4V plate produced using selective laser melting. *Metals*, 13 (2023) 575.
40. L. Lu,K. Guo, J. Sun, G. Zhang, S. Sui, C. Sun. Investigation on hot sizing process based on creep mechanism for laser cladded thin-walled titanium alloy components. *Journal of Laser Applications*, 31 (2019) 042002.
41. H.Y. Wan, Z.J. Zhou, C.P. Li, G.F. Chen, G.P. Zhang. Effect of scanning strategy on grain structure and crystallographic texture of Inconel 718 processed by selective laser melting. Journal of Mater*ials Science & Technology*, 34 (2018) 1799–1804.
42. C. Cui, M. Wu, R. He, Y. Gong, X. Miao. Investigation on the columnar-to-equiaxed transition and corrosion behavior in multi-track Stellite-6 coating fabricated by laser cladding. *Materials Chemistry and Physics*, 291 (2022) 126681.
43. P. Murugesan, V. Satheeshkumar, N. Jeyaprakash, G. Prabu, C.-H. Yang. Nano-level mechanical and tribological behavior of additively manufactured AlSi10Mg plates. *Archives of Civil and Mechanical Engineering* 23 (2023) 62.

44. R.E. Kim, E.S. Kim, G.M. Karthik, G.H. Gu, S.Y. Ahn, H. Park, J. Moon, H.S. Kim. Heterostructured alloys with enhanced strength-ductility synergy through laser-cladding. *Scripta Materialia*, 215 (2022) 114732.
45. M.R. Barnett. A rationale for the strong dependence of mechanical twinning on grain size. *Scripta Materialia*, 59 (2008) 696–698.
46. K. Li, X. Liu, Y. Liu, X. Wu, W. Zhang, J. Yang, W. Li, S. Zhou. Strengthening layer with nano-twins in titanium alloy induced by laser surface re-melting: Mechanism of high strength and ductility. *Materials Characterization*, 196 (2023) 112632.
47. J.W. Christian, S. Mahajan. Deformation twinning. *Progress in Materials Science*, 39 (1995) 1–157.
48. R.E. Kim, J. Moon, E.S. Kim, J. Lee, H.S. Kim. Surface heterostructuring of laser-clad 316L stainless steel through texture-driven deformation twinning. *Scripta Materialia*, 221 (2022) 114989.
49. G. Prabu, M. Duraiselvam. Tribological studies on AlCrFeCuCoNi high entropy alloy surface coated on Ti-6Al-4V using plasma transferred arc technique. *Archives of Metallurgy and Materials*, 67 (2) (2022) 409–420.
50. Y. Lin, Y. Lei, H. Fu, J. Lin. Mechanical properties and toughening mechanism of TiB2/NiTi reinforced titanium matrix composite coating by laser cladding. *Materials & Design*, 80 (2015) 82–88.
51. W. Chen, Q. Cai, Z. Luo, Y. Wu. Influence of heat treatment on strength and toughness of laser cladding iron-based coatings. *Materials Letters*, 335 (2023) 133733.
52. M. Masanta, S.M. Shariff, A. Roy Choudhury. Evaluation of modulus of elasticity, nano-hardness and fracture toughness of TiB2–TiC–Al_2O_3 composite coating developed by SHS and laser cladding. *Materials Science and Engineering: A*, 528 (16–17) (2011) 5327–5335.
53. N. Jeyaprakash, G. Prabu, C.-H. Yang. Microstructure and tribological behaviour of FeCoCrNiAlCuMn particles alloyed on Inconel-718 substrate using plasma transferred arc technique. *Intermetallics*, 139 (2021) 107360.
54. H. Zhang, Y. Pan, Y. Zhang, G. Lian, Q. Cao, L. Que. Microstructure, toughness, and tribological properties of laser cladded Mo2FeB2-based composite coating with in situ synthesized WC and La2O3 addition. *Surface & Coatings Technology*, 449 (2022) 128947.
55. P. Farahmand, S. Liu, Z. Zhang, R. Kovacevic. Laser cladding assisted by induction heating of Ni–WC composite enhanced by nano-WC and La2O3. *Ceramics International*, 40 (2014) 15421–15438.
56. J. Tuominen, J. Näkki, J. Poutala, J. Miettinen, T. Peltola, P. Vuoristo, I. Rasehorn, M.M. Alam, A.F.H. Kaplan. Fatigue behavior of laser clad round steel bars. *Journal of Laser Applications*, 27 (2015) 012006.
57. Y. Chew, J.H.L. Pang. Fatigue life prediction model for laser clad AISI 4340 specimens with multiple surface cracks. *International Journal of Fatigue*, 87 (2016) 235–244.
58. N. Jeyaprakash, G. Prabu, C.-H. Yang. The influence of different phases on the microstructure and wear of Inconel-718 surface alloyed with AlCuNiFeCr hard particles using plasma transferred arc. *Journal of Materials Engineering and Performance*, 31 (2022) 9921–9934.

59. M.Z. Ge, Y. Tang, Y.K. Zhang, Y. Wang. Enhancement in fatigue property of Ti-6Al-4V alloy remanufactured by combined laser cladding and laser shock peening processes. *Surface & Coatings Technology*, 444 (2022) 128671.
60. Y.C.J.H.L. Pang, G. Bi, B. Song. Effects of laser cladding on fatigue performance of AISI 4340 steel in the as-clad and machine treated conditions. *Journal of Materials Processing Technology*, 243 (2017) 246–257.

9

Residual Stress in Laser Cladding Material

B. Mallikarjuna, Aditya Kumar, N. Jayanth, and T. Sathies

9.1 Introduction

Laser stands for light amplification by stimulated emission of radiation [1]. The first ruby laser was invented by Theodore Maiman in 1960. Thereafter, various types of lasers were developed for different applications with improved laser beam efficiency and feasibility [2]. Lasers are classified into solid, liquid, fiber, and gas lasers. Each type of laser has specific applications for processing various materials. Hence, lasers are used in engineering applications such as cutting, drilling, welding, marking, bending, surface treatment, cladding, and additive manufacturing [3–7]. The application of laser cladding (LC) was widespread in industry in 1980. It has become a promising alternative to the electro spark process, tungsten inert gas (TIG) welding, plasma spray, or high-velocity oxy fuel (HVOF) [2,8]. LC is an advanced manufacturing and surface treatment technique that involves depositing a material, typically in the form of a powder or wire feedstock, onto a substrate using a focused high-energy laser beam, often a CO_2 or fiber laser. An illustration of the LC process with powder directed, wire fed, and pre-placed powder feedstock is shown in Figure 9.1a–c. The cladding process is called by various names, including LC and laser cladding deposition (LCD)/laser metal deposition (LMD). The LC is used to enhance the properties of the substrate, repair damaged components, or build up material layers with specific characteristics [9]. LC offers several advantages, including precision, minimal heat-affected zones, and the ability to work with a wide range of materials. It is noteworthy that a range of common clad and coating materials includes stainless steel, Ti6Al4V, titanium aluminide, H13 tool steel, 316L stellite 6, Al_2O_3, TiC, WC, cobalt-based alloy, aluminum-based alloy, Inconel 718, and more [2,10–15]. These materials are often applied onto substrate materials, which primarily encompass stainless steel, Ti6Al4V, low carbon steel, mild steel, medium carbon steel, and Inconel 718 [2,16–18]. Therefore, the right selection of clad materials with respect to substrate materials assumes significant importance in determining their compatibility for particular applications and desired properties. If there is a mismatch in thermal properties and a significant thermal gradient between the cladding and substrate, it can

DOI: 10.1201/9781032693576-9

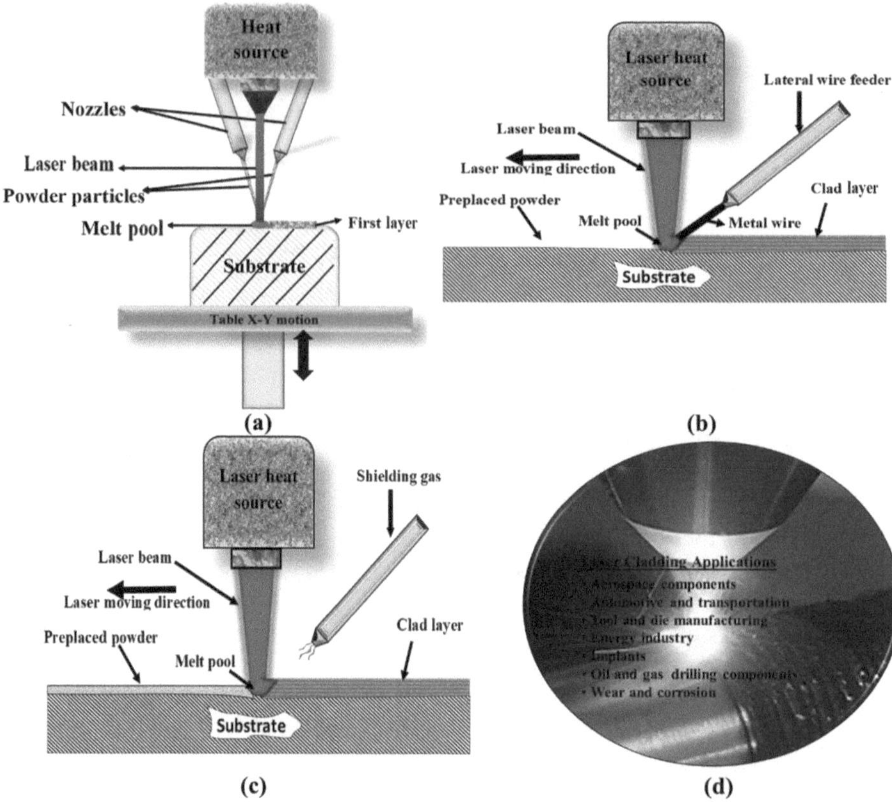

FIGURE 9.1
(a) laser cladding with powder injection, (b) laser cladding with wire injection (c) laser cladding with pre-placed powder, and (d) applications of laser cladding.

lead to the development of residual stresses in the deposited layer and the heat-affected zone (HAZ). The residual stress distribution and magnitude are influenced by process parameters as well. These stresses, in turn, result in issues such as dimensional inaccuracies, cracking, distortion, and premature failure of the components [7,10,19–21].

9.2 Classifications of Laser Cladding

The LC can be classified into three types based on their feedstock supply into the substrate surface [16]. The schematic of LC processes is shown in Figure 9.1a–c. Figure 9.1a is an illustration of the powder injection LC. In powder injection LC, a stream of metal powder is injected through a coaxial nozzle. The powder is directed into the path of a laser beam, which rapidly melts and fuses the powder onto the substrate. This method is suitable for applications where a fine control of the coating thickness and composition is needed. It can also be used for both additive manufacturing and repair processes [22,23]. Figure 9.1b is an illustration of the wire feed LC. In this method, a drum or spool is used to feed a continuous wire of coating material into the path of a laser beam. The laser beam is precisely focused on the substrate material where the coating is to be applied. As the laser beam melts the wire, it fuses with the substrate, creating a metallurgically bonded coating. This method is often used when a specific alloy composition or wire geometry is required for the coating. Figure 9.1c is an illustration of the pre-placed powder LC. In this method, a layer of coating powder is pre-deposited or spread onto the substrate surface [16]. The laser beam is then directed onto the pre-placed powder layer, which melts and fuses to form the coating. It allows for better control over the initial powder layer and can be used for creating large thick coatings. Each of these methods has its own advantages and is selected based on the specific application, including the material being used, the desired cladding/coating thickness, and the precision needed in the process [24]. However, LC with powder injection technique is preferred one compared to the other two.

9.3 Generic Steps in Laser Cladding Process

The generic process steps in the LC process are presented in Table 9.1 [25]. The LC process includes several key stages, starting with selecting the appropriate technique and feedstock material. Next, the substrate or the cladding surface undergoes preparation using different grades (100#, 240#, or 600#) of SiC papers or ultrasonic cleaning with ethanol/deionized water. It is expected to remove the oxide layer and to obtain a mirror finish. This meticulous preparation ensures strong material adhesion. Subsequently, the third step involves tasks like CAD geometry, process parameter selection, and substrate placement. In the fourth step, a highly focused laser beam is directed onto the substrate while continuously delivering the feedstock material to create a melt pool. The intense laser melts the material, fusing it with the substrate to form a bead and layer. Optionally, preheating the substrate can reduce temperature differences, leading to residual stresses. Additional layers can be added to reach the desired thickness or properties, with potential adjustments to laser parameters or feedstock material. In the fifth step, the molten material cools and solidifies, forming the clad layer. Solidification microstructure, influenced by factors such as temperature gradients (TG), solidification rate (R), composition (C), phase equilibrium reactions, nucleation and growth kinetics, crystallographic constraints, and

TABLE 9.1

Generic Processing Steps in Laser Cladding of Any Material System [2,12,28,90]

Feedstock	Substrate Preparation	Printing Process Preparation	Laser Set Up and Material Deposition	Melting and Solidification, Cooling	Post Processing	Final Inspection
• Powder • Wire • Pre-placed powder	• Substrate surface preparation using SiC and, after that, cleaning with ethanol • Degreasing • To ensure good adhesion of the cladding material	• CAD geometry (clad part) • Tool path • Printer preparation • Placement of substrate or part • Process parameters, such as power, spot size, and scanning speed • Desired thickness of clad	• Laser source, optics, and control parameters. A laser beam is precisely focused on the substrate surface, and material fed coaxial or Off-axis • Material introduced into the lasers path (powder/wire)	• The intense heat generated by the laser rapidly melts the cladding material • As the laser moves, the molten material solidifies, creating a solid, metallurgically bonded coating on the substrate • Proper cooling rates are essential to achieve the desired microstructures	• Defects, thickness measurement • Surface finish • Heat treatment • Machining • Quality checking • Clad thickness verification • Hardness measurement • Clad/coating surface evaluation	• A final inspection and testing of the coated component • To verify that it meets the intended specifications and quality standards.

rapid cooling rates (approximately 10^5 K/s), may introduce residual stresses and distortion. The sixth step involves removing the clad component from the machine, defect analysis, thickness measurement, hardness assessment, and surface quality evaluation. If needed, secondary operations can be performed to achieve specific surface properties. Finally, the last step includes a comprehensive inspection and testing of the clad component [23,25–27].

9.4 Applications of Laser Cladding

LC is a versatile and precise manufacturing process that offers various industrial applications. LC is an effective method for repairing damaged or worn parts, reducing the need for costly replacements in various industries, including aerospace, automotive, thermal plants, biomedical, and heavy machinery to marine equipment, as shown in Figure 9.1d. Often applies wear-resistant and corrosion-resistant coatings to critical components. It offers the ability to tailor material properties and extend the lifespan of critical components. In the aerospace industry, it is employed for repairing and refurbishing worn or damaged turbine blades, combustion chambers, and other high-value components [28,29]. LC is used in automotive parts to enhance the wear resistance of engine components, such as cylinder liners and camshafts [28,30]. Tool and die manufacturers use LC to improve the wear resistance of molds, dies, and cutting tools. This extends tool life and reduces downtime for maintenance [30]. In the energy sector, LC is applied to repair and refurbish components used in power generation, including turbine blades, boiler tubes, and steam generator tubes. It helps maintain the efficiency and reliability of power plants. LC is used to manufacture and repair biomedical implants and devices. It allows for the precise addition of biocompatible materials to improve the performance and longevity of medical components. LC is applied to oil and gas drilling and production components. It enhances the wear and corrosion resistance of drilling tools, downhole equipment, and pipelines, ensuring reliability in harsh environments [29]. Further, it has to find its application in electronics and microfabrication for depositing conductive materials on electronic components [2,31].

9.5 Defects in Laser-Cladded Parts

The various defects and flaws associated with the LC process are presented in Table 9.2. Laser-cladded parts witness the various defects. Because melting and solidification take place while deposition of material

TABLE 9.2

The Various Imperfections Associated with the Laser Cladding Process [32–36,78,91]

Defect	Reason	Assessment
Hot cracks	Hot cracks occur during solidification when the material is still in a semi-molten state. They can be caused by factors such as high TG, improper material selection, or cooling rates	Quality control measures, non-destructive testing, and scanning electron microscope
Stress cracks	Stress cracks develop after the cladding process when the material cools and contracts, leading to cracking	Quality control measures, non-destructive testing, and scanning electron microscope
Crater cracks	Crater cracks occur at the edges or boundaries of the cladding zone. They can be caused by insufficient material fusion or issues with the transition between the cladded layer and the substrate	Quality control measures, non-destructive testing, and scanning electron microscope
Oxidation, inclusions and segregation	Inclusions inside the clad part can range from segregation to impurities and oxidation. During the process, oxygen can react with the heated material and form unwanted oxides. This can lead to microscopic defects and irregularities	Thermo Gravimetric Analyzer (TGA), surface analysis methods, scanning electron microscope, and optical microscope
Dimensional inaccuracy	Shrinkage, warpage, distortion, poor selection of process properties	Coordinate measuring machine (CMM), calibration cube, Vernier calipers, and micrometers
Residual stress	Thermal stresses, solidification, cooling rate, reheating, difference between substrate and deposition, and metallurgical transformation stresses	X-ray diffraction, neutron diffraction, Contour diffraction, synchrotron radiation, ultrasonic testing, hole-drilling strain gauge method, ring-core method, Barkhausen noise measurement, etc.
Surface roughness	Unmelted powder particles, cracks, and inappropriate process parameters	Touch inspection, visual inspection, microscopic inspection, reflected light intensity, and surface photographs
Porosity, pores, voids, lack of fusion	Poor bonding between layers, inappropriate process parameters, oxygen pickup during the cladding, entrapped gas in powders, and insufficient melting of powder of the substrate	Optical microscope, scanning electron microscope, helium porosimeter, bulk density, core plugging test, density measurement, and gas sorption

on the substrate. Hence, each layer experiences rapid heating and cooling results in various defects including residual stress, distortion, porosity, and cracks. Also, improper selection of process parameters results in unmelted powder particles, poor surface integrity, poor interface bonding, voids, inclusions, segregation, etc. The residual stress and distortion are the common defects associated with the LC process due to the nature of deposition. It's important to measure and address these residual stresses (tensile or compressive in nature) because they can reduce part performance, and potentially lead to premature failure. Hence, understanding the state and magnitude of residual stress distribution in laser-cladded material is paramount [32–36].

9.6 Residual Stresses

Residual stresses, internal stresses existing within a material or component, even in the absence of external loads, persist after undergoing various mechanical, chemical, or thermal processes [13]. Understanding residual stress is crucial when working with various materials [37]. These stresses play a significant role in influencing the performance of components, impacting factors such as fatigue life, mechanical strength, and structural stability [38]. When combined with applied stress in structures and components, residual stress can alter load-carrying capacity. Moreover, residual stress may contribute to unforeseen failures, either due to a reduction in material strength or the initiation of stress-corrosion cracking [38]. Consequently, the following sections delve into the origin of residual stress, the various types of residual stress, the effects of residual stress on component integrity, and measurement techniques.

9.6.1 Origin of Residual Stresses in Laser-Cladded Materials

LC stands out as a remarkably versatile and advanced additive manufacturing method. However, it comes with the challenge of dealing with residual stress, which emerges from intricate interactions and transformations during the melting, solidification, and cooling of the deposited material [39]. Understanding the origins of these stresses holds paramount importance as it enables the optimization of the manufacturing process and guarantees the integrity of the final product [7,40]. During the LC process, both the substrate and the newly deposited material experience significant TG. The rapid heating and subsequent cooling generate thermal stresses within the component. Differences in materials coefficient of thermal expansion and elastic moduli, between the two materials contribute to these stresses. The materials undergo phase transformations as they cool. For example, austenitic stainless steel might transform into martensite, causing volume changes and transformation-induced stresses [41]. The magnitude of these stresses is influenced by factors such as material composition and cooling rate. Additionally, the bonding between the clad material and the substrate can introduce metallurgical stresses. Managing metallurgical stresses can be particularly challenging, as they may impact the adhesion of the clad layer and require careful control [34,42]. All of these factors collectively contribute to the development of residual stresses in both the clad and substrate regions of the component during and after the cladding process.

9.6.2 Types of Residual Stress

The residual stresses can be classified based on the different length scales, as shown in Figure 9.2a. Type I residual stress vary over the large distances of the component, $l_{0,I}$. Type I residual stresses typically refer to the residual stresses that arise due to thermal effects. These stresses that are result from thermal gradients and differences in thermal expansion or contraction within a material [43].

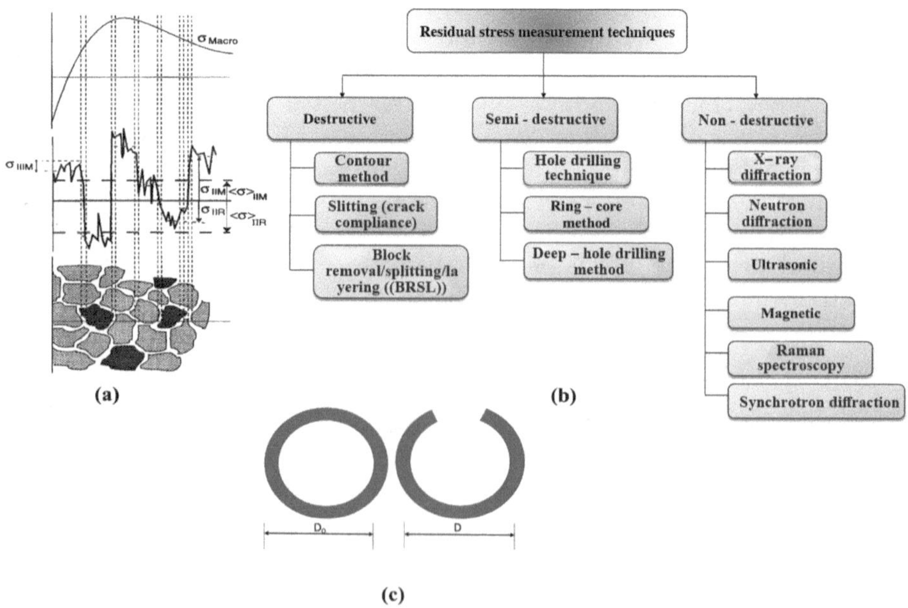

FIGURE 9.2
(a) Classification of residual stress on different length scales such as type I (over large scale of component), type II (over the grains), and type III (intergranular or atomic level). Reproduced from [43] and permission may not be needed. (b) Residual stress measurement techniques include destructive, semi-destructive, and non-destructive methods (c) tube splitting [47].

Type II residual stress varies across grains, extending over a distance equal to one grain, $l_{0,II}$. Type II residual stresses are often used to describe stresses related to phase transformations in materials. These stresses result from changes in material structure, such as the transformation from austenite to martensite in certain steels [43,44].

Type III micro residual stress is observed at the grain or intergranular level and spans multiple atomic distances denoted as, $l_{0,III}$ [43]. Conversely, type II and type III residual stresses have limited impact on the material's mechanical properties, rendering their detection challenging through conventional measurement techniques. Nevertheless, when addressing highly localized microscale or nanoscale residual stresses, advanced techniques such as the combination of focused ion beam milling, digital image correlation, or nano-indentation, along with transmission electron microscopy, can facilitate their measurement and analysis [45].

9.6.3 Residual Stress Measurement Techniques

There are several techniques available for measuring residual stresses including destructive, semi-destructive, and non-destructive techniques (refer to Figure 9.2b) [37,47,48].

9.6.3.1 Destructive Methods

Destructive methods are a fundamental category of residual stress measurement methods, delivering precise and acceptable results using cost-effective equipment. They are the preferred choice when specimen damage is acceptable, and Table 9.3 presents comparison of such measurement systems [47].

TABLE 9.3

Comparison between Destructive and Semi-Destructive Techniques for Measuring Residual Stress [46]

Destructive/ Semi-Destructive Method	Hole-Drilling Method	Ring-Core Method	Contour Method	Crack Compliance Method	Stripping Method
Depth	1–2 mm	0–5 mm	—	1–100 mm	10–100 mm
Accuracy	±50 MPa	±10 MPa	±20 MPa	0.01% × E	30–50 MPa
Applicable object	Isotropic elastic material	—	The sample stress gradient changes greatly	Low-stress amplitude sample	Hard film material
Operation situation	Complicated	More complicated	More complicated	More complicated	Complicated

TABLE 9.4

Comparison of the Non-Destructive Residual Stress Measurement Systems [37,47,48]

NDT Method	Ultrasonic	X-Ray Diffraction	Neutron Diffraction	Raman Spectroscopy	Magnetic
Lab based/portable	Both	Both	Lab-based method	Both	Both
Availability of equipment	GA	GA	Specialist	GA	GA
Speed of operation	Rapid	Fast/medium	Medium/slow	Fast	Rapid
Cost of equipment	Medium	Medium	Government facility	Medium	Medium
Level of expertise require	Medium	Medium	High	Medium	Low
Material type	Metal/ceramics	Metal/ceramics	Metal/ceramics	Polymers/ceramics	Ferromagnetic
Composites	Yes	Yes	Yes	Yes	No
Crystalline or amorphous	Crystalline	Crystalline	Crystalline	Either	Crystalline
Coated	Yes	Yes	No	Yes	No
Surface preparation	Not critical	Important	Not critical	Not critical	Not critical
Resolution (μm)	5,000	20 depth 1,000 laterally	500	0.5	1,000
Penetration	>100 mm along specimen	5 μm-Ti, 50 μm-Al, 1 mm-layer removal	100 mm for Al, 25 mm for Fe	Surface	20–300 μm (Barkhausen)
Stress type	Macro	Macro, micro	Macro, micro	Macro	Macro
Stress state	U, B	U, B	U, B, T	U, B	U, B

GA, generally available; U, unidirectional; B, bidirectional; T, tridirectional.

9.6.3.1.1 Excision

Excision, the most straightforward and destructive technique for assessing residual stress, entails affixing strain gauges to a specific location on a specimen. Following this, a portion of the material surrounding the strain gauges is precisely removed, leaving only a small fragment connected to the gauges. The excision process effectively relieves residual stresses within the remaining fragment, making it nearly stress-free. This distinctive approach ensures complete strain relaxation at the measurement sites, facilitating the straightforward computation of residual stresses directly using Hooke's law. When employing a rectangular rosette for measurements, stress values can be calculated using Equations (9.1) and (9.2) [48]:

$$\sigma_{\max}, \sigma_{\min} = -\frac{E}{2}\left(\frac{\varepsilon_3+\varepsilon_1}{1-\vartheta} \mp \frac{\sqrt{(\varepsilon_3-\varepsilon_1)^2+(\varepsilon_3+\varepsilon_1-2\varepsilon_2)^2}}{1+\vartheta}\right) \tag{9.1}$$

$$\beta = \frac{1}{2}\tan^{-1}\left(\frac{\varepsilon_3+\varepsilon_1-2\varepsilon_2}{\varepsilon_3-\varepsilon_1}\right) \tag{9.2}$$

In this context, σ_{max} and σ_{min} denote the maximum and minimum principal stresses, while ε_1, ε_2, and ε_3 are the strain gauge readings within the rosette. The angle β, measured anticlockwise from gauge 1 to the principal stress (σ_{max}) direction, is employed. Equation (9.1) transforms to Hooke's law when using one/two strain gauges instead of a rosette [47,48].

9.6.3.1.2 Sectioning

The sectioning technique resembles excision, involving strain gauge attachment and specimen removal. However, it differs in the way material is removed to induce partial strain relaxation at measurement sites. Cutting successive sections yields data for determining original residual stress magnitudes and positions [47].

9.6.3.1.3 Tube Splitting

Splitting is a fast method for the initial evaluation of circumferential residual stresses in thin-walled tubes, particularly in applications such as heat exchanger tubes (as specified in ASTM 1998). This approach begins with an initial measurement of the original outer diameter of a short tube sample. Subsequently, the tube is opened by creating a lengthwise cut, as depicted in Figure 9.2c. It provides only an approximate estimation of residual stresses, assuming a linear distribution of circumferential residual stresses across the tube wall thickness. The corresponding surface stresses can be explained as Equation (9.3):

$$\sigma_\theta = \mp\frac{Et}{1-\vartheta^2}\cdot\frac{D-D_i}{DD_i} \tag{9.3}$$

Here, D_i represents the initial outer diameter, D is the outer diameter after splitting, E stands for Young's modulus, and μ denotes Poisson's ratio. Tensile stress (+) leads to an expansion in the outer surface diameter, while an equivalent compressive stress (–) occurs on the tube inner surface.

9.6.3.2 Semi-Destructive Methods

Semi-destructive methods for measuring residual stresses involve techniques that may cause some damage to the specimen but are less destructive than completely destructive methods. These methods provide a balance between obtaining valuable stress information and preserving the specimen to some extent.

9.6.3.2.1 Hole Drilling and Ring Core

The hole-drilling method, as defined by ASTM 1997, stands out as the most commonly employed destructive approach to gauge residual stresses. Its popularity stems from its ease of use, consistent results, and the fact that it typically inflicts only minimal, and often acceptable, damage to the test specimen. Due to these advantages, it is often referred to as a "semi-destructive" technique. The hole-drilling technique involves attaching a specialized strain gauge rosette to the specimen, followed by the precise drilling of a shallow hole at the center of interest. The process of removing material from the stressed region leads to a redistribution of stress in the surrounding material. The strain gauges within the rosette record the ensuing strain relaxations. Following this, the initial residual stresses at the hole location can be calculated using Equation (9.4) from the measured strain relaxations:

$$\sigma_{max}, \sigma_{min} = -\frac{E}{2}\left(\frac{\varepsilon_3+\varepsilon_1}{(1+\vartheta)\bar{a}} \mp \frac{\sqrt{(\varepsilon_3+\varepsilon_1)^2+(\varepsilon_3+\varepsilon_1-2\varepsilon_2)^2}}{\bar{b}}\right) \tag{9.4}$$

where $\bar{a}$ and $\bar{b}$ are dimensionless calibration constants depending on the diameter and depth of the drilled hole [47].

9.6.3.2.2 Slotting

The slotting method, often called the crack compliance method, draws its name from its early development's analogy with changes in compliance due to deep crack propagation. In this method, strain gauges are affixed to the surface of the specimen, and strain relaxation is measured as a narrow slot is made [49]. These gauges can be placed on either side of the specimen, usually arranged perpendicular to the slot, reducing perpendicular stresses. The ring-core technique is responsive to all in-plane stresses resulting from the circular ring cut. Slotting is essentially a one-dimensional subset of ring coring, enabling the use of mathematical techniques for residual stress determination. Commonly employed methods include the differential strain method and the power series method, requiring careful selection of the appropriate measurement technique [48].

9.6.3.3 Non-Destructive Testing

Non-destructive testing techniques rely on the assessment of electromagnetic, optical, and various other physical phenomena within the residual stress domain. These methods are employed to residual stresses are measured in the tangential, axial, and radial directions.

9.6.3.3.1 X-Ray Diffraction Method (XRD)

The technique, initially proposed by Lester and Aborn in 1925, saw XRD methods emerging as competitors to mechanical techniques around 1930. XRD operates based on Bragg's law [50]. Turan et al. [51] explored residual stress in both flat and grooved rails through the use of XRD and strain gauge techniques. Employing the $\sin^2\psi$ method, they exhibited a notable degree of linearity in their data. Zhang et al. [52] utilized the $\sin^2\psi$ method to determine macroscopic in-plane stress based on measured strain values. The assessment of residual stress was carried out in a triaxial direction across the thickness of the turbine disk. In a study by Umapathi and Swaroop [53], laser peening was applied to TC6 material to investigate residual stress, revealing a transition from compressive to tensile residual stress with increased peening time. Measurements were conducted using angle-dispersive X-ray diffraction (AD-XRD) at the BL-11 beamline facility. Equation 9.5 presents the formulation for the $\sin^2\psi$ method:

$$\frac{d_{\phi\psi} - d_0}{d_0} = \frac{1+\upsilon}{E}\sigma_\phi \sin^2\psi - \frac{\upsilon}{E}(\sigma_{11} + \sigma_{22}) \quad (9.5)$$

9.6.3.3.2 Neutron Diffraction Technique

The basic principle of measuring residual strain through X-ray and neutron diffraction methods is analogous. Nevertheless, X-rays are utilized to measure the average strain in the immediate vicinity of the surface within a few microns, whereas neutron diffraction enables the assessment of residual stress across the material's various layers. Neutron diffraction relies on a source of neutrons, typically provided by a nuclear reactor or accelerator. Neutrons are used because they can penetrate deep into materials without damaging them, making them suitable for studying the internal structure of a component. When the neutron beam interacts with the crystal lattice of a material, it scatters in different directions. The extent and direction of this scattering are influenced by the arrangement of atoms in the lattice. Residual stresses within the material cause slight shifts in the positions of the atomic planes in the crystal lattice. These shifts are a result of the stress and strain present in the material. A detector records the scattered neutrons at various angles. By analyzing the diffraction pattern created by these scattered neutrons, it is possible to determine the spacing between atomic planes in the crystal lattice, which is influenced by the residual stresses. The data obtained from the neutron diffraction experiment are used to determine the residual stress in the specimen. This is done by comparing the measured lattice spacing to the lattice spacing in a stress-free reference material [47].

9.6.3.3.3 *Raman Spectroscopy Method*

Residual stress measurement via Raman spectroscopy relies on assessing stress-induced shifts in Raman frequencies within materials. To begin, prepare the sample meticulously, ensuring a clean and flawless surface. Set up the Raman spectrometer, calibrating it for the specific material and conditions. Choose an appropriate laser wavelength based on material properties. Precisely focus the laser beam on the area of interest. Acquire a baseline Raman spectrum from an unstressed region for reference. Collect Raman spectra from areas with expected residual stresses and analyze peak shifts. Residual stresses typically cause frequency shifts in Raman peaks. Analyze the data using suitable software or algorithms to determine stress values, commonly reported in megapascals (MPa). This non-destructive method provides insights into material stress levels [47,48].

9.6.3.3.4 *Synchrotron Method*

Synchrotrons produce powerful X-rays with enhanced penetration capabilities, reaching depths of approximately 50 mm in aluminum, surpassing traditional X-rays. This extended penetration enables synchrotron diffraction to create detailed three-dimensional maps of strain distribution within engineered components, even at depths up to a millimeter-a critical advantage over conventional XRD methods. Synchrotrons showcase a notable capability in producing focused, narrow beams ranging from 1 mm to 10 μm in size. This feature implies that spatial resolutions rely on the crystallite size within the sample rather than the instrument itself. Additionally, measurements are notably faster compared to conventional XRD methods. With measurement times reduced to a fraction of a second, comprehensive strain maps of components can be efficiently generated within just a few hours of beam time [37,47,48].

9.6.3.3.5 *Magnetic Method*

The magnetic technique is based on the principle of magnetostriction, where a material expands or contracts when magnetized. Ferromagnetic materials inherently possess magnetism, but an external magnetic field can also induce magnetization in them. Residual stress is assessed using two magnetic techniques: magnetic Barkhausen noise and acoustic Barkhausen emission [37,47,48].

9.6.3.3.6 Eddy Current Method

To use this method, prepare the sample, ensuring a clean, superconductive surface. Choose a suitable eddy current probe based on material characteristics, coil type, frequency, and size. Position the probe over the area to assess residual stresses. When the probe is energized with alternating current, it induces eddy currents within the material. The interaction between these currents and the material's electrical conductivity, influenced by residual stresses, changes the probe's response. The instrument records impedance or phase shift data, reflecting stress-induced conductivity changes. Analyze this data to detect variations related to residual stresses. Impedance or phase changes can be correlated with stress levels if a calibration curve or known relationship exists for the material. Interpret the results and report the estimated residual stress levels, often in megapascals (MPa). This non-destructive method is valuable for materials analysis and quality control [37,47,48].

9.6.4 Mitigation of Residual Stresses

Residual stresses present in laser-cladded materials can result in various implications. Elevated residual stresses might trigger the development of cracks in both the cladding layer and the underlying substrate material. Such cracks pose a threat to the structural integrity of the component. Additionally, these stresses can induce distortion or warping in the laser-cladded part, causing inaccuracies in dimensions and potentially causing issues during assembly. Particularly, high tensile residual stresses can curtail the component's fatigue life, rendering it more catastrophic to failure under cyclic loading [36,54,55]. To ensure the structural soundness of laser-clad components, various approaches can be employed to mitigate residual stresses [56]. One strategy is to preheat the substrate before LC, which serves to reduce the thermal gradient and the subsequent thermal stresses. This method is especially valuable for materials susceptible to cracking due to rapid cooling. An appropriate preheating temperature of 150°C yields favorable mechanical properties. As the preheating temperature increases, the surface's residual stress on each specimen diminishes. The rear-end deposition section experiences higher longitudinal stress, while the lateral residual stress is more pronounced along both sides in the direction of scanning [57]. Another effective technique involves post-heat treatment, wherein the laser-clad component is subjected to controlled heating and cooling cycles after the cladding process. This allows for the relaxation of residual stresses and can enhance the material's properties [56]. Employing a multi-pass cladding method, where the cladding is split into multiple stages with intermittent cooling and stress relief, is effective for managing residual stresses. Adjusting laser power, travel speed, scan direction, and layer thickness can also reduce stress in laser-cladded components. Furthermore, using computational modeling of the LC process aids in minimizing residual stress formation. It allows for a detailed analysis of stress distribution under various process parameters, offering valuable insights for LC material processing [10,58].

9.7 Experimental Analysis of Residual Stresses in Laser Cladding Materials

Hurtado et al. in 2016 [56], the clad layer was made from AISI 316L stainless steel powder, and the base material was also AISI 316L stainless steel. Residual stresses in samples that underwent LC, laser shock peening, and post-weld heat treatment were evaluated using the hole-drilling method. Tensile residual stresses were observed near the surfaces of the laser-clad region, arising from the rapid localized thermal expansion and subsequent contraction. In Ding et al.'s [57] study, 12CrNi2 alloy steel powder was laser-clad onto a camshaft for a nuclear power plant. They employed a PG YLS-6000 laser device and varied process parameters, including preheating temperatures (20°C and 150°C), scan speed (8 mm/s), powder feed rate (8 g/min), and beam diameter (3 mm). Residual stress in the clad area was assessed using a KJS-3 indentation stress tester. The results showed that increasing the preheating temperature from 20°C to 150°C led to a decrease in residual stress on the surface of the test specimens, which was linked to tensile stress. In a study by An et al. [59], residual stresses in thin-walled curved structures made of Ni-based superalloys were evaluated. The welding of the material was accomplished using laser powder bed fusion additive manufacturing, and the obtained data were compared with numerical simulations derived from finite element modeling. Caporali et al. [60] applied neutron diffraction techniques to statistically analyze both type I and type II residual stress in three different directions. The ENGIN-X neutron beamline was utilized to investigate how variations in grain size and composition impact lattice displacement. Fan et al. [61], through the application of neutron diffraction techniques, investigated the microstructure, residual stress, and crystallographic properties of a ceramic composite specimen processed via a green method. Masoomi et al. [62] analyzed the influence of heat treatment on the distribution of residual stress in powder bed fusioned stainless steel 17-4 PH. Wang et al. [36] explored the utilization of TiC and SiC powders for cladding on a Ti6Al4V alloy substrate. The cladding process involved a laser power of 1,100 W and a travel speed of 6 mm/s to create a functionally graded material (FGM). The distribution of residual stress on both the upper and lower surfaces of the TC4 sheet, coated with FGM layers, was evaluated using a portable X-ray residual stress analyzer via the XRD method. Along the longitudinal direction, significant tensile residual stresses were observed.

Liu et al. [63] identified a relationship between the movement of domain walls (DWs) and magnetic Barkhausen noise in the presence of tensile stress. The influence of the material's microstructure on both DW motion and magnetic Barkhausen noise was examined. Gupta et al. [64] established a correlation between DW motion and magnetic Barkhausen noise for tensile stress. Neslusan et al. [65] studied magnetic Barkhausen technique was utilized to assess tool steel X210Cr12 after semi-solid forging, enabling the simultaneous monitoring of both microstructure and stress conditions. Das et al. [66] studied residual stresses in plasma-sprayed coatings, and individual splats of zirconia laser melt and coating were assessed using both Raman spectroscopy and XRD techniques. The results indicate that Raman spectroscopy is a rapid and precise method for measuring residual stresses, including depth profiling. Matsumoto et al. [67] studied the assessment of residual stress in low carbon steels was conducted using eddy current magnetic signature (EC-MS) testing. This investigation involved measuring residual stresses across various conditions, including within the elastic limit, Luders-strain range, and work-hardening region. Additionally, Bassam et al. [68] applied a range of frequency-dependent apparent eddy current conductivity (AECC) to analyze residual stresses present on the surfaces of metals subjected to shot peening. Emamian et al. [69] reported the LC of lateral powder injection (Fe-TiC) on AISI 1030 carbon steel hardened substrate. It was found that the TiC particle distribution was influenced by cooling rates during deposition. The micro-cracks in the first track are reduced by subsequent layers of deposition due to lower cooling rates and thermal stresses.

Eo et al. [70] have used the DED technique for cladding FGM material onto MN steel. While depositing at high laser intensities, the first and second layers exhibited hot cracks and solidification cracks. The majority of cracks were identified at the grain boundaries formed during solidification, particularly in regions where dendrites exhibited varying growth directions, as shown in Figure 9.3c. This occurrence was attributed to the solidification shrinkage and the narrowing effect that took place when two grains with different crystallographic orientations came into contact. As a result, the tensile strain became concentrated at these grain boundaries [70]. Hot cracks (as shown in Figure 9.3a) occur above the solidification line, lacking oxidation color and metal luster. Uneven distribution of coarse and brittle compound phases, impurities, and disordered grain growth are the main factors leading to the formation of these cracks [71]. The cold cracks were shown in Figure 9.3b are relatively straight and have a metallic luster, unlike hot cracks, and appear after the solidification temperature line. Cold crack formation is primarily attributed to the inappropriate choice of process parameters and the presence of excessive TG in the melt pool [72]. The cracks observed in the cladding layer, as depicted in Figure 9.3c, result from the rapid cooling during the solidification process, inducing significant thermal stress. Additionally, an uneven distribution of stress, associated with irregular mixing of cladding material or the presence of impurities, can contribute to the formation of cracks within the cladding layer [73]. Cracks at the substrate interface result from significant differences in thermal properties between the substrate and the clad layer. These differences induce thermal stresses during solidification and cooling, exacerbated by substantial TG. Additionally, the molten metal undergoes restraint stress due to thermal expansion and contraction against

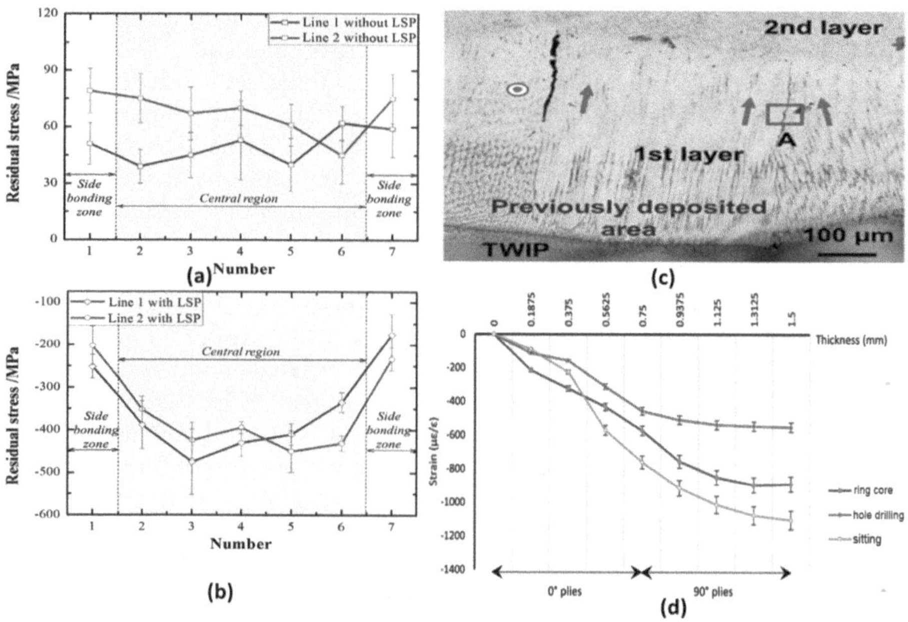

FIGURE 9.3
(a) XRD measured surface residual stress in (a) the cladded sample, (b) laser shock peened sample [76], (c) optical micrograph of the cladded region with dendritic structure and cracks in the FGM layers [70], and (d) comparison of different techniques measured strains in a longitudinal direction for samples [77].

the substrate. The combined thermal and restraint stresses ultimately lead to the formation of interface substrate cracks that propagate to the cladding layer's surface [74]. Figure 9.3d cracks in the overlap zone primarily emerge due to an imprudent selection of the overlap rate. This choice results in the accumulation of heat, elevated TG, and increased thermal stress in the lap zone. Consequently, the grain growth duration is extended, leading to the formation of coarser grains, along with the presence of pores or impurities. The expansion of these defects ultimately culminates in the development of cracks in the lap zone [75]. Table 9.2 presents the various cracks and their reasons.

Luo et al. [76] conducted the LC using a YLS-4000eCT-Y11 ytterbium-doped fiber laser of 304 Stainless steel onto a 316L austenite stainless steel substrate. The process used were P=1,800 W, laser diameter d=3mm, 22.5% of laser overlapping ratio, the speed was 4 mm/s, and the shielding gas velocity was 5 L/min. The stresses were measured using XRD sin2 ψ technique. Residual stress measurements (Figure 9.4a–c) clearly indicate that the tensile stresses in the cladded region are in the range of 36–81 MPa. The average tensile stress 56 MPa of the same sample after the laser shock peening (LSP) turned to compressive residual stresses –183 MPa [76]. Chen et al.'s [79] study involved

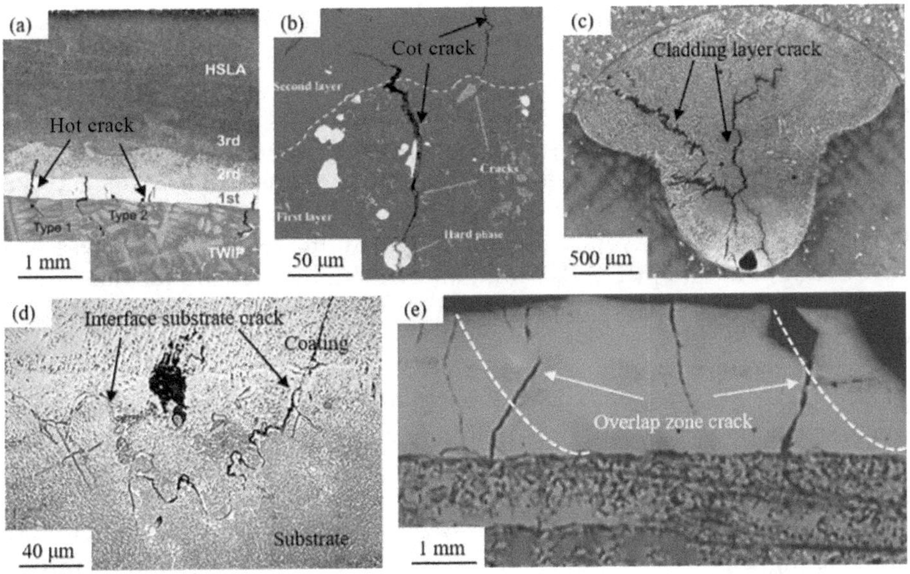

FIGURE 9.4
Various cracks are associated with the cladding: (a) hot cracks at P=2,900 W and V=10 mm/s, (b) cold cracks at P=5,000 W and V=30 mm/s, (c) crack in cladding layer at P=600 W and V=9 mm/s, (d) crack in interface at P=576 W and V=4 mm/s, and (e) crack in overlap zone at P=3,200 W and V=416.7 mm/s. Reproduced from open access journal [78].

the cladding of IN 625 powder onto a wrought IN 625 substrate. Residual stresses within the clad material were evaluated using the hole-drilling strain gauge method, following ASTM E837-95 guidelines. In the clad region, biaxial tensile stresses were identified, measuring σ_L and σ_T at 387 MPa and 265 MPa, respectively, as shown in Figure 9.4d. Subsequent post-clad annealing at 760°C for 1 hour resulted in a reduction of these biaxial tensile stresses, decreasing from 387 and 265 MPa to 337 and 229 MPa. Further annealing led to a decrease in stresses, reducing them from 337 and 229 MPa to 334 and 167 MPa. In Ghaedamini et al.'s [77] study, hole-drilling, ring core, and slitting residual stress measurement techniques were used to assess the layered composite (glass/epoxy composite) specimen's stresses. There is an increase in strain with greater depth across all three methods. However, variations in strain values are evident among the different methods. In general, when working with laminated composite samples, the choice of method can significantly impact the magnitude of released strains. Figure 9.4e illustrates that the slitting method yields the highest released strains, followed by the ring-core method, while the hole-drilling method results in the lowest released strains. Based on the results obtained from varying sample depths, the slitting method produces approximately 15% more released strains than the ring-core method, which, in turn, generates approximately 40% more released strains than the hole-drilling method [77]. Roy et al. [80] have investigated the residual stress magnitude and distribution in the vicinity of clad, HAZ, and substrate of the laser-cladded rail (A 600-mm long rail piece and 400mm length). The Kowari neutron diffractometer at ANSTO was used to measure the residual stresses. In the cladding layer, the most substantial compressive residual stress was observed, whereas the highest tensile residual stress was identified in the HAZ along the longitudinal (X) directions. In the case of 410L two layers, the maximum compressive residual stress was -396 ± 10 MPa at a depth of 0.75mm, and the highest tensile residual stress was 578 ± 8 MPa at a depth of 3.5mm, both occurring in the longitudinal (X) direction.

From the above literature review, it can be concluded that several papers have reported on residual stress measurement using destructive, semi-destructive, and non-destructive techniques in laser-cladded materials. The techniques are well reasonable for assessing the stresses within the cladded and substrate materials. The state and magnitude of residual stresses are influenced by various process parameters used during cladding process. Right process parameters will give low-stress levels and distortion components.

9.8 Computational Modeling of Laser Cladding Process to Predict the Residual Stress and Distortion of the Materials

Hao et al. [81] carried out transient thermal analysis of LC process of the Ti6Al4V alloy. The predicted melt pool and temperature were validated with the experimental ones. Tamanna et al. [10] have reported on thermomechanical analysis of LC process of the Ceramic clad (Al_2O_3 and TiC) on substrate (Ti6Al4V). They have predicted the melt pool, thermal cycles, residual stress, and distortion. The predicted model showed that the Al_2O_3 induced compressive stress and TiC induced the tensile stress. It was concluded that the Al_2O_3 and Ti6Al4V substrates are a good combination for the LC, unlike TiC and Ti6Al4V substrates. Ding et al. [57] reported on temperature and residual stress model of LC process with substrate preheating and non-preheating. The assessment of residual stress on the surface of every test specimen was conducted, comparing non-preheated specimens to those preheated at 150°C and 300°C. As the preheating temperature increased, both longitudinal and lateral residual stresses exhibited a significant reduction, with longitudinal residual stress slightly surpassing lateral residual stress. This behavior aligns closely with the numerical simulation results depicted in Figure 9.5, with a

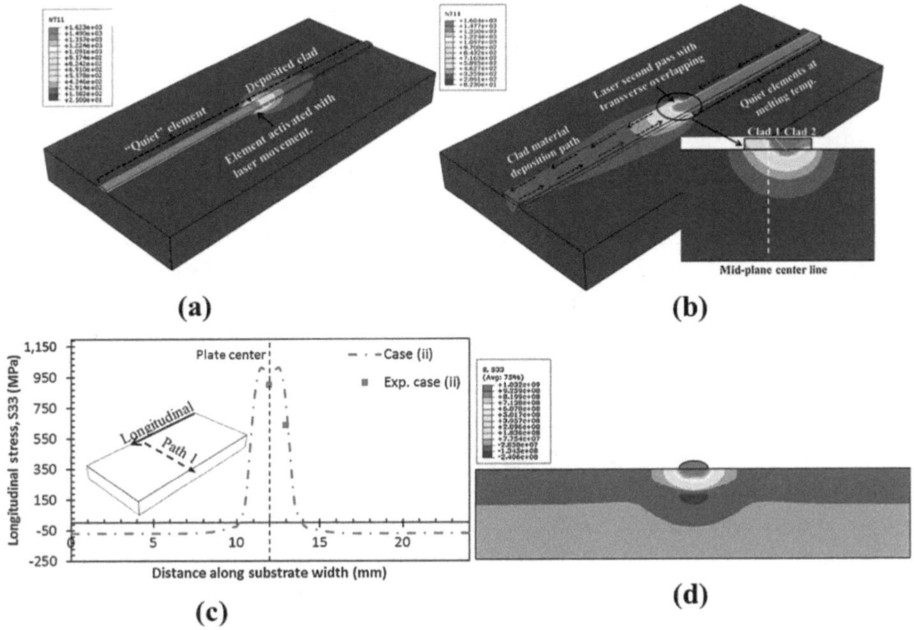

FIGURE 9.5
Temperature contours of (a) single-clad, (b) double-clad, (c) comparison of XRD measurement with S33 longitudinal stress, and (d) S33 stress contour. Reproduced with permission from [83].

numerical average deviation of 7.93%, indicating strong agreement between the simulation and the actual test values. Kumar and Roy [82] have investigated the blown powder LC of two-dimensional thermal conduction equation via finite volume method. The preheating of powder enhances the faster scan speed and lower dilution rate. Wang et al. [36] investigated the computational modeling of LC of TiC and SiC powders on Ti6Al4V substrate. A laser power of 1,100 W and a travel speed of 6 mm/s were used in the cladding of FGM. Plati et al. (2006) investigated the 3D transient thermomechanical analysis of LC process. It was considered the non-linear material properties and elastoplastic model in the simulation. It was found that the large thermal gradients typically arise in the through-thickness direction. This leads to differential thermal contraction and frequently creates significant stresses, which can cause yielding in the hot, near-surface areas. This plastic deformation results in the formation of residual stresses and also contributes to distortion of the workpiece. The predicted thermal history is around 260°C in accordance with the measured ones.

Heigel et al. [84] carried out thermomechanical modeling of LC of feedstock with powder and wire reported. It was found that the single layer of wire deposition yields lower temperature and distortion compared to the powder injection [24]. In Chew et al.'s [83] study, thermomechanical modeling and simulation of LC process of AISI 4340 steel carried out to predict the thermal history and residual stresses. The experimentally cladded samples residual stress measured using XRD technique. Thermal history of the laser-cladded contours for single-clad and double-clad, as shown in Figure 9.5a and b. The predicted temperature is between 1,600°C and 1,800°C and validated against the published experimental value. On the other hand, the predicted residual stress is 727.7 and 953 MPa, matched with the measured residual stresses of 635.1 ± 72.7 and 898.6 ± 137.7 MPa, as shown in Figure 9.5c. Figure 9.5d shows the maximum stress found in the cladded region.

Nian et al. [85] used computational modeling to predict parameters in the LC process, including cooling rates, thermal gradients, melt pool characteristics (from thermal analysis), and residual stress and distortion (from structural analysis). They applied specific parameters: 2,400 W laser power, 20 m/min travel speed, and 20.32 g/min powder injection rate. The tensile residual stress was observed in both substrate and interface. Tensile residual stress was also detected in the clad region itself, which was mainly oriented in the LC direction. Kang et al. [86] carried out thermomechanical analysis of LC process of 316L stainless steel powder cladding on 316L stainless steel substrate to understand the thermal gradients and residual stress magnitude and distribution. The residual stress in the laser-cladded sheet was greater along the scanning direction than perpendicular to it. However, the RS values in both the LC sheet and the substrate remained below their respective yield strength values. Hence, the residual stress model introduced in this paper is well-suited for the efficient and precise prediction of residual stresses in laser-cladded sheets used for the repair of circular steel tubular columns via LC. Wu et al. [87] studied LC using Ni60A-25% WC, and the authors conducted a 3D transient thermomechanical analysis. The melt pool took on an ellipsoidal shape, with temperatures exceeding the material's melting point, reaching 3,111.8 K during deposition. Temperature cycles indicated that laser power directly affected temperature magnitude. The substrate temperature (2,500 K) was lower than the clad region (3,700 K). They observed that increasing the scan speed decreased compressive residual stresses (CRS). With a 2,800 W laser power, residual stresses reached up to 130 MPa (x), 88 MPa (y), and 72 MPa (z). In summary, CRS was influenced by the laser and indirectly by the scan speed. Interestingly, higher LC power led to a gradual increase in residual stresses within the HAZ.

From the literature discussed above, it can be concluded that several researchers have conducted numerical modeling and simulations of the LC process using various materials. The objective was to predict parameters such as the melt pool, cooling rates, thermal gradients, residual stress, and distortion during the LC process by considering various process parameters like laser power and travel speed. The predicted results are then compared to experimental measurements for validation. These findings align with experimental results obtained through techniques such as XRD, blind hole drilling, and neutron diffraction. As a result, 3D transient thermomechanical modeling of the LC process proves valuable in reducing the number of experiments needed for cladding or repairing expensive components.

9.9 Case Study 1: Residual Stress in Laser Cladding Rail

This study concentrates on examining the impact of cladding high-value track components, constructed from standard rail steel UIC 900A/grade 260, with a layer of premium martensitic stainless steel, aiming to mitigate wear and fatigue [88].

LC is essential for enhancing the surface properties of various components. It helps increase wear resistance, repair damaged surfaces, and improve performance. LC can introduce residual stresses into the component, affecting its mechanical integrity and long-term performance. This small case study aims to understand residual stress in LC components through the paper by Narayanan et al. [88] titled "Residual stress in LC rail," published in *Tribology International*. The martensitic steel has a composition of 0.04% C, 13% Cr, 4.1% Ni, and Fe balance used as clad material, and the substrate is pearlite steel (UIC 900A/grade 260) [88].

The case study investigates the residual stress in LC rail components, focusing on how these stresses influence the rail's wear resistance and fatigue performance. The study was carried out on railway rails, employing LC as a surface modification technique. The aim was to increase rail durability by enhancing wear resistance, which is essential for ensuring railway systems' safety and reliability. Residual stress measurements were a core part of this study. The researchers utilized the deep hole-drilling method, a well-established technique for evaluating material residual stresses. Neutron diffraction was used to investigate the distribution of residual stress across the surface of the rail. By employing these measurement techniques, the researchers obtained comprehensive data on the magnitude and distribution of residual stress in the LC rail components [88].

The findings of the study can be helpful for future use. The LC process led to compressive and tensile residual stresses in the rail. The distribution of these stresses through the rail's cross-section was found to be non-uniform because of the nature of the LC process. The CRS significantly enhanced the rail's wear resistance. This is a crucial factor for extending the lifespan of the rail and reducing maintenance costs. The presence of tensile residual stresses near the surface of the rail raised concerns regarding its fatigue life. Fatigue life is needed for any structural component subjected to cyclical loading. Tensile residual stresses can contribute to the initiation and propagation of fatigue cracks, eventually leading to component failure. In railway systems, fatigue failure is a safety concern, and understanding how residual stress impacts it is essential [88].

9.10 Case Study 2: Residual Stress Investigation of Co-Based Laser Clad by X-Rays and Synchrotron Diffraction Methods

This study concentrates on the evaluation of residual stress in clad layers of Co-W-Cr (Eutroloy and Stellites) applied to a substrate made with C45 steel. The analysis includes assessing surface strain tensors using the $\sin^2\Psi$ technique with the X-ray tube of a laboratory Cu-Kα radiation source. Residual stress mapping was performed on both unaltered and cross-sectional samples using synchrotron and XRD analysis [89].

In this work, the powder of Co-W-Cr alloy, namely Eutrolloy 16012 and Stellite 20 were used. These powders are deposited onto a substrate made of C45 low-alloyed steel. The LC process is conducted using the Sulzer Metco Twin 10C system, which consists of nozzles of side and coaxial feeding, CNC table with an XYZ-Rotation, and Nd: YAG continuous wave Rofin Sinar laser (a 2KW) [89].

Figure 9.6c and d shows the strains and residual stress calculated using empirical relation. The strains were computed in the reflection mode by focusing on the central tracks of the Eutroloy 16012 cladded layer at scanning

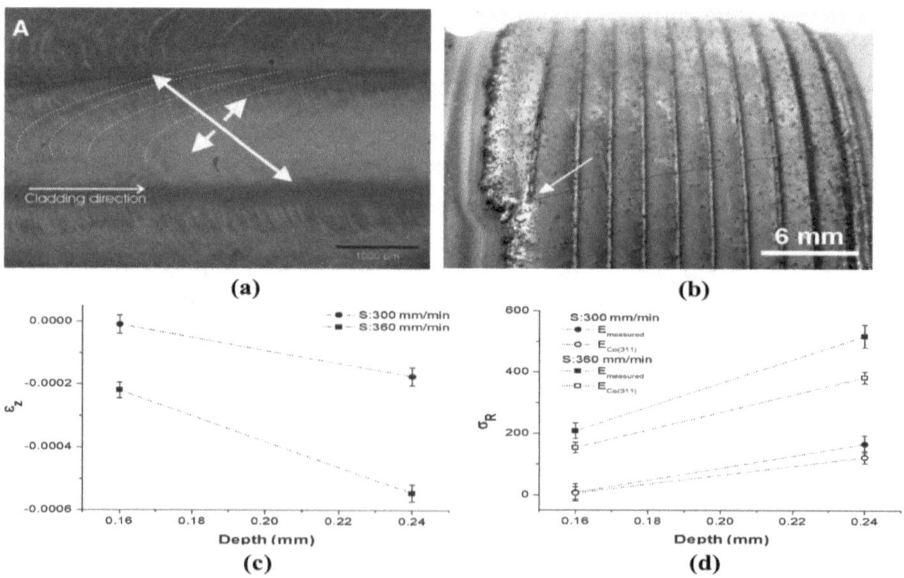

FIGURE 9.6
(a) The stress tensor's principal components at the measurement location, denoted by arrows, on the central tracks of the Stellite 20 cladding on a plane substrate. (b) Cracks that on a round steel substrate coating of Eutroloy 16012. The arrow points to the site where crack initiation occurred. (c), (d) Residual strains along the Z-axis and residual stresses after computation using empirical equation. Reproduced with permission from [89].

speeds of 300 and 360 mm/min. The diffraction took place at uniform depths in both samples, highlighting the uniformity of the grain structure within the cladding. Residual stresses in both instances were minimal at the surface but exhibited a swift escalation with increasing depth (Figure 9.6d).

XRD was used to evaluate residual stresses in Co-based laser-deposited claddings. These findings were combined with microstructural data, phase analysis, Nano hardness measurements, and elastic modulus calculations. The formation of larger grains and the development of a strong texture during laser track solidification pose challenges for stress analysis using synchrotron radiation diffraction in the reflection mode throughout the entire coating depth. Residual stresses generally increase with depth, reaching a magnitude close to the yield stress at approximately half the coating's thickness. Near the coating-substrate interface, internal stresses shift toward compression. Residual stress levels are influenced by scanning speed, track overlap degree, and the transition from a flat to a round substrate shape (Figure 9.6a and b).

9.11 Conclusions

Residual stresses in LC result from rapid heating and cooling rates, and understanding their origins and effects is essential for ensuring the quality of manufactured components. Various methodologies, including destructive and non-destructive techniques, can be employed for residual stress measurement, with the choice influenced by factors such as material, geometry, and the intended use of laser-clad parts and materials. Effectively managing residual stresses allows LC to consistently ensure the quality, reliability, performance enhancement, and cost-efficiency of critical components across various industrial domains.

In a case study of residual stress in laser-clad rail, the stresses across the rail's cross-section were non-uniform due to the nature of the LC process. The presence of CRS significantly improved the rail's wear resistance, a critical factor in extending the rail's lifespan and reducing maintenance costs. However, the existence of tensile residual stresses near the surface raised concerns regarding its fatigue life.

References

1. Tarasov LV. *Laser Physics*. Moscow: Mir Publishers; 1986.
2. Fayaz G, Zakeri SS. *Laser Cladding of Metals*. 2020. Springer. https://doi.org/10.1007/978-3-030-53195-9_3.

3. Wang Z, Denlinger E, Michaleris P, Stoica AD, Ma D, Beese AM. Residual stress mapping in Inconel 625 fabricated through additive manufacturing: Method for neutron diffraction measurements to validate thermomechanical model predictions. *Mater Des* 2017;113:169–77. https://doi.org/10.1016/j.matdes.2016.10.003.
4. Yilbas BS, Karatas C, Arif AFM, Abdul Aleem BJ. Laser control melting of alumina surfaces and thermal stress analysis. *Opt Laser Technol* 2011;43:858–65. https://doi.org/10.1016/j.optlastec.2010.10.009.
5. Balichakra M, Bontha S, Krishna P, Balla VK. Laser surface melting of γ-TiAl alloy an experimental and numerical modeling study. *Mater Res Express* 2019;6:046543. https://doi.org/10.1088/2053-1591/aafc89.
6. Balichakra M, Bontha S, Krishna P, Das M, Balla VK. Understanding thermal behavior in laser processing of titanium aluminide alloys. *Proc. 6th Int. 27th All India Manuf. Technol. Des. Res.* Conf., Pune, India, 2016, pp. 73–7.
7. Mallikarjuna B, Reutzel EW. Reclamation of intermetallic titanium aluminide aero-engine components using directed energy deposition technology. *Manuf Rev* 2022;9:1–22. https://doi.org/10.1051/mfreview/2022024.
8. Valente C, Teresa M, Sharma N. LASER cladding - A post processing technique for coating, repair and re-manufacturing. In Kapil Gupta (ed.), *Materials Forming, Machining and Post Processing*, 2019, pp. 231–49. Switzerland AG: Springer Nature.
9. Prabu NJG, Hua C, Manikanta Y, Banda P. Structural and nanohardness of laser melted CM247LC nickel - based superalloy. *Trans Indian Inst Met* 2023;76:287–95. https://doi.org/10.1007/s12666-022-02734-y.
10. Tamanna N, Kabir IR, Naher S. Thermo-mechanical modelling to evaluate residual stress and material compatibility of laser cladding process depositing similar and dissimilar material on Ti6Al4V alloy. *Therm Sci Eng Prog* 2022;31:101283. https://doi.org/10.1016/j.tsep.2022.101283.
11. Kathuria YPU. Some aspects of laser surface cladding in the turbine industry. *Surf Coat Technol* 2000;132:262–9.
12. Maliutina IN, Si-Mohand H, Piolet R, Missemer F, Popelyukh AI, Belousova NS, et al. Laser cladding of γ-TiAl intermetallic alloy on titanium alloy substrates. *Metall Mater Trans A Phys Metall Mater Sci* 2016;47:378–87. https://doi.org/10.1007/s11661-015-3205-9.
13. Mallikarjuna. *Effect of Process Variables on Residual Stress and Microstructure in Laser Additive Manufacturing of γ-TiAl Alloy*. National Institute of Technology Karnataka; 2020.
14. Safari M, Moradi M, Joudaki J, Babaei AM. Experimental investigation of laser cladding of H13 hot work steel by Stellite 6 powder. *Sādhanā* 2022;47:1–10. https://doi.org/10.1007/s12046-022-01909-z.
15. Aqida BNSN, Brabazon ECD. Surface modification of AISI H13 tool steel by laser cladding with NiTi powder. *Appl Phys A* 2016;122:1–6. https://doi.org/10.1007/s00339-016-9937-6.
16. Liu Y, Ding Y, Yang L, Sun R, Zhang T, Yang X. Research and progress of laser cladding on engineering alloys: A review. *J Manuf Process* 2021;66:341–63. https://doi.org/10.1016/j.jmapro.2021.03.061.
17. Farotade GA, Adesina OS, Kolesnikov A, Popoola API. Computational analysis of heat transfer within a Ti-6Al-4V alloy substrate during laser cladding process. *Mater Res Express* 2019;6:046516. https://doi.org/10.1088/2053-1591/aafab9.

18. Xia M, Gu D, Yu G, Dai D, Chen H, Shi Q. Influence of hatch spacing on heat and mass transfer, thermodynamics and laser processability during additive manufacturing of Inconel 718 alloy. *Int J Mach Tools Manuf* 2016;109:147–57. https://doi.org/10.1016/j.ijmachtools.2016.07.010.
19. Hornberger H, Virtanen S, Boccaccini AR. Biomedical coatings on magnesium alloys - A review. *Acta Biomater* 2012;8:2442–55. https://doi.org/10.1016/j.actbio.2012.04.012.
20. Nowotny S, Scharek S, Beyer E, Richter K. Laser beam build-up welding-Precision in repair, surface cladding, and direct 3D metal deposition. *J Therm Spray Technol* 2007;16:344–8. https://doi.org/10.1007/s11666-007-9028-5.
21. Pathania A, Kumar SA, Nagesha BK, Barad S, Suresh TN. Reclamation of titanium alloy based aerospace parts using laser based metal deposition methodology. *Mater Today Proc* 2021;45:4886–92.
22. Jeyaprakash N, Yang CH, Ramkumar KR. Microstructure and wear resistance of laser cladded Inconel 625 and Colmonoy 6 depositions on Inconel 625 substrate. *Appl Phys A Mater Sci Process* 2020;126:1–11. https://doi.org/10.1007/s00339-020-03637-9.
23. Jeyaprakash N, Yang CH, Sivasankaran S. Laser cladding process of Cobalt and Nickel based hard-micron-layers on 316L-stainless-steel-substrate. *Mater Manuf Process* 2020;35:142–51. https://doi.org/10.1080/10426914.2019.1692354.
24. Heigel JC, Gouge MF, Michaleris P, Palmer TA. Selection of powder or wire feedstock material for the laser cladding of Inconel® 625. *J Mater Process Tech* 2016;231:357–65. https://doi.org/10.1016/j.jmatprotec.2016.01.004.
25. Liu B, Deng Z, Liu D. Preparation and properties of multilayer Ca/P bio-ceramic coating by laser cladding. *Coatings* 2021;11:1–17.
26. Song L, Bagavath-Singh V, Dutta B, Mazumder J. Control of melt pool temperature and deposition height during direct metal deposition process. *Int J Adv Manuf Technol* 2011;58:247–56. https://doi.org/10.1007/s00170-011-3395-2.
27. Shepeleva L, Medres B, Kaplan WD, Bamberger M, Weisheit A. Laser cladding of turbine blades. *Surf Coat Technol* 2000;125:45–8.
28. Toyserkani E, Khajepour A, Corbin S. *Laser Cladding*. Florida: CRC Press; 2005.
29. Ocylok S, Weisheit A, Kelbassa I. Increased wear and oxidation resistance of titanium aluminide alloys by laser cladding. *Adv Mater Res* 2011;278:515–20. https://doi.org/10.4028/www.scientific.net/AMR.278.515.
30. Yao JH, Zhang QL, Kong FZ. Laser remanufacturing to improve the erosion and corrosion resistance of metal components. In Chi Tat Kwok (ed.), *Laser Surface Modification of Alloys for Corrosion and Erosion Resistance*, 2012, pp. 320–54. Woodhead Publishing, Sawston, United Kingdom.
31. Val J, Comesaña R, Riveiro A, Lusquiños F, Quintero F, Boutinguiza M, et al. Laser direct writing of Co-superalloy lines for micro-fabrication applications. *Surf Coat Technol* 2018;345:76–88.
32. Haldar B, Saha P. Identifying defects and problems in laser cladding and suggestions of some remedies for the same. *Mater Today Proc* 2018;5:13090–101. https://doi.org/10.1016/j.matpr.2018.02.297.
33. Borges B, Quintino L, Miranda RM, Carr P. Imperfections in laser clading with powder and wire fillers. *Int J Adv Manuf Technol* 2010;50:175–83. https://doi.org/10.1007/s00170-009-2480-2.

34. Bendeich P, Alam N, Brandt M, Carr D, Short K, Blevins R, et al. Residual stress measurements in laser clad repaired low pressure turbine blades for the power industry. *Mater Sci Eng A* 2006;437:70–4. https://doi.org/10.1016/j.msea.2006.04.065.
35. Withers PJ, Bhadeshia HKDH. Residual stress. Part 2 – Nature and origins. *Mater Sci Technol* 2001;17:366–75. https://doi.org/10.1179/026708301101510087.
36. Wang Q, Shi J, Zhang L, Tsutsumi S, Feng J, Ma N. Impacts of laser cladding residual stress and material properties of functionally graded layers on titanium alloy sheet. *Addit Manuf J* 2020;35:101303.
37. Kandil FA, Lord JD, Fry AT, Grant PV. *A Review of Residual Stress Measurement Methods - A Guide to Technical Selection*. 2001. Middlesex: NPL Publications.
38. Zhan Y, Liu C, Kong X, Lin Z. Experiment and numerical simulation for laser ultrasonic measurement of residual stress. *Ultrasonics* 2017;73:271–6. https://doi.org/https://doi.org/10.1016/j.ultras.2016.08.013.
39. Gmb HL. Laser cladding. *Laserline* 2021:1–6.
40. Mallikarjuna B, Bontha S, Krishna P, Balla VK. Characterization and thermal analysis of laser metal deposited γ-TiAl thin walls. *J Mater Res Technol* 2021;15:6231–43. https://doi.org/10.1016/j.jmrt.2021.10.133.
41. Manvatkar VD, Gokhale AA, Reddy GJ, Venkataramana A. Estimation of melt pool dimensions, thermal cycle, and hardness distribution in the laser-engineered net shaping process of austenitic stainless steel. *Metall Mater Trans A* 2011;42:4080–4087. https://doi.org/10.1007/s11661-011-0787-8.
42. Liu Y, Yang Y, Wang D. A study on the residual stress during selective laser melting (SLM) of metallic powder. *Int J Adv Manuf Technol* 2016;87:647–56. https://doi.org/10.1007/s00170-016-8466-y.
43. Withers PJ, Bhadeshia HKDH. Residual stress Part 1- Measurement techniques. *Mater Sci Technol* 2001;17:355–65. https://doi.org/10.1179/026708301101509980.
44. Mercelis P, Kruth JP. Residual stresses in selective laser sintering and selective laser melting. *Rapid Prototyp J* 2006;12:254–65. https://doi.org/10.1108/13552540610707013.
45. Li C, Liu ZY, Fang XY, Guo YB. Residual stress in metal additive manufacturing. *Procedia CIRP* 2018;71:348–53. https://doi.org/10.1016/j.procir.2018.05.039.
46. Guo J, Fu H, Pan B, Kang R. Recent progress of residual stress measurement methods : A review. *Chinese J Aeronaut* 2021;34:54–78. https://doi.org/10.1016/j.cja.2019.10.010.
47. Schajer GS. *Practical Residual Stress Measurement Methods*, Vol. 1. West Sussex, United Kingdom: A John Wiley & Sons, Ltd; 2013. https://doi.org/https://doi.org/10.1002/9781118402832.
48. Schajer GS. *Residual Stresses: Measurement by Destructive Testing*; 2001. Canada: Encyclopaedia of Materials: Science and Technology.
49. Prime MB. *Residual Stress Measurement by Successive Extension of a Slot- The Crack Compliance Method*. Los Alamos: American Society of Mechanical Engineers; 1997.
50. Balichakra M, Bontha S, Krishna P, Balla VK. Prediction and validation of residual stresses generated during laser metal deposition of γ titanium aluminide thin wall structures. *Mater Res Express* 2019;6:106550. https://doi.org/10.1088/2053-1591/ab38ee.
51. Turan ME, Aydin F, Sun Y, Cetin M. Residual stress measurement by strain gauge and X-ray di ff raction method in di ff erent shaped rails *Eng Fail Anal* 2019;96:525–9. https://doi.org/10.1016/j.engfailanal.2018.10.016.

52. Zhang Z, Feng Y, Tan Q, Zou J, Li J, Zhou X, et al. Residual stress distribution in Ni-based superalloy turbine discs during fabrication evaluated by neutron / X-ray diffraction measurement and thermomechanical simulation. *Mater Des* 2019;166:107603. https://doi.org/10.1016/j.matdes.2019.107603.
53. Umapathi A, Swaroop S. Measurement of residual stresses in titanium alloys using synchrotron radiation. *Measurement* 2019;140:518–25. https://doi.org/10.1016/j.measurement.2019.04.021.
54. Saboori A, Gallo D, Biamino S, Fino P, Lombardi M. An overview of additive manufacturing of titanium components by directed energy deposition: Microstructure and mechanical properties. *Appl Sci* 2017;7:883. https://doi.org/10.3390/app7090883.
55. Rangaswamy P, Griffith ML, Prime MB, Holden TM, Rogge RB, Edwards JM, et al. Residual stresses in LENS ® components using neutron diffraction and contour method. *Mater Sci Eng A* 2005;399:72–83. https://doi.org/10.1016/j.msea.2005.02.019.
56. Hurtado AM, Francis JA, Stevens NPC, Hurtado AM, Francis JA, Stevens NPC. An assessment of residual stress mitigation strategies for laser clad deposits. *Mater Sci Technol* 2016;32:1484–94. https://doi.org/10.1080/02670836.2016.1192766.
57. Ding C, Cui X, And JJ, Zhu P. Effects of substrate preheating temperatures on the microstructure, properties, and residual stress of 12CrNi2 prepared by laser cladding deposition technique. *Materials (Basel)* 2018;11:2401. https://doi.org/10.3390/ma11122401.
58. Caiazzo F, Alfieri V, Bolelli G. Residual stress in laser-based directed energy deposition of aluminum alloy 2024: simulation and validation. *Int J Adv Manuf Technol* 2022;118:1197–211. https://doi.org/10.1007/s00170-021-07988-2.
59. An K, Yuan L, Dial L, Spinelli I, Stoica AD, Gao Y. Neutron residual stress measurement and numerical modeling in a curved thin-walled structure by laser powder bed fusion additive manufacturing. *Mater Des* 2017;135:122–32. https://doi.org/10.1016/j.matdes.2017.09.018.
60. Caporali S, Pratesi G, Kabra S, Grazzi F. Type I and type II residual stress in iron meteorites determined by neutron diffraction measurements. *Planet Space Sci* 2018;153:72–8. https://doi.org/10.1016/j.pss.2017.12.015.
61. Fan K, Ruiz-hervias J, Ygnacio J, Gurauskis J, Baudín C. Residual stress and diffraction line-broadening analysis of Al 2O3/Y-TZP ceramic composites by neutron diffraction measurement. *Int J Refract Met Hard Mater* 2017;64:122–34. https://doi.org/10.1016/j.ijrmhm.2017.01.011.
62. Masoomi M, Shamsaei N, Winholtz RA, Milner JL, Gnäupel-herold T, Elwany A, et al. Residual stress measurements via neutron diffraction of additive manufactured stainless steel 17-4 PH. *Data Br* 2017;13:408–14. https://doi.org/10.1016/j.dib.2017.06.027.
63. Liu J, Yun G, Gao B, Zeng K, Zheng Y, Chen J. Micro-macro characteristics between domain wall motion and magnetic Barkhausen noise under tensile stress. *J Magn Magn Mater* 2020;493:165719. https://doi.org/10.1016/j.jmmm.2019.165719.
64. Gupta B, Ducharne B, Uchimoto T, Sebald G, Miyazaki T, Takagi T. Non-destructive testing on creep degraded 12% Cr-Mo-W-V ferritic test samples using barkhausen noise. *J Magn Magn Mater* 2020;498:166102. https://doi.org/10.1016/j.jmmm.2019.166102.

65. Neslu M, Minárik P, Čillikova M, Kola K, Rube K. Barkhausen noise emission in tool steel X210Cr12 after semi-solid processing. *Mater Charact* 2019;157:109891. https://doi.org/10.1016/j.matchar.2019.109891.
66. Das B, Brodard P, Bandyopadhyay PP. Raman spectroscopy assisted residual stress measurement of plasma sprayed and laser remelted zirconia splats and coatings. *Surf Coat Technol* 2019;378:124920. https://doi.org/10.1016/j.surfcoat.2019.124920.
67. Matsumoto T, Uchimoto T, Takagi T, Dobmann G. Investigation of electromagnetic nondestructive evaluation of residual strain in low carbon steels using the eddy current magnetic signature (EC-MS) method. *J Magn Magn Mater* 2019;479:212–21. https://doi.org/10.1016/j.jmmm.2019.01.103.
68. Abu-nabah BA, Nagy PB. Iterative inversion method for eddy current profiling of near-surface residual stress in surface-treated metals. *NDT E Int* 2006;39:641–51. https://doi.org/10.1016/j.ndteint.2006.05.001.
69. Emamian A, Alimardani M, Khajepour A. Correlation between temperature distribution and in situ formed microstructure of Fe – TiC deposited on carbon steel using laser cladding. *Appl Surf Sci* 2012;258:9025–31. https://doi.org/10.1016/j.apsusc.2012.05.143.
70. Eo DR, Chung SG, Yang JH, Cho WT, Park SH, Cho JW. Surface modification of high-Mn steel via laser-DED: Microstructural characterization and hot crack susceptibility of clad layer. *Mater Des* 2022;223:111188. https://doi.org/10.1016/j.matdes.2022.111188.
71. Chen Y, Lu F, Zhang K, Nie P, Elmi Hosseini SR, Feng K, et al. Dendritic microstructure and hot cracking of laser additive manufactured Inconel 718 under improved base cooling. *J Alloys Compd* 2016;670:312–21. https://doi.org/10.1016/j.jallcom.2016.01.250.
72. Liu S, Sun Y, Zhai P, Fan P, Zhang Y, Li M, et al. Microstructure and properties of nickel-based gradient coatings prepared using cold spraying combined with laser cladding methods. *Materials (Basel)* 2023;16:1627. https://doi.org/10.3390/ma16041627.
73. Dmitrieva A, Klimova-Korsmik O, Gushchina M, Korsmik R, Zadykyan G, Tukov S. Effect of the laser cladding parameters on the crack formation and microstructure during nickel superalloy gas turbine engines repair. *Metals (Basel)* 2023;13:393. https://doi.org/10.3390/met13020393.
74. Zhang Z, Zhao Y, Chen Y, Su Z, Shan J, Wu A, et al. The role of the pulsed-wave laser characteristics on restraining hot cracking in laser cladding non-weldable nickel-based superalloy. *Mater Des* 2021;198:109346. https://doi.org/10.1016/j.matdes.2020.109346.
75. Meng L, Sheng P, Zeng X. Comparative studies on the Ni60 coatings deposited by conventional and induction heating assisted extreme-high-speed laser cladding technology: Formability, microstructure and hardness. *J Mater Res Technol* 2022;16:1732–46. https://doi.org/10.1016/j.jmrt.2021.12.110.
76. Luo KY, Jing X, Sheng J, Sun GF, Yan Z, Lu JZ. Characterization and analyses on micro-hardness, residual stress and microstructure in laser cladding coating of 316L stainless steel subjected to massive LSP treatment. *J Alloys Compd* 2016;673:158–69. https://doi.org/10.1016/j.jallcom.2016.02.266.
77. Ghaedamini R, Ghassemi A, Atrian A. A comparative experimental study for determination of residual stress in laminated composites using ring core, incremental hole drilling, and slitting methods. *Mater Res Express* 2019;6:025205. https://doi.org/10.1088/2053-1591/aaee46.

78. Li M, Huang K, Yi X. Crack formation mechanisms and control methods of laser cladding coatings: A review. *Coatings* 2023;13:1117. https://doi.org/10.3390/coatings13061117.
79. Chen J, Wang SH, Xue L. On the development of microstructures and residual stresses during laser cladding and post-heat treatments. *J Mater Sci* 2012;47:779–92. https://doi.org/10.1007/s10853-011-5854-4.
80. Roy T, Paradowska A, Abrahams R, Law M, Mutton P, Soodi M, et al. Residual stress in laser cladded heavy-haul rails investigated by neutron diffraction. *J Mater Process Tech* 2020;278:116511. https://doi.org/10.1016/j.jmatprotec.2019.116511.
81. Hao M, Sun Y. A FEM model for simulating temperature field in coaxial laser cladding of Ti6Al4V alloy using an inverse modeling approach. *Int J Heat Mass Transf* 2013;64:352–60. https://doi.org/10.1016/j.ijheatmasstransfer.2013.04.050.
82. Kumar S, Roy S. Development of theoretical process maps to study the role of powder preheating in laser cladding. *Comput Mater Sci* 2006;37:425–33. https://doi.org/10.1016/j.commatsci.2005.10.005.
83. Chew Y, Hock J, Pang L, Bi G, Song B. Thermo-mechanical model for simulating laser cladding induced residual stresses with single and multiple clad beads. *J Mater Process Tech* 2015;224:89–101. https://doi.org/10.1016/j.jmatprotec.2015.04.031.
84. Heigel JC, Michaleris P, Reutzel EW. Thermo-mechanical model development and validation of directed energy deposition additive manufacturing of Ti-6Al-4V. *Addit Manuf* 2015;5:9–19. https://doi.org/10.1016/j.addma.2014.10.003.
85. Nian L, Wang M, Ge X, Wang X, Xu Y. Thermo-mechanical coupling numerical simulation for extreme high-speed laser cladding of chrome-iron alloy. *Coatings* 2023;13:1–15.
86. Kang L, Zhang C, Bradford MA, Liu X. Residual stresses in circular steel tubular columns repaired by laser-cladding additive manufacturing. *Thin Walled Struct* 2023;193:111275.
87. Wu S, Liu Z, Gong Y, Liang X, Wu Y. Analysis of the sequentially coupled thermal–mechanical and cladding geometry of a Ni60A-25%WC laser cladding composite coating. *Opt Laser Technol J* 2023;167:109595.
88. Narayanan A, Mostafavi M, Pirling T, Kabra S, Lewis R, Pavier MJ, et al. Residual stress in laser cladded rail. *Tribol Int* 2019;140:105844.
89. De Oliveira U, Ocelík V, De Hosson JTM. Residual stress analysis in Co-based laser clad layers by laboratory X-rays and synchrotron diffraction techniques. *Surf Coat Technol* 2006;201:533–42. https://doi.org/10.1016/j.surfcoat.2005.12.011.
90. Nie Z, Wang G, McGuffin-Cawley JD, Narayanan B, Zhang S, Schwam D, et al. Experimental study and modeling of H13 steel deposition using laser hot-wire additive manufacturing. *J Mater Process Technol* 2016;235:171–86. https://doi.org/10.1016/j.jmatprotec.2016.04.006.
91. Ni C, Composite M. Defect formation mechanism and performance study of laser cladding Ni/Mo composite coating. *Coatings* 2021;11:1460.

10

Impact of Alloying Element on Wear Resistance through Laser Cladding

T. Satish Kumar and Titus Thankachan

10.1 Introduction

Surface modification of materials via different techniques has been adopted to achieve modified surface properties without compromising their bulk properties based on the application. When it comes to surface modification of metals, laser technology is widely used due to the high energy density of laser along with their coherence and directionality. Mainly two methods, laser remelting and laser cladding, have proven to be effective techniques for improving the surface characteristics of different metals [1–4]. However, owing to the high production efficiency, interface bonding and deposition process control, laser cladding has established itself as an excellent technique for coating process or surface modification. During laser cladding process, the surface modification of the substrate metal is carried out by melting the pre-deposited powder and then rapidly solidifying it. This would form a coating which will be having a metallurgical bond with the metal. Through the laser cladding process, the substrate metal will be having a fine microstructure with excellent tribological and corrosion characteristics. However, the alloying powder deposited onto the surface of the substrate metal has a great impact on defining the characteristics of the modified surface [5–8]. When melted at the surface of the metal, alloying powders undergo a metallurgical reaction with the substrate metal, leading to the formation of non-equilibrium phases based on which the characteristics of the modified surface tend to vary. Considering the various factors that affect the quality of a cladded surface as shown in Figure 10.1, the material characteristics for both the substrate and alloying powders play a major role in defining the clade quality along with fixturing design, process parameters and component design.

Most commonly used laser clad materials in research include molybdenum, chromium, cobalt, nickel and so on, which have been found to enhance the

DOI: 10.1201/9781032693576-10

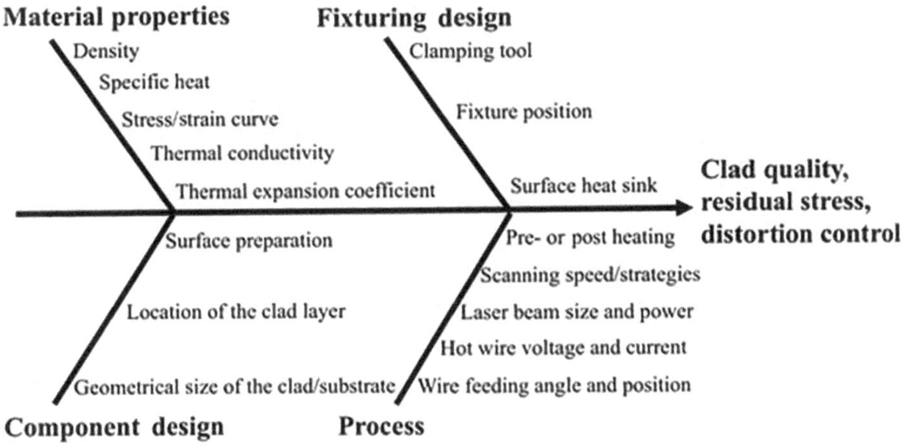

FIGURE 10.1
Factors influencing the clad quality [9].

surface properties of the substrate metal either through the formation of metal ceramic composites on the surface or by altering the phases in coating and interface. Considering the commercially used laser clad materials, the most commonly used are iron-based, nickel-based and cobalt-based alloy powders. Iron-based alloy powders, when introduced to the steel-based substrate, create hard phases, thereby enhancing the abrasive wear resistance of the substrate metal [10–13]. Nickel-based alloy powders, due to their good wettability, corrosion and wear resistance along with moderate price and availability, make them a candidate material for the laser cladding technique. Alloy element used in the laser cladding process will be identified on the basis of where the particular substrate metal is going to be used along with its cost and availability. Addition of each alloying element has different purposes while laser cladding like grain refining, wear resistance, corrosion resistance and thermal characteristics, proper alloy powders has to be identified for the process. When it comes to enhancing the thermal characteristics of a substrate metal, the most commonly used laser clad materials include tungsten (W), molybdenum (Mo), chromium (Cr), vanadium (V), boron (B), silicon (Si), nickel (Ni), cobalt (Co), niobium (Nb), titanium (Ti) and so on. This chapter provides a detailed explanation of these alloy powders in defining the wear resistance of a substrate metal while laser cladding is provided. A discussion on the different elements used for the laser cladding purpose, including high-entropy alloys, is shown in Figure 10.2.

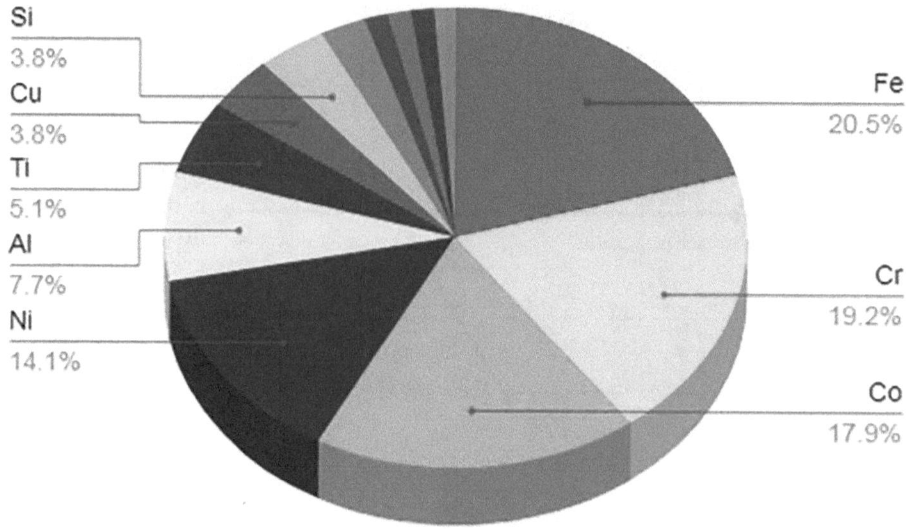

FIGURE 10.2
Elements used in laser cladding based on literature [14].

10.2 Effect of Chromium or Tungsten Addition on Wear Resistance

Chromium as an alloying element, when laser cladded on an iron-based alloy, enhances the mechanical and corrosive characteristics of the material. The formation of a chromium atom-based solid solution promotes the formation of a chromium oxide layer, which enhances the corrosive and tribological characteristics of materials. The introduction of chromium initially enhances the hardness of the substrate metal, followed by a decreasing trend; further addition of chromium leads to an increased hardness value. This initial trend in hardness enhancement with respect to increase in chromium can be ascribed to serious lattice distortion that happens in the alpha iron as chromium atoms dissolve in the alpha iron and the presence of residual austenite present on the surface. As with further increase in chromium, residual austenite tends to increase in the laser cladded surface, reducing the hardness until chromium composition gets increased such that formation to carbides happens increasing hardness of cladded layer. The wear resistance of a material, which has an inverse relationship with the hardness of the material, follows an opposite trend in wear rate with chromium addition.

Tungsten, when added as an alloying element onto a steel substrate metal, combines with carbon element and forms tungsten carbide, which enhances the hardness of the cladded surface. With the addition of tungsten, tungsten atoms dissolve in the substrate matrix, leading to solid solution strengthening and thereby enhancing material's hardness. Grain boundary refinement also occurs with tungsten addition, inhibiting dislocation motion and enhancing grain boundary density; this also enhances surface hardness of the cladded surface. As the presence of tungsten enhances the hardness of substrate, wear resistance of laser cladded material tends to decrease. Along with this, formed tungsten carbide forms a layer with low shear strength, which in turn helps in reducing the wear rate during the wear process.

10.3 Effect of Molybdenum or Vanadium Addition on Wear Resistance

Considering the addition of molybdenum (Mo) onto steel via laser cladding, the formation of martensite or bainite phase occurs, enhancing the hardness of the modified surface. This upsurge in hardness can be ascribed as a major factor for the increase in the wear resistance of the laser cladded surface with Mo addition, as shown in Figure 10.3. Along with the formation of martensite

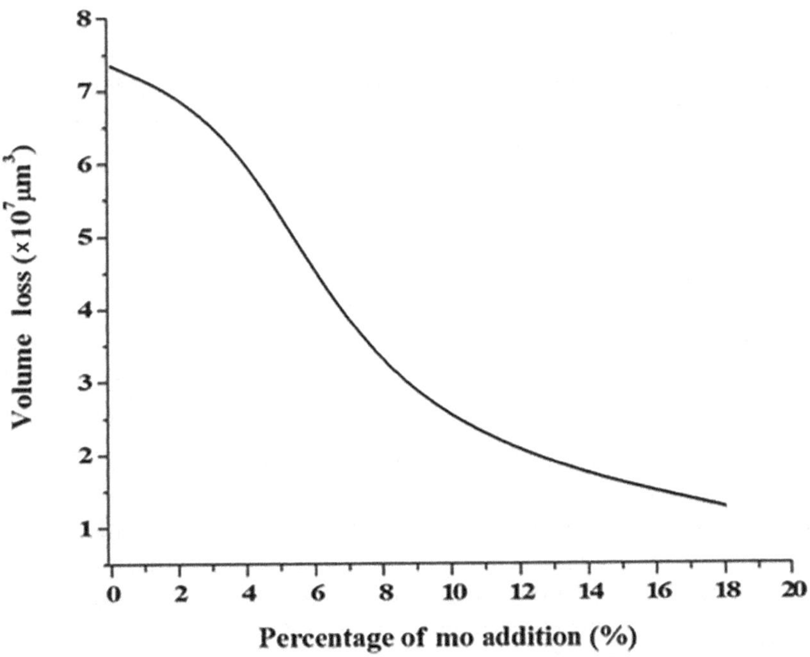

FIGURE 10.3
Impact of molybdenum content on wear resistance.

or bainite phases, the formation of molybdenum carbide, a hard phase on the surface of the laser cladded surface, can also be proposed as a major reason for reduction in volume loss with respect to increase in molybdenum addition. During laser cladding, addition of Mo refines the grain size of the coated surface, which in turn enhances wear resistance of the modified surface.

Laser cladding was carried out by Wang et al. [15] on a 45-grade steel substrate with dimensions of 150×150×12 mm, onto which molybdenum particles are deposited. Laser cladding was carried out at 2000 W laser power at a scanning rate of 240 mm/min and the powder fed at a frequency of 15 g/min. Microstructural evaluation proposed eutectic structures between dendrites and a reduction in dendrite size was observed with an increase in molybdenum addition. Molybdenum addition have an impact on the laser cladded surfaces nucleation process, hinder austenite growth thereby refining the structure and leads to lattice distortion by dissolving in cladded surface. Studies proved that $M_{23}(C,B)_6$ content increases with increase in Mo addition, while $M_7(C,B)_3$ content decreases. The wear resistance of the cladded surface tends to increase with respect to an increase in molybdenum addition. This increase in wear resistance can be credited to grain refinement that occurs with respect to molybdenum addition. Additionally, the formation of molybdenum carbide also tends to enhance wear characteristics of cladded surface. It was also identified that wear mechanism shifted from abrasive wear to adhesive wear with respect to addition of molybdenum element, which can also be a major reason for reduction in wear rate.

With the addition of vanadium to the surface of a substrate metal, the hardness of the laser cladded surface tends to decrease initially and then increase. This reduction in hardness initially can be attributed to the burning out of vanadium element reducing the hardness of the substrate metal. With further increase in vanadium addition, the hardness of the laser clades surface tends to increase correspondingly. This increase in hardness with further addition of vanadium can be due to the solid solution strengthening and particle strengthening. Along with this, the presence of many intermetallic compounds would enhance the hardness of the laser cladded surface by hindering the movement of dislocations. Furthermore, the addition of vanadium leads to more refined microstructure and inhibits the segregation of elements, thereby changing the Laves phase morphology, leading to property enhancement. This variation in hardness directly affects the wear characteristics of the modified laser cladded surface as with an increase in hardness; the wear resistance of the materials tends to increase.

A high-entropy alloy (HEA) coating with varying vanadium content was used by Liao et al. [16] on a 904 L stainless steel substrate through laser cladding to augment the wear characteristics and service life of an agitator blade. AlCoCrMoVx HEA was considered in this work with varying x(0, 0.2, 0.4, 0.6, 0.8, 1.0) at different atomic percentages, and the chemical conformation of substrate (50×10×10 mm) is provided in Table 10.1. This HEA with varying vanadium preposition was prepared via a RC-LMS-6000R fiber laser

TABLE 10.1

Chemical Composition of Substrate

Cr	Ni	Mo	C	Mn	Si	P	S	Cu	Fe
19–23	23–28	4–5	≤0.02	≤2	≤2	≤0.045	≤0.035	1–2	Bal

machine, wherein the powder was placed above the substrate at a thickness of 1.5 mm. Laser cladding was carried out at a 3.5 kW laser output power, with a scanning rate of 5 mm/s and argon flow rate of 5 L/min.

Microstructural evaluation of the developed AlCoCrMoV HEA coating with 0% vanadium content showed a needle-like microstructure with cellular dendrites and crack particles. With the addition of vanadium particles into the composition, the needle-like microstructure tends to decrease, forming irregular particles with an average size of 2–3 μm. Reduction in hardness value was observed with respect to increase in vanadium addition initially followed by an increase in hardness subsequently. This trend of reduction and increase in hardness value with respect to alloy addition can be attributed to the burning down of vanadium initially during laser cladding, thereby reducing the hardness value. Trend in the increase of microhardness value later is because of solid solution and particle strengthening. The presence of second-phase particles hinders the motion of dislocation, thereby enhancing the hardness value. With the addition of vanadium during laser cladding, the wear rate initially increases and then decreases. This variation in wear rate can be attributed to the incidence of alpha phase in HEA coating, which enhances the wear resistance of the cladded surface.

10.4 Effect of Boron and Silicon Addition on Wear Resistance

The addition of boron as laser clad powder onto an alloy steel substrate enhances the hardness of metal. This surge in hardness of substrate metal with boron introduction can be credited to the formation of hard boride-based phases. For instance, the formation of boron reacts with iron to form iron boride, which restrains grain growth and enhances its hardness, wear resistance and corrosion resistance. Similarly, boron reacts with other alloying elements of the steel, such as chromium, manganese and molybdenum, to form its borides, which in turn alter the characteristics of the modified surface. The wear resistance of a laser cladded alloy steel with respect to the increase in boron content is shown in Figure 10.4.

It is evident from Figure 10.4 that with a rise in boron addition on surface coating, the wear characteristics of the substrate metal tend to increase. This increase in wear resistance can be credited to the presence of boride phase

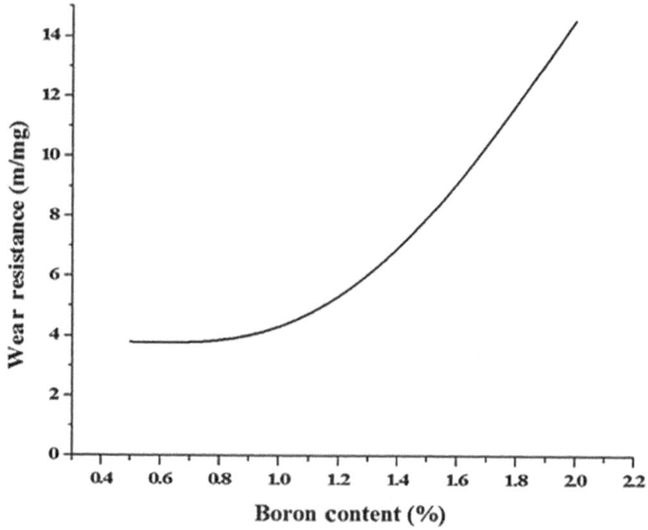

FIGURE 10.4
Impact of boron content on wear resistance.

present on the substrate along with its distribution and morphology. With an increase in the boride's volume fraction on the surface of the substrate metal, the hardness tends to intense, which in turn enhances the wear resistance of the modified surface. As the borides formed are distributed uniformly on the laser cladded surface, they form a network that resists the eradication of the substrate metal during the wear process. As the boride phase being brittle, with increase in of boron content increases the boride phases tend to get cracked and peeled off from matrix phase thereby dipping wear resistance of laser cladded substrate.

FeCoCrNiB HEA with different boron content was deposited onto SA515 Gr60 steel alloy of dimensions 120×60×10mm through laser cladding process by Liu et al. [17]. Microstructural evaluation proposed that with an increase in boron content, a reduction in its dendritic and interdendritic structure was observed. M_2B phase tends to increase with the addition of boron element, while the dendrite phase tends to decrease. This increase in the hard boride phase tends to enhance the hardness of the cladded surface. With an increase in boron content, the hardness of the cladded surface tends to increase, which can also be ascribed to the incidence of nano-precipitates and grain refinement occurrence. Wear volume being inversely proportional with hardness value of cladded surface, wear rate of cladded surface tends to decrease with respect to boron addition.

As the silicon content increases, refinement of microstructure happens, and at the same time, a linear relationship between hardness and silicon content can be observed in different research. The increase in hardness value with

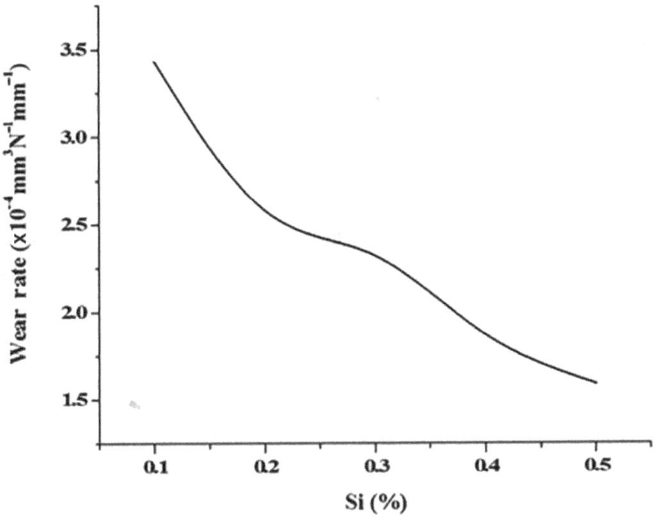

FIGURE 10.5
Impact of silicon content on wear resistance.

respect to silicon addition can be attributed to the outcome of dislocation strengthening apart from fine-grain refinement. This enhancement in hardness leads to increased wear resistance of the laser cladded surface, and the depiction of wear rate with respect to silicon addition is shown in Figure 10.5. Along with this, the presence of silicon dioxide and silicon oxide film acts as a lubricant, which reduces the wear rate of the modified surface. Silicon compounds present in the modified surface can also be proposed as one of the major factors that carry away load during wear, thereby reducing the wear rate of laser cladded surface.

Liu et al. [18] conducted a noteworthy work in which they used laser cladding technology to apply HEA coatings on the surface of AISI 304 stainless steel. These coatings featured variations in the silicon (Si) content, with x values ranging (x=0, 0.1, 0.2, 0.3, 0.4, 0.5), allowing for a comprehensive exploration of the alloy's properties across the Si spectrum. For this experiment, AISI 304 grade stainless steel plates as the chosen substrate material. These plates possess dimensions measuring 200, 300 and 15 mm as its length, width and thickness, respectively. High-quality elemental powders, each with a purity level exceeding 99.5%, were considered as primary raw materials in this study. These elemental powders consisted of aluminum (Al), chromium (Cr), iron (Fe), cobalt (Co), nickel (Ni) and silicon (Si). To create powders with desired nominal compositions of AlCoCrFeNiSix, denoted by the molar ratio x, various mixtures were meticulously prepared. The laser cladding experiments were executed with the assistance of a high-performance fiber laser, specifically the IPG YLS-4000. Several critical parameters were meticulously controlled during the process, including a laser spot diameter, scanning speed, powder feed rate, overlap rate and laser power output, which were maintained at 4 mm, 5 mm/s, 3.3 g/min, 50% and 2000 W, respectively. To ensure a controlled environment, high-purity argon gas was employed at a flow rate of 4 L/min as shielding gas throughout the laser cladding process. A comprehensive examination was undertaken to thoroughly investigate the coating's microhardness, wear behavior and microstructure evolution. The central aim was to gain insight into the influence of the silicon (Si) element on these aspects.

The results of the study unveiled several noteworthy findings. The HEA coatings were found to comprise two solid solution phases: ordered Al-Ni and disordered Fe-Cr, each with a body-centered cubic (BCC) structure. The introduction of the silicon (Si) element into the mix led to its dissolution into these solid solutions, causing lattice distortion. With an increase in Si content, a noticeable refinement of the microstructure occurred, accompanied by the precipitation of a limited extent of $Cr_{23}C_6$ phase along grain boundaries. Interestingly, microhardness of coatings exhibited a linear correlation with the Si content. Particularly, noteworthy is the fact that the average microhardness of HEA coating with a Si content of 0.5 reached its peak, registering an impressive 8.19 GPa value. A detailed study on the mechanisms behind the enhancement in microhardness revealed consequence of dislocation strengthening as the primary attribute, other than the contributions of solution strengthening or fine-grain strengthening. Moreover, as Si content increased, there was a substantial reduction in both the wear rate and average friction coefficient of the HEA coatings.

A notable observation from the study is the substantial reduction in friction coefficients in the HEA coatings that included silicon (Si) when compared to the Si-free HEA coating (Si0). As Si content increased, a minimal decrease in the wear rate was noted. Notably, the wear rate exhibited a more significant reduction when the concentration of Si rose from 0 to 0.2, after which the rate of decrease slowed down. The wear rate and friction coefficient of the coatings both significantly decreased with an increase in silicon (Si) content. Adhesive, abrasive and delamination wear were the main wear mechanisms seen in the Si0 and Si0.1 HEA coatings. However, in coatings with Si concentration ranging from 0.2 to 0.5, the predominant wear mechanisms shifted toward oxidative wear, with a minor presence of abrasive wear. What's particularly intriguing is the composition of the wear products generated during oxidative wear, which included Fe_2O_3, Fe_3O_4, Cr_2O_3, SiO and SiO_2. Among these, SiO and SiO_2 within the oxide film appeared to perform as effective lubricants during wear progression. This lubricating effect contributed to the enhanced wear resistance and a reduction in the friction coefficient observed in Si-containing HEA coatings. In essence, the presence of silicon compounds played a pivotal role in mitigating wear and improving the overall performance of these coatings.

10.5 Effect of Nickel or Cobalt Addition on Wear Resistance

Ni-based alloying powder is mainly used on parts where wear resistance, corrosion resistance and fatigue resistance are required. Nickel exhibits excellent wettability with high corrosion resistance and self-lubricating quality, making it an excellent alloying element for enhancing the wear resistance of substrate while laser cladded. The addition of hard nickel powder into steel substrate would considerably increase the hardness of the laser cladded surface. This increase in hardness with respect to the addition of nickel onto the steel substrate can be ascribed to the establishment of hard gamma phase and intermetallic compounds. The addition of nickel into steel through laser cladding has enhanced the wear rate of the modified surface by about 35 times. This enhancementin hardness can be credited to mixing of gamma nickel with the borides in the metal substrate to forming strong network of interdendritic eutectic structure enhancing the surface characteristics.

Feng et al. [19] conducted successful laser cladding of two different coatings, namely 55 NiTi coating and 55 NiTi+5 Ni coating, onto an SS316L substrate. The research aimed to investigate the impact of nickel introduction on the NiTi coating and understand the association between its microstructure and surface characteristics. To achieve this, raw materials in the form of Ni and 55 NiTi powders with a purity of 99.9% were chosen. These powders were combined by taking pure Ni powder and 55 NiTi alloy powder at a weight ratio of 1:8 and then mixed so that a total weight ratio of 60% Ni and 40% Ti was attained. The mixing process was carried out in a planetary ball mill operated for a time period of 8 hours at 250 rpm. As the substrate, a 100×100×10 mm 316 L stainless steel was utilized.

Specific parameters were used to build the laser cladding coatings on the substrate: a powder feeding rate of 50 rpm, a laser scanning speed of 2 mm/s and a power of 2.0 KW. A ball-on-plate conformation was employed for the wear experiment, utilizing Si balls of diameter 5 mm at room temperature, with a 15 N load and 224 r/min rotational speed. The wear test samples were sliced into 30 mm by 30 mm pieces. Energy dispersive X-ray spectroscopy (EDS) and X-ray diffraction (XRD) studies showed that NiTi and Ni3Ti were the main phases in the coatings. Additionally, other phases such as Ni_4Ti_3, Fe_2Ti and $NiTi_2$ were detected, with $NiTi_2$ being exclusive to the 55 NiTi coating. The cladding layer's hardness was approximately four times higher than the substrate's, with a range of 770–830 HV. Moreover, 55 NiTi+5 Ni coating exhibited hardness approximately 8% greater than the 55 NiTi coating.

Furthermore, the wear resistance characteristics of 55 NiTi+5 Ni coating outperformed the 55 NiTi coating, as indicated by a decrease in wear loss of about 13% and friction coefficient at a minimal level. These improvements were accredited to the solid solution strengthening outcome brought about by Ni introduction and second-phase strengthening consequence resulting from an increased content of the Ni_3Ti phase within the cladding layer.

Cobalt-based alloy powders, due to their low melting point and wettability, are considered a prominent alloying element in laser cladding, as they melts first and reacts with other elements in the substrate metal to form new phases during the solidification process, thereby enhancing the properties of the laser clades surface. Cobalt-based alloy powders exhibit excellent corrosion resistance, wear resistance and high temperature properties, making them a prominent alloying element for laser cladding in metallurgical, aerospace, petroleum, chemical industries, etc.

A decrease in wear was observed with an increase in the addition of cobalt as an alloying powder during laser cladding process, as shown in Figure 10.6. This increase in wear resistance with respect to the increase in cobalt content can be attributed to increase in hardness. Cobalt tends to dissolve in the gamma phase resulting in the solid solution strengthening of matrix metal enhancing wear resistance of substrate metal. The addition of cobalt into the substrate through laser cladding process also leads to enhanced high temperature characteristics of the substrate, thereby reducing the wear rate of the substrate metal after heating up of the cladding wear while wear. Studies have also proven that the addition of cobalt through laser cladding shifted the wear mechanism from abrasive to adhesive wear, and the adhesive wear debris tends to get oxide during wear tests. These oxide residues act as solid lubricants, which attribute to the reduction in the wear rate of the cladding layer during tribological tests.

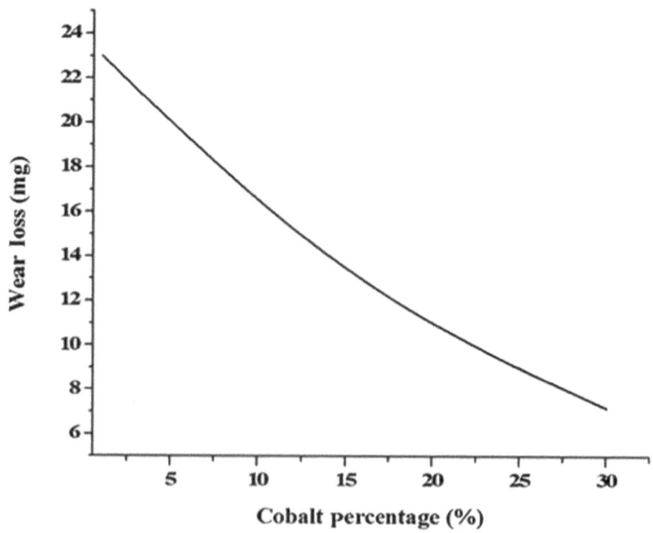

FIGURE 10.6
Impact of cobalt content on wear resistance.

Various nickel-based alloy powders, featuring varying cobalt concentrations, were tested using the fiber laser method on a 42CrMo steel substrate surface by Kaiming Wang et al. [20]. The study delved into how the cobalt concentration influenced the composition, hardness, wear and microstructure characteristics of these coatings. Three distinct Ni-based alloy powders were employed, each comprising chromium of 7.01 weight percentage, carbon of 0.30 weight percentage, boron of 1.59 weight percentage, silicon of 3.05 and iron of 3.24 weight percentage with the remainder as Ni. The alloy powders had differing Co proportions of 0 wt. % (denoted as NiCo-00), 15 wt. % (NiCo-15) and 30 wt. % (NiCo-30) in this investigation. The selected substrate material was 42CrMo steel, which has the following composition: 0.43 weight percent C, 1.17 weight percent Cr and 0.21 weight percent Mo. The substrate specimens were 150 × 50 × 10 mm in size.

The findings reveal that predominant phases within the cladding layers consist of $M_{23}(C, B)_6$, M_2B, $M_7(C, B)_3$ and γ. With increase in cobalt concentration, there is a gradual reduction in the quantities of M_2B, $M_7(C, B)_3$ and $M_{23}(C, B)_6$, along with a gradual decrease in eutectic structure width observed in the cladding layer. Notably, there is no distinct cobalt phase observed within cladding layers following the introduction of cobalt. This can be attributed to the similarity in atomic radii between Co and Ni, which facilitates the easy dissolution of Co into the Ni matrix. The microhardness experiences a decrease; however, when the Co concentration increases, the cladding layer's resistance to wear gets better over time. In particular, the NiCo-30 cladding layer has 3.6 times more wear resistance than the NiCo-00 cladding layer. The cladding layer experiences an abrasive wear process that changes to an adhesive wear mechanism as the Co concentration increases. Notably, weight loss of the cladding layer exhibits a gradual decline as the Co content increases. The wear resistance is most pronounced in the NiCo-30 cladding layer, surpassing that of the NiCo-00 cladding layer by approximately 3.6 times. Evidently, incorporation of Co into Ni-based alloy brings about a substantial enhancement in - wear resistance of - cladding layer.

To investigate the worn morphologies of cladding layers with varying Co contents, SEM studies were carried out. In these observations, deep furrows were noted on the worn surface of the cladding layer, which is a distinctive indicator of abrasive wear. However, with the introduction of the cobalt element into the Ni-based cladding layer, these furrows on the worn-out surface became noticeably shallower. Furthermore, worn surfaces exhibited the presence of debris, and the quantity of debris increased significantly with the escalating Co content. This observation points to an adhesive wear mechanism replacing abrasive wear as the main cladding layer wear mechanism. In conclusion, it was discovered that an increase in Co content affected the cladding layer's wear mechanism, causing it to shift from abrasive to adhesive wear.

10.6 Effect of Niobium or Titanium Addition on Wear Resistance

Niobium as an alloying element in steel enhances the mechanical properties of the metal due to its capability in grain size refinement and precipitation strengthening process. Niobium, a carbide-forming element, enhances the hardness of the substrate metal when laser cladded. Studies proposed that an increase in hardness occurs with an increase in niobium element, as a BCC crystal structure yields more hardness than a face-centered cubic structure. Studies also stated that the larger atomic radius of niobium induces a lattice distortion, which leads to solid solution strengthening. Based on various research, it was also indicated that the addition of niobium will stimulate the formation of martensite nuclei, thereby increasing the proportion of martensite in the cladded layer. With the presence of martensite, it can be proposed that the hardness and hardenability of the cladded material enhances. Along with this, the formation of the Laves phase with niobium addition will change the phase morphology of the laser cladded surface, thereby inducing strength and hardness of the substrate metal. It is evident that the addition of niobium would enhance the hardness of the substrate metal, which in turn enhances the wear resistance of the laser cladded surface.

Qi et al. [21] conducted a study involving cobalt-based alloys with varying levels of niobium (Nb) content. These alloys were prepared using a laser cladding technique assisted with an alternating magnetic field. Importantly, the presence of this magnetic field was found to have a momentous impact on the performance of the laser cladding layer. The researchers chose 42CrMo steel as the substrate for the laser cladding process. Different powder sizes were used for various materials: Cr powders with 100–200 mesh, Nb powders with 300–500 mesh, boron powders were about 800 mesh and carbon powders were roughly 200 mesh. During the laser cladding process, specific parameters were employed, including a 1300 W laser power at 3 mm/s scanning speed and a 3 mm spot size. They fabricated four samples, labeled C_1, C_3, C_5 and C_7, without using a magnetic field (0 mT). Additionally, four more samples, designated C_2, C_4, C_6 and C_8, were created in the presence of magnetic field, which is about 20 mT.

In this study, the investigation focused on the wear behavior of the cladding layer under consistent friction experimental conditions. Average coefficients of friction for the various laser cladding samples, namely C_1, C_2, C_3, C_4, C_5, C_6, C_7 and C_8, were found to be about 0.58, 0.56, 0.60, 0.55, 0.72, 0.65, 0.56 and 0.63, correspondingly. An observation was made regarding the coefficient of friction values of the laser cladding layer when assisted by an alternating magnetic field of 20 mT. It was noted that this coefficient of friction was lesser compared to the coefficient of friction of the laser cladding layer without alternating magnetic field assistance. Furthermore, the C_8 laser cladding layer exhibited the highest hardness among all the samples. However, it is worth highlighting that the coefficient of friction for C_8 was not the lowest among all the samples. This suggests that hardness alone is not the sole determinant of the wear property of the laser cladding layer. Another important factor in determining wear performance is the elastic modulus of the laser cladding layer. Several theoretical and experimental studies have indicated that wear properties are related to the ratio of hardness (H) and elastic modulus (E), with higher H/E values often indicating better wear-resistant performance. The H/E values for the C_1, C_2, C_3, C_4, C_5, C_6, C_7 and C_8 laser cladding layers were determined to be 0.051, 0.053, 0.042, 0.057, 0.037, 0.041, 0.051 and 0.056, respectively. Surprisingly, the C1 cladding layer, which had the highest H/E values compared to C_3 and C_5, displayed the lowest wear resistance. This observation was attributed to the presence of a significant spalling crater on the wear surface of C_1, which increased the friction of the cladding layer. Consequently, it was concluded that H/E values alone may not accurately characterize the wear resistance of the cladding layer in this context.

Among all the samples, it was observed that the C_4 laser cladding layer consistently displayed the lowest friction coefficient. This was attributed to the fact that C_4 had the highest H/E ratio among all the cladding layers. The magnetic field's stirring effect increased the hardness of the cladding layer while reducing the material's elastic modulus, resulting in reduced frictional behavior. When examining the wear morphologies of the C_1, C_3 and C_5 cladding layers, common features included deep plow grooves, severe spalling and plastic deformation on the wear surface of the laser cladding layer. In particular, the wear morphology of the C_5 cladding layer exhibited numerous microcracks, which could be attributed to its low hardness and high elastic modulus. These factors led to severe wear under cyclic contact stresses, resulting in significant plastic deformation. In contrast, the C_6 laser cladding layer showed an improved wear profile compared to C5, tending to be flatter. However, both C_6 and C_8 exhibited peeling pits in the NbB (niobium boride) material, which increased the coefficient of friction. Conversely, the C_2 and C_4 laser cladding layers exhibited more favorable wear morphology due to their reduced elastic modulus and increased hardness. The above proposed mechanical characteritics effectively reduced cutting when in contact with Si_3N_4 ceramic balls, consequential in a shallower wear surface with substantial decline in spalling crater presence. C_7 cladding layer also displayed a flat wear profile and can be attributed to the presence of a ceramic phase within the clad layer. This ceramic phase helped to diminish the contact area between the metal part of the cladding layer and Si_3N_4, subsequently reducing the coefficient of friction at the laser cladding layer.

The addition of titanium as alloying powder through laser cladding enhances the hardness of the substrate metal. It was also observed that titanium incorporation retarded the growth of dendrites, forming fine and equiaxed grains. Titanium addition also hinders cross-slip dislocation and jog formation. Considering the wear resistance of the laser cladded substrate

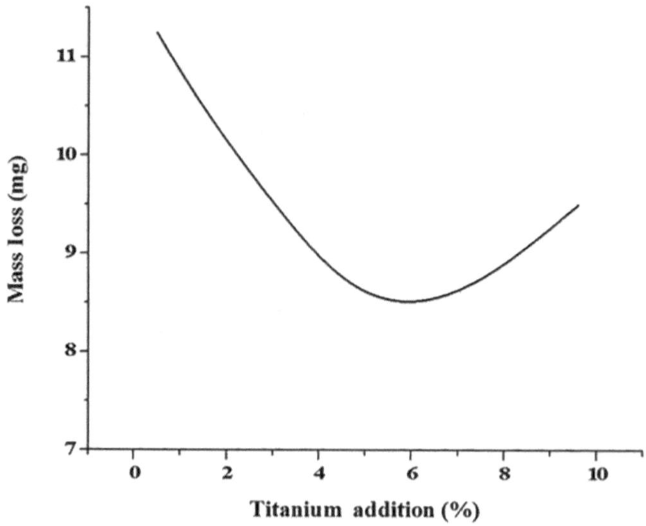

FIGURE 10.7
Impact of titanium content on wear resistance.

with titanium as the major alloying element, studies proved that an enhancement in wear resistance was experiential with the addition of titanium, and after a certain percentage of titanium addition, the wear resistance tends to decrease, which is represented in Figure 10.7. The abovementioned increase in wear resistance with the introduction of titanium content can be ascribed to the formation of carbides and other secondary phases, which would enhance the hardness and wear resistance of the material. The formation of martensitic or bainitic structure during the laser cladding process can also be proposed as a major factor for hardness and wear resistance enhancement. However, studies portrayed a reduction in wear resistance was observed for the substrate metal after a certain percentage of titanium addition. This reduction can be attributed to the increase in spalling and pores present on the laser cladded surface, which leads to the peeling away of the metal.

Jing et al. [22] conducted a study on Ni-35/tungsten carbide coatings with varying titanium content prepared through laser melting in different environments. The coating material considered in this research consisted of a blend of Ni-35 alloy powder and spherical-shaped tungsten carbide powder. The ratio of mass fraction was 6.5:3.5, indicating that Ni35 alloy powder accounted for 65% of the mixture and WC powder accounted for 35%. Pure Ti powder was added to the Ni-35/WC mixture with varying mass fractions. Three different Ti content levels were studied: 0%, 4% and 8%. The substrate used for coating deposition was 65Mn alloy steel. Specimens considered in this research had measurements of 50×50×10mm. The Ni-35 powder had a particle size distribution in the range of 89–104μm. The WC powder of 99% purity with a 53–74μm particle size range was used in this research. Similarly, the Ti powders used were 99% pure with an aggregate size of 53–74 μm. Coatings were prepared under two different working conditions: an oxygen-free environment and a normal atmospheric environment. The presence of titanium in the coating fabrication process can lead to a splash phenomenon, which affects the coating's quality. This splash phenomenon is more pronounced at atmospheric working conditions and results in a high dilution rate and obvious defects on the coating's exterior. Increasing titanium content in the coating leads to an intensification in the degree of spalling and the number of pores. This indicates that controlling the titanium content is crucial for achieving a high-quality coating. The splash effect that titanium causes during the coating process is lessened when laser cladding is done in an oxygen-free atmosphere. This results in a minimal dilution rate and a smoother coating surface with no outward defects. The Ni-35/35%WC coating chiefly consists of γ-(Ni,Fe) solid solution and carbides such as $Cr_{23}C_6$, M_6C, WC and W_2C. The addition of Ti in the coating synthesis leads to the in-situ synthesis of TiC in large numbers. Some titanium also exists as (Ti, W)C1–*x*. This addition of Ti inhibits the formation of other carbides and refines WC particles within the coating, resulting in a further uniform distribution. Laser cladding carried out in oxygen-free conditions results in a coating with an obvious lath martensite microstructure. This modification enhances the coating's hardness and toughness. While the microhardness of the coating with titanium addition decreases slightly compared to the coating without Ti, the distribution of microhardness becomes more uniform. In friction and wear tests, Ni-35/35%WC coating experiences alternating oxidative wear and spalling. Ti incorporation lessens coating spalling and delamination brought on by oxidative wear and micro flaws. As a result, the wear resistance of the coating gets improved. Laser cladding carried out in an oxygen-free environment leads to a denser coating and maintains further actual titanium concentration. Coatings fabricated using this method exhibit a minimal wear rate with superior friction stability.

10.7 Conclusion

A detailed study on the role of each commonly used alloying element on the wear characteristics of the substrate is carried out in this study. The role of chromium, tungsten, molybdenum, vanadium, boron, silicon, nickel, cobalt, niobium and titanium is explained in this chapter. Even though certain deviation like reduction in wear rate happens at the initial stage, research has proved that the addition of the abovesaid alloying element would enhance the mechanical and tribological characteristics of the cladded surface. This increase in wear resistance is observed in most of the scenarios as the incorporation of the abovesaid alloying elements led to grain refinement of the cladded surface, and along with that, solid solution strengthening and particle strengthening were also observed. The presence of intermetallic compounds and secondary phases performed an inevitable part in defining the hardness enhancement of the cladded surface and thereby reducing the wear rate of the developed cladded substrate.

References

1. Zhang, Z., Farahmand, P. and Kovacevic, R., 2016. Laser cladding of 420 stainless steel with molybdenum on mild steel A36 by a high power direct diode laser. *Materials & Design*, 109, pp. 686–699.
2. Kaiming, W., Yulong, L., Hanguang, F., Yongping, L., Zhenqing, S. and Pengfei, M., 2018. A study of laser cladding NiCrBSi/Mo composite coatings. *Surface Engineering*, 34(4), pp. 267–275.
3. Wang, C., Gao, Y., Zeng, Z. and Fu, Y., 2017. Effect of rare-earth on friction and wear properties of laser cladding Ni-based coatings on 6063Al. *Journal of Alloys and Compounds*, 727, pp. 278–285.
4. Yin, Y., Pan, C., Zhang, R., Zhao, C. and Qu, Y., 2018. The effect of Ti addition on the microstructure and properties of high chromium iron-based coatings. *Journal of Alloys and Compounds*, 765, pp. 782–790.
5. Gao, W., Zhang, Z., Zhao, S., Wang, Y., Chen, H. and Lin, X., 2016. Effect of a small addition of Ti on the Fe-based coating by laser cladding. *Surface and Coatings Technology*, 291, pp. 423–429.
6. Tao, X.P., Zhang, S., Zhang, C.H., Wu, C.L., Chen, J. and Abdullah, A.O., 2018. Effect of Fe and Ni contents on microstructure and wear resistance of aluminum bronze coatings on 316 stainless steel by laser cladding. *Surface and Coatings Technology*, 342, pp. 76–84.
7. Lai, Q., Abrahams, R., Yan, W., Qiu, C., Mutton, P., Paradowska, A. and Soodi, M., 2017. Investigation of a novel functionally graded material for the repair of premium hypereutectoid rails using laser cladding technology. *Composites Part B: Engineering*, 130, pp. 174–191.
8. Liu, S. and Pang, M., 2021. Effect of TiB2 Content on properties of nickel-coated graphite self-lubricating coating prepared by laser cladding. *Coatings*, 11(12), p. 1501.
9. Yao, M., Kong, F. and Tong, W., 2023. A 3D finite element analysis of thermally induced residual stress distribution in stainless steel coatings on a mild steel by laser hot wire cladding. *The International Journal of Advanced Manufacturing Technology*, 126(1–2), pp. 759–776.
10. Jeyaprakash, N., Yang, C.H. and Sivasankaran, S., 2020. Laser cladding process of Cobalt and Nickel based hard-micron-layers on 316L-stainless-steel-substrate. *Materials and Manufacturing Processes*, 35(2), pp. 142–151.
11. Wang, C., Zhang, S., Zhang, C.H., Wu, C.L., Zhang, J.B. and Abdullah, A.O., 2018. Phase evolution and wear resistance of in situ synthesized V8C7 particles reinforced Fe-based coating by laser cladding. *Optics & Laser Technology*, 105, pp. 58–65.
12. Zhang, H., Zou, Y., Zou, Z. and Shi, C., 2014. Effects of chromium addition on microstructure and properties of TiC–VC reinforced Fe-based laser cladding coatings. *Journal of Alloys and Compounds*, 614, pp. 107–112.

13. Domitner, J., Silvayeh, Z., Buzolin, R.H., Krisam, S., Achterhold, K., Povoden-Karadeniz, E., Sommitsch, C. and Mayr, P., 2022. Microstructure characterization of nickel matrix composite reinforced with tungsten carbide particles and produced by laser cladding. *Advanced Engineering Materials*, 24(11), p. 2200463.
14. Arif, Z.U., Khalid, M.Y., urRehman, E., Ullah, S., Atif, M. and Tariq, A., 2021. A review on laser cladding of high-entropy alloys, their recent trends and potential applications. *Journal of Manufacturing Processes*, 68, pp. 225–273.
15. Wang, K., Chang, B., Chen, J., Fu, H., Lin, Y. and Lei, Y., 2017. Effect of molybdenum on the microstructures and properties of stainless steel coatings by laser cladding. *Applied Sciences*, 7(10), p. 1065.
16. Liao, T., Wang, Z., Wu, X., Liu, Q., Guo, Y., Ding, K. and Shang, X., 2023. Effect of V on microstructure, wear and corrosion properties in AlCoCrMoVx high entropy alloy coatings by laser cladding. *Journal of Materials Research and Technology*, 23, pp. 4420–4431.
17. Liu, D., Zhao, J., Li, Y., Zhu, W. and Lin, L., 2019. Effects of boron content on microstructure and wear properties of FeCoCrNiBx high-entropy alloy coating by laser cladding. *Applied Sciences*, 10(1), p. 49.
18. Liu, H., Sun, S., Zhang, T., Zhang, G., Yang, H. and Hao, J., 2021. Effect of Si addition on microstructure and wear behavior of AlCoCrFeNi high-entropy alloy coatings prepared by laser cladding. *Surface and Coatings Technology*, 405, p. 126522.
19. Feng, Y., Du, Z. and Hu, Z., 2021. Study on the effect of Ni addition on the microstructure and properties of NiTi alloy coating on AISI 316L prepared by laser cladding. *Materials*, 14(16), p. 4373.
20. Wang, K., Chang, B., Lei, Y., Fu, H. and Lin, Y., 2017. Effect of cobalt on microstructure and wear resistance of Ni-based alloy coating fabricated by laser cladding. *Metals*, 7(12), p. 551.
21. Qi, K. and Yang, Y., 2022. Microstructure, wear, and corrosion resistance of Nb-modified magnetic field-assisted Co-based laser cladding layers. *Surface and Coatings Technology*, 434, p. 128195.
22. Jing, P., Wang, H., Chen, W., Chen, L., Yin, H., Wu, H. and Li, D., 2022. Effect of Ti addition on microstructure and tribological properties of laser cladding Ni35/WC coating in an oxygen-free environment. *Surface and Coatings Technology*, 440, p. 128480.

11

Effect of Laser Cladding on Fatigue Property

Natarajan Jeyaprakash, Sundara Subramanian Karuppasamy, and Che-Hua Yang

11.1 Introduction

Metallic alloys derived from iron (Fe), titanium (Ti), aluminum (Al), nickel (Ni), copper (Cu), and cobalt (Co) have been extensively used in the fabrication of metallic components that are employed in most industrial and service sectors [1–3]. These alloys have independent mechanical, physical, thermal, electrical, electrochemical, and chemical characteristics, which made these alloys as a mandatory manufacturing material in a wide range of applications including nuclear reactors, automobiles, boilers, fuel tanks, pressurized water reactors, power plants, and so on. These applications are said to work in severe harsh environments such as high temperature, high pressure, cyclic loading, dissolved corrosive salts such as chlorides, iodide, fluorides, and sulfides, dissolved oxygen such as marine environments, and so on [4,5]. After a long-term usage, it is reported that the previously mentioned factors could serve as the root cause for the reduction in the mechanical characteristics exhibited by these metallic alloys.

Among the above-mentioned factors, cyclic loading turns out to be a major issue that will lead to a decrease in the life span offered by the components. Under cyclic loading, the tubes that are used to carry hot steam in power plants and boilers experience crack formation on their surface. This might be due to the effect of high-temperature service environment which reacts with the tube's surface and renders its tensile properties thereby inducing crack formation [6,7]. On the other hand, the rotatory components of automobile and aerospace engines exhibit wear and tear due to cyclic motion (loading) after prolonged usage. This wear phenomenon initiates crack formation which leads to catastrophic failure [8,9]. The components that work in service environments containing high pressure and temperature experience stress-induced cracking which will minimize the fatigue behavior offered by those components. The residual stress developed in the working environments is reported as the root cause for diminishing the fatigue resistance offered [10,11]. The periodical (cyclic) loading between the rail tracks and rail wheels will pave the way for wear in the rail tracks and often cause major accidents in the railways. In power

DOI: 10.1201/9781032693576-11

plants, the blades of steam turbines experience erosion due to the high pressure and high temperature of the hot gas (steam) which is reported to diminish the fatigue resistance of the blades [12,13]. Moreover, the dies used in the casting of metallic parts are reported to exhibit failures because of the cyclic load, which initiates thermal cracking and affects its thermal fatigue resistance [14]. Hence, the reduction in fatigue resistance serves as the key issue in all the industrial sectors and this issue could lead to fatal accidents and property loss. Therefore, there is a need to find a solution to improve the fatigue behavior offered by the components/metallic alloys [15].

Surface engineering is a branch of engineering disciplines that focuses on treating worn-out or damaged surfaces thereby improving its service life. This discipline employs several advanced surface treatment techniques that could be able to repair or rebuild the surfaces with high precision [16]. These techniques can be categorized into many types based on the material type (treating surface), service environment, energy source, and so on. The major surface treatment techniques are briefed as follows. Vapor deposition belongs to the category of surface treatment techniques where a coating is obtained on the surface to be treated by depositing the coating materials in vapor form. The surface to be treated serves as the substrate and the coating particles are directed on the substrate's surface via vapors [17]. This technique is quite common and can be further classified into two types, namely, physical vapor deposition (PVD) and chemical vapor deposition (CVD). In PVD, the coating particles are deposited in a physical manner whereas in CVD, chemical methods are used to coat the coating particles on the substrate. Though PVD and CVD are quite common techniques, however, researchers reported that these techniques lack in repairing complex surfaces and requirement of skilled laborers, are time-consuming in approach, could emit harmful gases, which are highly volatile, and so on [18]. Thermal spraying is another surface treatment technique, which is employed in most industries to repair the damaged surface. Here, a thermal spray gun is used where the feedstock is preheated and directed on the substrate's surface in order to obtain a fine coating. The substrate can be either preheated or used at normal temperature to obtain the coating. However, only thin films can be obtained via thermal spraying, which are expensive and the coated layer is greatly influenced if high temperature is a key factor in its service environments [19,20].

The coatings prepared by the surface treatment techniques such as PVD, CVD, hardfacing, and thermal spraying are reported to have poor metallurgical bonding between the coating and the substrate. Hence, after a long time, the service environment makes these coatings brittle and minimizes the surface characteristics offered. In addition, the lifespan of these coatings tends to decrease due to poor metallurgical bonding [21]. Hence, the researchers work on several advanced surface treatment techniques that could improve the metallurgical bonding that could improve the service life of the coatings even in severe aggressive working environments.

The above-mentioned limitations faced by the traditional surface treatment techniques could be overcome by the laser-based surface treatments. The laser-based surface treatments are able to fabricate highly efficient coatings on the substrate's surface with good metallurgical bonding between the coated layer and the substrate [22–24]. It employs a high-energy laser, which is responsible for elevating the mechanical characteristics of the coating when compared to those coatings obtained by vapor deposition and thermal spraying methods. This laser-based surface treatment can be categorized into many types based on the laser energy and heat flow in the substrate's surface. Out of the various laser-based surface treatment techniques, laser cladding serves as the predominantly used technique for treating the damaged surfaces. It employs a high-energy laser beam which is used to melt the substrate's surface thereby forming a molten metal pool on the surface [25,26]. The coating particles were directed on this molten metallic pool and with the combined effect of heat, a fine coating was obtained on the substrate's surface. The clad layer obtained after the laser cladding process is reported to exhibit better metallurgical bonding between the substrate and the coating and also this layer exhibits better mechanical characteristics such as wear resistance, hardness, anti-corrosion behaviors, fatigue resistance, and so on. This chapter is focused on discussing and elaborating on the laser cladding principle, types of fatigue, fatigue failure mechanism, process parameters influencing the fatigue property, fatigue resistance improvement through laser cladding via cladding powders along with relevant literatures.

11.2 Laser Cladding

As mentioned, the laser-based surface treatments use the laser as the energy source to treat the repaired or worn-out surfaces in order to improve the service life offered by them. Since the heat transferred from the laser to the substrate's surface is so high, this could melt or vaporize the substrate's surface. Depending on the demanded substrate's surface temperature, the laser-based surface treatment can be broadly classified into three categories as shown in Figure 11.1. The laser cladding process belongs to the second category where the laser power is in the range between the substrate melting temperature and substrate vaporizing temperature. Here, a high-energy laser beam is employed to irradiate the sample's surface.

This irradiation could lead to the formation of the molten metallic pool on the substrate but not vaporized. The coating materials are directed to this molten melt pool where they get uniformly deposited on this pool and solidify to form the coating layer [27,28]. Some of the advantages of this process over traditional surface treatment techniques are as follows [29,30]:

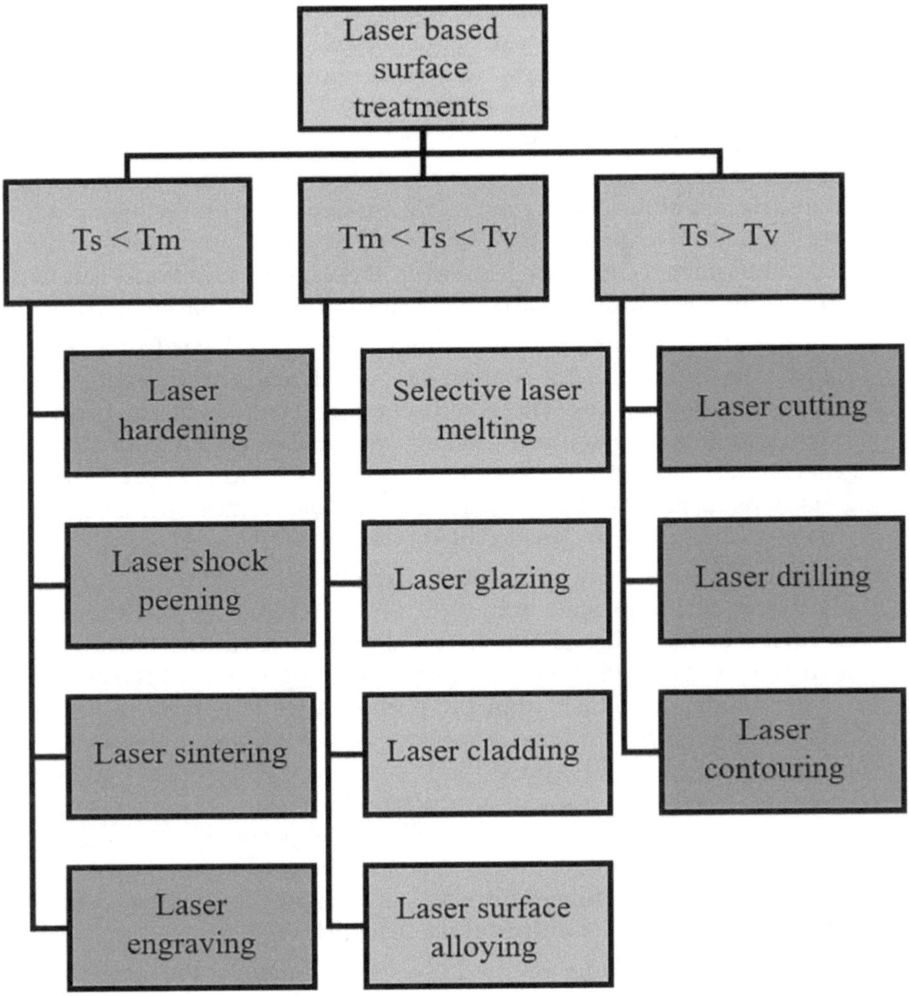

FIGURE 11.1
Classification of laser-based surface treatments where Ts is the surface temperature of the substrate, Tm is the melting temperature of the substrate, and Tv is the vaporizing temperature of the substrate.

i. Substrate/coating materials – Unlike thermal spraying or vapor deposition techniques, the laser cladding could be adapted to a wide range of substrates and a vast range of coating materials [31]. Some of the substrate and coating materials that get benefits via the laser cladding process are given below.

   a. Substrates – Metallic alloys and superalloys (iron-based superalloys, cobalt-based superalloys, nickel-based superalloys, aluminum alloys, titanium alloys, magnesium alloys, cobalt-chromium alloys, copper alloys), polymers, thermoplastics, glasses, and so on.

b. Coating materials – Metallic alloys and superalloys (iron-based superalloys, cobalt-based superalloys, nickel-based superalloys, aluminum alloys, titanium alloys, magnesium alloys, cobalt-chromium alloys, copper alloys), polymers, thermoplastics, glasses, and so on.

ii. Surface defects – Since the laser cladding process involves heating the surface and the coating materials are deposited on the melt pool, very few surface defects such as pores and voids are present rather than the other techniques. Moreover, these pores and voids can be minimized by optimizing the laser cladding process parameters [32].

iii. Mechanical behaviors of the coatings – The clad layer (coating) is fabricated under a higher cooling rate, which leads to segregation of secondary precipitates. These secondary precipitates or the phases that formed during solidification are reported to enhance mechanical characteristics such as hardness, wear resistance, fatigue behaviors, tensile properties, creep resistance, and corrosion resistance thereby improving the service life of the substrates [33].

iv. Metallurgical bonding – The major drawback of traditional surface treatment techniques such as PVD, CVD, hardfacing, and thermal spraying is the existence of poor metallurgical bonding between the substrate and coated layer. Particularly, it is reported that the coated layer initiates to detach from the surface after long-term usage or under high-temperature service environment. However, in the cladding process, the clad layer exhibits superior metallurgical bonding between the clad layer and the substrate. This is due to the laser irradiation followed by the deposition of coating particles in the molten melt pool. This combined effect leads to the intermixing of phases present in both substrate and clad particles at the interface region ensuring superior metallurgical bonding [34,35].

v. Microstructure – Under optimal process parameters, the laser cladding process is reported to produce a fine coating with better surface quality. The lower distortion rate leads to a better deposition rate of coating particles thereby forming fine grains in the clad region. Moreover, via better cooling rate, in some metallic alloys, the primary grains in the microstructure are governed by the secondary precipitates which will aid in improving the mechanical characteristics offered by the clad layers [36].

vi. Costs – The maintenance costs of the equipment and the production costs of the coatings were comparatively lower than the other surface treatment techniques [37].

vii. Multilayers – Unlike the other surface treatment processes, in cladding, the coating layer can be fabricated as a single layer or multilayers depending upon the demands. In addition, this multilayer could exhibit outstanding characteristics, which could improve the service life of the substrates [38–40]. Thus, the laser cladding serves as the unavoidable technique to enhance the surface characteristics of the substrates.

11.3 Fatigue in Materials – Definition and Case Studies

Generally, the components made from metallic alloys such as iron-based, copper-based, aluminum-based, nickel-based alloys are employed in severe aggressive service environments like high temperature, high pressure, marine, dissolved salts, and so on. Mostly, these components are exposed to cyclic loading for a prolonged time. The mechanical, thermal, electrical, electrochemical, and physical characteristics offered by those components are reported to diminish in this prolonged service lifetime. In due course, other than the above-mentioned characteristics, the cyclic loading for a prolonged interval could lead to the formation of cracks and voids in the components. These cracks and voids act as the nucleation sites thereby reducing the mechanical strength offered by the components and causing catastrophic failure followed by property loss. Thus, the failure of components caused by cyclic loading is termed as fatigue [41]. In other words, the cyclic loading leads to the residual stresses thereby initiating minute crack formation in the components. Due to repeated loading, these minute cracks are reported to develop and lead to sudden failure, which is called as fatigue failure [42]. Some examples of fatigue failure are as follows: (i) Connecting rod failure in internal combustion (IC) engines – The petrol and diesel engine has numerous components that are reported to operate in cyclic conditions. The connecting rod, which serves as the connecting tool between the crankshaft and piston, is exposed to numerous cyclic loadings due to the rapid motion for generating power to drive the vehicles. This cyclic loading initiates the crack formation and leads to sudden failure (fatigue) thereby collapsing the entire engine. (ii) Turbine blade failure – The turbine blades of jet and aircraft engines are reported to operate at high speeds in a range of 1,000–1,500 revolutions per second. At these high speeds, the turbine blades were subjected to extreme thermal and mechanical loads in order to generate the necessary trust to move the aircraft. These extreme loads will render its mechanical strength thereby inducing cracks and notches in them. In the meantime, these cracks and notches will develop resulting in catastrophic fatigue failure followed by major accidents. The collapse of a fan disk in the center tail-mounted engine of United Airlines Flight 232, a DC-10 airliner, in July 1989 is a well-known and tragic example of a severe fatigue-related jet engine failure. The majority of the hydraulic flight controls in the tail assembly were destroyed by the shrapnel explosion that followed the disk crack, making the aircraft almost hard to fly. The large jet was successfully navigated to the Sioux Falls, Iowa, airport by the pilots, but they were unable to maintain adequate control over the aircraft to make a safe landing. 112 of the 269 occupants of the airplane made it out alive. According to the accident investigation, a catastrophic fatigue crack in the titanium fan disk was caused by an undiagnosed material fault. The passenger compartment of the De Havilland DH-106, the world's first commercial jetliner, was initially intended to have huge square windows. Three different aircraft broke apart in mid-air within a few years after going into service, killing every person on board. All contemporary jet aircraft employ small, oval-shaped windows because, upon closer examination, it was discovered that two of these failures were caused by metal fatigue of the aluminum aircraft skin, which resulted in splitting at the corners of square windows and faulty riveting. Thus, fatigue failure is considered a major issue in most of the industrial sectors.

11.4 Types of Fatigue

Fatigue failure can occur due to various reasons such as thermal, mechanical, electrical, electromechanical, thermomechanical, electrochemical, and so on. This section briefly explains the types of fatigue (Figure 11.2) that occur in materials.

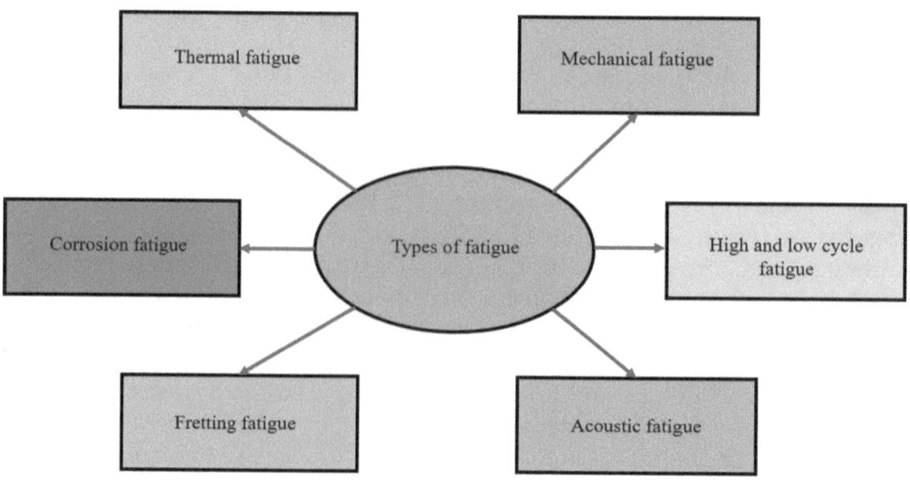

FIGURE 11.2
Types of fatigue.

11.4.1 Thermal Fatigue

As the name implies, thermal fatigue occurs due to the influence of thermal cycles that take place in materials/components. Certain components or materials operate in aggressive environments that are exposed to various heating and cooling cycles in a broad range. In such environments, the continuous thermal cycles lead to the formation of minute cracks that might develop into large cracks which results in failure. Thus, the thermal cycles lead to catastrophic failure of the components and are termed as thermal fatigue. Thermal fatigue might also occur when two or more materials with different thermal constraints such as the coefficient of thermal expansion are combined to form the final product [43,44]. Examples: turbine blades of gas and steam turbines, boilers, and nuclear reactor components experience thermal fatigue due to extreme thermal cycles.

11.4.2 High-Cycle and Low-Cycle Fatigue

Certain types of fatigue occur due to the alternating stress over a broad range of cycles (roughly 12,000 cycles) and are termed cyclic fatigue. Cyclic fatigue is broadly classified into two types, namely, high-cycle and low cycle fatigue. In a material, the high-cycle fatigue results from the alternating stress that takes place in the elastic and slightly inelastic phase (range) which exhibit lower strains [45]. On the other hand, the low cycle fatigue is reported to take place in the plastic phase of the material, thereby exhibiting higher strains [46].

11.4.3 Mechanical Fatigue

The most common type of fatigue experienced by all components irrespective of their working environment is mechanical fatigue. Mechanical fatigue arises due to mechanical constraints such as load and force over a large number of cycles. The force and load can be either internal or external. Under cyclic loading/force, the materials start to deteriorate due to the formation of cracks and voids thereby causing sudden failure followed by losses [47]. In recent years, a combination of thermal and mechanical fatigue termed thermomechanical fatigue occurs in certain materials that are exposed to extreme thermal and mechanical cycles but in the out-of-phase, both mechanical and thermal constraints are independent of each other [48].

11.4.4 Corrosion Fatigue

Certain materials experience fatigue failure in terms of electrochemical phenomena. Corrosion is a type of electrochemical phenomenon that alters the mechanical characteristics offered by the materials. This corrosion is responsible for diminishing the tensile strength, toughness, hardness, wear resistance, and other characteristics of the material. When the material is exposed to corrosive media for a prolonged interval, a chemical reaction takes place between the material's surface and the corrosive media. This reaction leads to the formation of minute holes called pits and pores on the material surface thereby inducing corrosion to a greater extent. In due course, the cyclic loading along with the corrosion results in corrosion fatigue where the pits and pores act as the nucleation sites for forming minute cracks at the end of pits and pores. These minute cracks will further propagate inside the material resulting in corrosion fatigue failure and property loss. In other words, the combined effect of cyclic loading and corrosion by-products (pores and pits) leads to corrosion fatigue failure in components [3,49].

11.4.5 Fretting Fatigue

A special type of fatigue that occurs due to the relative motion is the fretting fatigue. A tiny relative movement between two solid surfaces in contact is referred to as fretting. The surface deterioration brought on by this motion as fretting damage. Fretting damage is expected to develop on contacting surfaces of tightly fitting mechanical assemblies that vibrate and cause minor relative displacements. The gradual deterioration of a material due to little rubbing that results in abrasion and fracture is known as fretting fatigue [50]. Engineering assemblies may experience fretting fatigue as a result of oscillatory loads passing across a contact interface between neighboring components. In other words, the occurrence of crack initiation and propagation from the area of concentrated stress in reciprocating frictional contacts is known as fretting fatigue. It has been determined to be the root source of significant harm in aircraft engine's dovetail connections [51].

11.4.6 Vibration or Acoustic Fatigue

An ongoing problem for aircraft maintenance since the development of the turbojet engine is acoustic fatigue, also known as sonic fatigue. The problem has affected both military and civilian aircraft, and as noted by Rudder and Plumblee, it has only gotten worse as a result of the "evolution" of high-performance military aircraft. Resonant response in aircraft structures with narrow gauges and minimal damping is indicative of acoustic fatigue. Because of the stochastic nature of the jet and/or boundary-layer noise, the structural response is frequently non-linear and random. High-cycle fatigue known as "acoustic fatigue" develops as a result of acoustic loadings that are applied randomly and have the potential to harm panel structures. Regarding airplanes, a lack of knowledge regarding flow characteristics across panel structures limits the design of aircraft that operate at high speeds under atmospheric circumstances. It is still very difficult to explain and predict acoustic fatigue, despite substantial theoretical, computational, and experimental attempts in the past [52,53].

11.5 Fatigue Failure Mechanism

As mentioned earlier, fatigue failure occurs due to the effect of repetitive cycles of loading and alternating stress. The mechanism of fatigue failure can be categorized into three stages, namely, crack initiation, crack growth, and failure (fracture) [54]. These three stages are briefly discussed as follows:

i. Crack initiation

   This stage is considered as the initial stage of fatigue failure. When the material or components is exposed to periodic loading cycles and alternating stresses, very minute pores might be formed inside the material. These minute pores were joined together to initiate the crack formation in the materials. In this early stage, the material or component still exhibits better mechanical characteristics since the cracks remain smaller.

ii. Crack growth

   Crack growth is termed as the second or middle stage in the fatigue failure process. In this stage, minute cracks, which are distributed over the material's inner cross-section, come closer and join together under repetitive cyclic loads. Thus, the joining of minute cracks leads to the formation or growth of large cracks thereby reducing the mechanical characteristics offered.

iii. Fracture

   The last stage of fatigue failure is the fracture. In this stage, at extreme thermomechanical loads, the cracks further propagate deeper inside the material and at one stage, the material cannot withstand the cyclic loads and ruptures along the crack pathways causing sudden failure of the component. The schematic illustration of the stages of fatigue failure is shown in Figure 11.3.

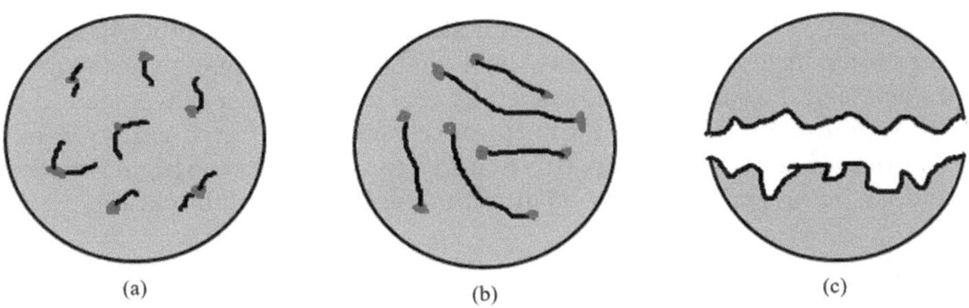

FIGURE 11.3
Schematic illustration of the stages in fatigue failure, (a) Crack initiation where minute cracks develop from the nucleation sites (red color), (b) Crack growth and propagation occurs due to repetitive cyclic loading, and (c) Fracture (failure).

11.6 Laser Cladding Process Parameters Influencing the Fatigue Property

Compared to the traditional surface treatment techniques, laser cladding is reported to produce a clad layer (coating) with excellent metallurgical bonding and significant mechanical characteristics. This process involves the combined effect of high-energy lasers, feedstocks, inert gases, and so on to produce a fine clad layer over the substrate's surface. To achieve a better clad layer with an adequate deposition rate of the coating particles and fine microstructures without pores or with fewer pores, there is a need to control the process parameters involved in this process. Several researchers reported that the mechanical characteristics offered by the clad layer vary with the cladding process parameters. In addition, the optimization of the laser cladding parameters varies with the type of substrate and coating materials [55–57]. This section details the process parameters used while performing the laser cladding process.

11.6.1 Laser Power

The main difference between the traditional surface treatment techniques and the laser cladding is the usage of lasers for fabricating the highly efficient coatings on the substrate. This continuous beam laser serves as the soul of the laser cladding process. The heat generated during the laser irradiation on the substrate's surface decides the formation of phases in the clad region. This heat is responsible for heating the surface to a temperature between the melting and vaporizing temperature of the substrate. During irradiation, the high-energy laser beam melts the substrate's surface, which leads to the formation of the localized metallic melt pool. Moreover, sufficient power is required to achieve his molten melt pool without vaporizing the substrate [58,59]. The coating alloy particles were deposited on this pool and with the effect of heat, the coating particles are reported to form various phases that will improve the mechanical characteristics offered. On the other hand, the laser power decides the thickness of heat-affected zone formation on the interface between the substrate and coated layer. Thus, the laser power should be in the nominal range so that the substrate remains unaffected with minimal heat-affected zone. In addition, this nominal range results in a higher cooling rate so that fine grains are formed after the solidification of the clad layer [60].

11.6.2 Laser Scanning Speed

In the laser cladding process, the clad layer is fabricated on the substrate's surface with the combined effect of high heating and cooling rates. Since, higher heating and cooling rates lead to the formation of surface defects such as pores, voids, segregation of unmelted coating particles, and so on.

In real time, these defects will serve as the nucleation zones for corrosion and erosion, which will affect the mechanical characteristics offered by the clad layer. To avoid these issues and to obtain coatings with better surface finish, care should be taken to optimize the scanning speed of the laser [61,62]. This scanning speed decides the microstructural and morphological evolution happening in the clad layer. By gradually increasing the laser scanning speed, researchers reported that there is a notable difference in the grain structure formed on the clad layer. Particularly, the columnar grains formed at the clad layer change to equiaxed and cellular grains at higher scanning speeds. These changes could effectively influence the mechanical behaviors offered by the clad layers. In some metals, the segregation of secondary precipitates is quite higher at higher scanning speeds. In addition, at higher scanning speeds, better grain refinement and better distribution of elements were achieved which further enhances the mechanical characteristics such as hardness, wear resistance, and anti-corrosion behaviors. Thus, higher scanning speeds could result in better deposition rates, effective grain refinements, finer microstructures, and a higher amount of segregation of secondary precipitates at the clad layer [63,64].

11.6.3 Powder-Feeding Rate

This parameter applies to the co-axial powder-based laser cladding system and not to preplaced powder laser cladding and wire feedstock-based laser cladding processes. In the co-axial powder-feed system, the gas-atomized powder (coating) particles are directed on the molten melt pool, which is formed at the substrate's surface due to the combined effect of laser power and laser scanning speed. Once, the powder particles reach the molten melt pool, the interactions will begin between the powder particles and the hot pool. The powder particles were influenced by the heat present in the molten melt pool and started to melt. The melted particles intermix with the molten melt pool, which will pave the way for the formation of metallic phases followed by the solidification to form the clad layer. This intermixing could result in better grain refinement and distribution at the clad layer thereby enhancing the mechanical characteristics offered by them [65,66]. Several researchers reported that at a lower powder-feeding rate, the clad region lacks in the formation of phases, that is, the clad layer has a huge contribution to the substrate's phases than the powder phases. Hence, the clad layer formed at a lower powder-feeding rate exhibits poor mechanical characteristics. On the other hand, a higher powder-feeding rate could result in insufficient heat distribution to the deposited particles. This insufficient heat distribution results in improper melting of powder particles that will affect the surface finish of the clad layer. Moreover, in real time, these improper melted particles could contribute to the embrittlement of the clad layers in the service environments. Hence, depending upon the laser power and laser scanning speed, the powder-feed rate is adjusted to obtain the clad layer with better structural integrity. Different substrates have various melting temperatures and various heat distribution patterns, which would have a greater influence on the coating materials. Therefore, the optimization of these three process parameters entirely depends on the type of substrate material and the preferred coating material for that substrate [67–69].

Some of the relevant literature regarding the optimization of process parameters for producing clad layers on different substrates are as follows. Reddy et al. [70] proposed mathematical models for optimizing the laser cladding process parameters for fabricating Fe-Cr-B clad layer on the 15Mo3 type steel substrate. They established three different models based on optimizing the deposition rate (efficiency) on 15Mo3 steel, the effect of dilution, and the effect of porosity by varying the process parameters. They concluded that the deposition rate is dependent on the laser power that is used to form the molten melt pool. In other words, the deposition rate of Fe-Cr-B particles attains maximum at higher laser power. Further, on analyzing the rate of dilution, the dilution rate is influenced by both laser power and Fe-Cr-B powder-feed rate. Other than these two factors, the dilution is dependent on the melting point of the Fe-Cr-B powder particles and its specific heat capacity. On the other hand, regarding the porosity that occurs on the clad layer, they concluded that the porosity is an independent phenomenon that occurs due to outgassing. And, the process parameters such as laser power and Fe-Cr-B powder-feeding rate do not have any influence on the formation of the pores on the clad layer. Xu et al. [71] examined the effect of laser power, powder-feeding rate, and laser scanning speed on the microstructure, hardness, and yield strength exhibited by the Inconel 718 clad layer obtained on the A-286 gas turbine alloy (stainless steel). For fabricating an efficient Inconel 718 clad layer, the laser power, Inconel 718 powder-feeding rate, and scanning speed were varied in the following ranges: laser power –1,200 to 2,400 W, powder-feed rate –200 to 350 mg/s, and scanning speed –3 to 7.5 mm/s, respectively. Based on the Taguchi analytical models and experimental results, the major outcomes of their work are as follows. The laser power is found to have a significant influence on the tensile properties offered by the Inconel 718 clad layer. At very high laser power (above 2,000 W), the clad region exhibits a coarser microstructure like the microstructure present in the cast Inconel alloy. This coarser microstructure is reported to diminish the mechanical behaviors offered by the Inconel 718 clad layer. They concluded that at 1.2 kW (laser power), 250 mg/s (Inconel 718 feed rate), and 4.5 mm/s (laser scanning speed) are the optimal parameters for coating Inconel 718 on the A-286 steel substrate. The clad region obtained from the mentioned parametric values exhibits finer microstructure with outstanding hardness and yield strength. Moreover, the formation of laves phase is much reduced in this clad layer and this reduction is responsible for enhancing the mechanical characteristics of the Inconel 718 clad layer. Ma et al. [72] established the empirical model based on a multi-objective quantum behaved particle swarm optimization (PSO) algorithm for optimizing the laser cladding process parameters in order to fabricate the cobalt alloy-based clad layer. On varying the process parameters such as laser power and scanning speed were optimized based on the dilution rate and residual stress developed during the clad deposition. Their studies reported that the amount of residual stress that developed during the cladding process is directly proportional to the laser power and powder-feed rate. Higher laser power and higher powder-feed rate result in the maximum amount of residual stress developed during the laser cladding process. In addition, the defocus distance from the laser serves as the important parameter that decides the dilution rate of the clad layer. On considering the laser power and laser scanning speed as independent parameters, the laser scanning speed is found to be a prominent parameter that has a significant effect on the induced residual stress. Sun et al. [73] investigated the process parameters for producing Ti6Al4V coatings via laser cladding. By means of ANOVA modeling, the relationship between each process parameter was analyzed and their output responses were recorded. Moreover, this research is focused on the clad profile, geometry, and optimization of process parameters based on response surface methodology. It is reported the clad layer width increases with increasing the laser power. At maximum laser power (700 W), the width and height of the Ti6Al4V clad layer are reported to decrease due to the high dilution rate obtained. In addition, a reduction is noticed in the respective responses obtained at peak laser power. On the other hand, the powder-feeding rate greatly influences the clad bead height and width. A lower powder-feeding rate diminishes the height and width of the Ti6Al4V clad bead, respectively. Yu et al. [74] proposed the analytical model for cladding the Fe-based alloy and established the relationship between the various process parameters via response surface methodology. Their major findings regarding the Fe-based clad layer are as follows. The laser power is found to be the most influential parameter that decides the clad layer's width. On the other hand, the combined parametric effect of laser power and laser scanning speed serves as the deciding criteria for the dilution rate and Fe-based clad layer's height. They mentioned that the powder-feeding rate does not have a significant influence on the dilution rate, clad layer's height, and thickness. For cladding Fe-based coating, the optimized values of laser power, powder-feeding rate, and laser scanning speed are as follows: laser power –365 W, scanning speed –7 mm/s, and powder-feeding rate –0.65 r/min. Under these values, they claimed that the obtained Fe-based clad layer is free from pores and voids with a good surface finish. Thus, there is a need to optimize the laser cladding process parameters for obtaining clad layer with fewer pores along with better surface finish.

11.7 Fatigue Resistance Improvement via Laser Cladding

Since fatigue failure is unavoidable in any industrial circumstance, the laser cladding technique aids in fabricating fatigue resistance coating on various metallic substrates that are exposed to extreme cyclic loads. The rapid

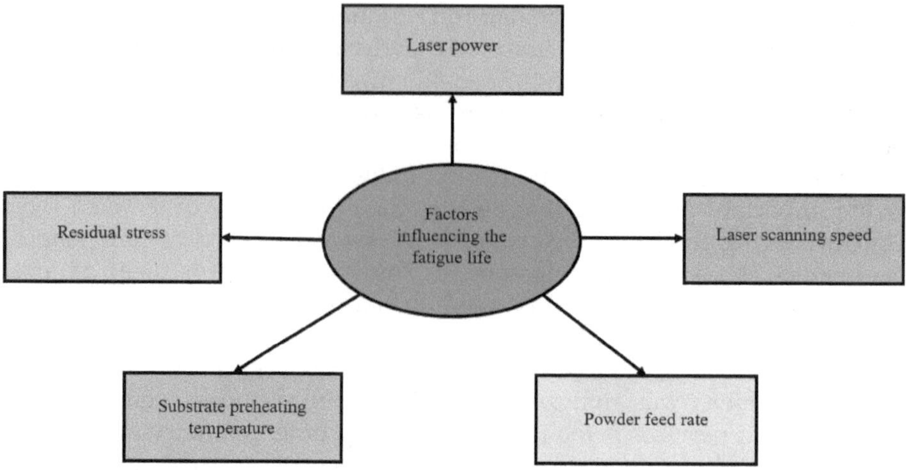

FIGURE 11.4
Factors influencing the fatigue behaviors of the clad layers.

solidification rate of the molten metallic pool results in the formation of fine grains without any surface defects. The clad layer with higher thickness can be obtained and the precipitation rate of the secondary and intermetallic precipitates that aids in improving the fatigue resistance property tends to be higher in laser cladding when compared with the other surface treatment techniques. Moreover, the fatigue resistance characteristics could be improved by the following effects in laser cladding (Figure 11.4).

11.7.1 Effect of Preheating on Fatigue Resistance

In the laser cladding process, the substrate (base metal) plays a vital role in deciding the mechanical characteristics offered by the clad layer. The substrate ensures adequate mechanical bonding between the clad layer and the substrate's surface to avoid the clad layer peeling off from the surface. For this purpose, the substrate material is pretreated via the sandblasting process to make sure that adequate metallurgical bonding exists between the substrate surface and clad layer. Moreover, it is reported that heat treatment will alter the mechanical characteristics offered by the materials to a greater extent. During the heat treatment process, the coarsening of grains takes place in a larger manner followed by uplifting the mechanical behaviors such as tensile strength, toughness, hardness, wear resistance, and fatigue resistance offered by the materials. Pasquale Cavaliere et al. [75] examined the effect of preheating the substrate (AISI 316L stainless steel) on the fatigue resistance behavior offered by the MoB/Co-Cr and MoB/CoTi clad layers obtained via laser cladding. They performed preheat treatment on the AISI 316L substrate at different temperatures such as 150°C and 300°C in order to compare the fatigue resistance behavior before and after preheating. By performing the fatigue failure tests, the results are concluded as follows. The non-preheated sample withstands only for shorter loading cycles (300,000 cycles). When the range is beyond 300,000 cycles, the non-preheated sample fractures because of the non-uniform microstructure exhibited by the clad layer. On the other hand, the microstructure of the clad layer tends to be uniform in the preheated substrate samples. Particularly, the preheated sample obtained at 300°C experiences greater fatigue resistance even up to the cyclic loads of 1,000,000 cycles. This excellent fatigue resistance is due to the effect of preheating the AISI 316L substrate, which aids in obtaining finer microstructure of the clad layer and exhibiting excellent fatigue resistance when compared with the non-preheated sample. Liu et al. [76] investigated the effect of preheating the copper alloy substrate at various temperatures and analyzed the mechanical behaviors of the Inconel clad layers obtained on the preheated substrate. Their study concluded that the clad layer obtained on the preheated substrate results in better microstructure of grains and also the precipitation rate of secondary intermetallic precipitates was quite higher due to the effect of preheating. The 200°C preheated sample has a better hardness, wear resistance, and fatigue resistance behavior than the other samples. Thus, preheating of the substrate results in elevating the fatigue resistance characteristics offered by the clad layers.

11.7.2 Effect of Residual Stress on Fatigue Resistance

Other than the substrate preheating, residual stress is also considered an important factor in influencing the fatigue resistance offered by the clad layers produced via laser cladding. Notably, the compressive residual stress is reported to proliferate the fatigue resistance characteristics of the clad layers. In addition, the effect of substrate preheating is more beneficial in inducing the residual stress in both circumferential and axial directions thereby improving the fatigue performance of the clad layers. By performing a rotary bending fatigue test, Hutasoit et al. [77] analyzed the fatigue behavior of Stellite 6 and Deloro 40G clad layers obtained on the stainless steel 4130 substrate. They clarified that, in comparison to substrates of the same size, the fatigue life was shortened by the existence of tensile residual stresses in the clad zone. In comparison to the specimen laser clad with Deloro 40G, the specimen laser clads with Stellite 6 exhibited a longer fatigue life due to the existence of higher compressive residual stress in the substrate region and lower tensile residual stresses in the coating region, all at similar coating thickness levels. Because of a tendency for the coating zone to have higher tensile residual stress in the axial stress direction, increasing the coating thickness decreased the fatigue life of laser-clad specimens with the same final structure size. In a four-point bending and torsion fatigue test, Alam et al. [78] examined the fatigue life of 42CrMo4 steel bar coated with Inconel 625 and S355 structural steel clad with Stellite

21 alloy. With the exception of the Inconel 625 clad specimen under greater applied stress, the clad specimen generally showed lower fatigue strength. Defects including inclusions and porosity, as well as excessive tensile residual stress determined by hole drilling, were blamed for the reduced fatigue lifetimes. Additionally, it was discovered that post-weld heat treatment had no appreciable effect on the fatigue performance of either pair of clad–substrate materials. Barr et al. [79] analyzed the effect of fatigue performance of various clad layers fabricated via the laser cladding process. They discovered some degree of as-clad mechanical property degradation in their investigation. Nonetheless, the fatigue properties were enhanced by the clad layer's material. As-clad samples had much longer fracture propagation lifetimes from a common beginning depth of 0.25 mm than baseline samples, according to test data, by a ratio of 3 to 4. The beneficial compressive residual pressures brought on by the repair process are responsible for the prolonged life.

11.8 Fatigue-Resistant Cladding Powders

As previously mentioned, the fatigue behavior offered by different substrates is said to be greatly influenced by the cyclic loading conditions. During cyclic loading, the material is under deformation via wear and erosion thereby minimizing the service life of the substrates. In particular, it is reported that the fatigue behaviors could be improved by the segregation of hard phases and secondary precipitates at the clad region. The cladding material should be wisely chosen to obtain such precipitates [80,81]. Superalloys are regarded as high-performance alloys that are reported to deliver excellent mechanical characteristics even near their melting point. While coated on substrates via laser cladding process, the clad region made from these alloys has superior hardness, higher wear resistance, better anti-corrosion behaviors, excellent fatigue resistance, and so on. This is due to the segregation of hard phases and secondary precipitates at their clad layer and offering excellent mechanical behaviors [82,83]. In most cases, the substrates were coated with these superalloys to improve their fatigue performance. The superalloys can be broadly classified into three major categories, namely, nickel-based, cobalt-based, and iron-based superalloys. Each type has a series of alloys that could be adapted as the coating feedstock in laser cladding. Nickel-based superalloys (Inconel and Colmonoy series) contain significant nickel and chromium composition along with other refractory elements. These are precipitate-hardened alloys that are well known for their wear and fatigue resistance. While cladding, these alloys form a hard lave phase along with secondary precipitates in their microstructure. The chromium-rich carbides and borides (secondary precipitates) along with the hard laves phase offer superior hardness thereby improving the fatigue and wear resistance offered by the clad layer. These alloys are quite suitable for improving the fatigue resistance exhibited by the components of nuclear reactors and marine equipment. Moreover, fine dendrites with equiaxed and cellular grain structures were obtained in the clad region of these alloys which contributes to a better surface finish thereby reducing the surface roughness [84,85]. On the other hand, the cobalt-based superalloys (Stellite series) are mostly preferred as the coating material where high-temperature fatigue resistance is the major key factor. As mentioned, these alloys are reported to enhance thermal fatigue behaviors and are usually employed in high-temperature service environments. During solidification of the clad layer, these alloys are reported to segregate metal carbides, which will improve the fatigue resistance and act as the potential barrier against crack formation under cyclic loading [86]. Other than superalloys, based on the working environment, normal metallic alloys made from aluminum, titanium, iron, etc., can be preferred to improve the fatigue behaviors offered by the substrates [87,88].

Shi Da Sun et al. [89] examined the fatigue characteristics offered by the clad layers obtained from the AISI 4340 steel powder and AerMet 100 steel powder, which were fabricated on the AISI 4340 substrate. The major outcomes of their work are as follows. The AISI 4340 clad region and the AerMet 100 clad region contain quite similar phases, which are reported to be rich in columnar dendrites. Under heat treatment, there are no changes in the AISI 4340 clad layer but the transition of columnar dendrites to equiaxed martensite grains takes place in the AerMet 100 clad layer. Further, the hardness studies revealed that higher hardness is observed on both AerMet 100-as-clad and post-heat treated AerMet 100 clad regions when compared to the AISI 4340-as-clad and post-heat treated AISI 4340 clad regions. This increase in hardness value is due to the martensite formation on the AerMet 100 clad regions. Moreover, the tensile properties such as yield strength, and ultimate tensile strength were compared between the AISI 4340 and AerMet 100 clad regions. It is claimed that the AerMet 100 clad region has the maximum yield strength and ultimate tensile strength values than the AISI 4340 clad region. After heat treatment, the tensile properties of the AerMet 100 clad region are further enhanced than the AISI 4340 substrate and the AISI 4340 clad regions. Also, on comparing the fatigue behaviors offered by both AISI 4340 and AerMet 100 clad regions, the post-heat treated AerMet 100 clad region has the highest fatigue life withstanding 12,851 cycles. Thus, they concluded that the AerMet 100 serves as the suitable cladding material for fabricating the clad layer on the AISI 4340 substrate to improve the fatigue behaviors of the substrate. Tuominen et al. [90] compared the fatigue behavior offered by the two clad layers made from cobalt-based and nickel-based superalloys on S355 steel and 42CrMo4 bar, respectively. Stellite 21 alloy is used to coat the S355 steel whereas Inconel 625 is preferred as the coating material for the 42CrMo4 substrate, respectively. Their observations are as follows. The Inconel 625 clad region was rich in the hard laves phases which were distributed in the gamma nickel matrix. These hard lave phases are rich in Ni which further enhances the hardness and wear characteristics offered by the clad layer. Moreover, there is an adequate segregation of chromium-rich carbides that aid in improving the fatigue life of the clad region. On the other hand, the Stellite 21 clad region exhibits primary dendrites with secondary eutectic phases, which were distributed in the Co-Cr matrix. Further, the clad region was decorated with cobalt and chromium-rich carbides. They concluded that both materials (Stellite 21 – cobalt-based alloy and Inconel 625 – nickel-based alloy) can be used in industries as suitable cladding materials. These two cladding materials are reported to form adequate secondary precipitates that will improve the hardness, wear resistance, thermal fatigue resistance, and tensile properties (yield and tensile strength) of the steel substrates used in industries. Taposh Roy et al. [91] compared the influence of fatigue by depositing Stellite 6, stainless steel 420 and 410L materials on the rails. The major outcomes of their work are as follows. The microstructures of the 410L and stainless steel 420 clad regions consist of columnar martensite dendrites whereas the Stellite 6 clad region contains eutectics phases, which were distributed in the Co-Cr matrix. Further, on heat treatment, the columnar martensite phases were transitioned to equiaxed grains whereas cellular eutectics were reported in the case of the Stellite 6 clad region after heat treatment. From the microhardness analysis, the hardness offered by the stainless steel 420 clad layer is higher than the other two clad regions. In addition, the hardness value at the Stellite 6 clad region is quite higher than the 410L clad layer. The maximum hardness at the stainless steel 420 clad layer is due to the segregation of equiaxed martensite phases in its microstructure. Moreover, all three clad regions obtained from stainless steel 420, Stellite 6 and 410L exhibit higher wear resistance than the rail substrate. Regarding the cracking tendency under cyclic loading conditions, the Stellite 6 clad layer shows higher fatigue resistance than the other two layers and thus, they concluded that the Stellite 6 material is found to be the most suitable material for improving the hardness, wear resistance, and fatigue behaviors of the rails. On the other hand, the wear and thermal fatigue resistance of the Stellite 6 alloy can be further enhanced by doping or adding tungsten carbide particles. Wang et al. [92] investigated the effect of adding tungsten carbide powder to Stellite 6 powder and analyzed the obtained microstructure, wear behaviors, and thermal fatigue resistance offered by the Stellite 6 + tungsten carbide clad layer. They found that the addition of tungsten carbide results in better dissolution of Stellite 6 particles in the molten melt pool. The clad region was rich in Co-Cr-WC phases with the segregation of carbides in the grain boundaries. Further, they concluded that the addition of WC to Stellite 6 results in better resistance against wear and thermal fatigue in high-temperature service environments [93].

11.9 Conclusion

Laser-based surface treatments have more advantages than traditional surface treatment techniques. Among the various laser-based surface treatment techniques, laser cladding is considered as the most optimal technique for treating the repaired surfaces in industries. It involves a high-power laser beam, which is directed on the sample to irradiate the sample's surface leading to the formation of the molten melt pool. The materials to be coated were deposited on this molten pool and the heat interactions between the coating particles and the molten pool initiates the dissolution of coating particles in the hot pool. After solidification, a fine clad layer is obtained which is reported to enhance the service life of the substrate. This chapter elaborated on the laser cladding principle, fatigue failure case studies, types of fatigues that occur in materials, fatigue-resistant cladding powders, and the effect of residual stress and preheating in elevating the fatigue performance of the clad layers along with relevant literature works.

Conflicts of Interest

The authors declare that there are no conflicts of interest.

Acknowledgment

The authors like to acknowledge the editors of this book for providing an opportunity to write this section.

References

1. Santos MC, Machado AR, Sales WF, Barrozo MA, Ezugwu EO. Machining of aluminum alloys: a review. *The International Journal of Advanced Manufacturing Technology*. 2016;86:3067–80.
2. Francis A, Yang Y, Virtanen S, Boccaccini AR. Iron and iron-based alloys for temporary cardiovascular applications. *Journal of Materials Science: Materials in Medicine*. 2015;26:1–6.
3. Jeyaprakash N, Yang CH, Karuppasamy SS, Radhika N. Microstructural evolution and anti-corrosion characteristics of as-cast and LPBF printed Inconel plates in seawater environment. *Surface Review and Letters*. 2023;31:2450032.

4. Li L, Sagues AA. Chloride corrosion threshold of reinforcing steel in alkaline solutions—open-circuit immersion tests. *Corrosion*. 2001;57(1):19–28.
5. Jeyaprakash N, Karuppasamy SS, Yang CH. 1 Application of wear-resistant laser claddings. In Hitesh Vasudev, Chander Prakash (Eds), *Handbook of Laser-Based Sustainable Surface Modification and Manufacturing Techniques*, 2023 (pp. 1–26). CRC Press.
6. Botha M, Hindley MP. One-way fluid structure interaction modelling methodology for boiler tube fatigue failure. *Engineering Failure Analysis*. 2015;48:1–10.
7. Karuppasamy SS, Jeyaprakash N, Yang CH. The sensors in future vehicles. In Jeyaprakash Natarajan, Mahendra Babu Kantipudi, Che-Hua Yang, Yaojung Shiao (Eds) *The Future of Road Transportation*, 2023 (pp. 261–283). CRC Press, USA.
8. Merati A. A study of nucleation and fatigue behavior of an aerospace aluminum alloy 2024-T3. *International Journal of Fatigue*. 2005;27(1):33–44.
9. Karuppasamy SS, Jeyaprakash N, Yang CH. Alternative propulsion systems. In Jeyaprakash Natarajan, Mahendra Babu Kantipudi, Che-Hua Yang, Yaojung Shiao (Eds), *The Future of Road Transportation*, 2023 (pp. 47–68). CRC Press, USA.
10. Khatibi G, Horky J, Weiss B, Zehetbauer MJ. High cycle fatigue behaviour of copper deformed by high pressure torsion. *International Journal of Fatigue*. 2010;32(2):269–78.
11. Karuppasamy SS, Jeyaprakash N, Yang CH. 27 Application of corrosion-resistant laser claddings. *Handbook of Laser-Based Sustainable Surface Modification and Manufacturing Techniques*. 2023;51(51):27.
12. Hou J, Wicks BJ, Antoniou RA. An investigation of fatigue failures of turbine blades in a gas turbine engine by mechanical analysis. *Engineering Failure Analysis*. 2002;9(2):201–11.
13. Jeyaprakash N, Yang CH, Sivasankaran S. Laser cladding process of cobalt and nickel based hard-micron-layers on 316L-stainless-steel-substrate. *Materials and Manufacturing Processes*. 2020;35(2):142–51.
14. Wegener T, Krochmal M, Möller TR, Le MT, Czap A, Marianek F, Fakesch H, Niendorf T. On the low-cycle fatigue behavior of a novel high strength mold steel. *International Journal of Fatigue*. 2023;175:107754.
15. Bhat S, Patibandla R. Metal fatigue and basic theoretical models: a review. In Eduardo Valencia Morales (Ed), *Alloy Steel-Properties and Use*, 2011 (Vol. 22, pp. 203–238). InTech Europe.
16. Karuppasamy SS, Jeyaprakash N, Yang CH. Battery technology for hybrid and battery vehicles. In Jeyaprakash Natarajan, Mahendra Babu Kantipudi, Che-Hua Yang, Yaojung Shiao (Eds), *The Future of Road Transportation*, 2023 (pp. 136–158). CRC Press, USA.
17. Schalk N, Tkadletz M, Mitterer C. Hard coatings for cutting applications: physical vs. chemical vapor deposition and future challenges for the coatings community. *Surface and Coatings Technology*. 2022;429:127949.
18. Murugesan P, Satheeshkumar V, Jeyaprakash N, Yang CH, Karuppasamy SS. Effect of α-Al and Si precipitates on microstructural evaluation and corrosion behavior of laser powder bed fusion printed AlSi10Mg plates in seawater environment. *Metals and Materials International*. 2023;29:2515–32.
19. Pierlot C, Pawlowski L, Bigan M, Chagnon P. Design of experiments in thermal spraying: a review. *Surface and Coatings Technology*. 2008;202(18):4483–90.

20. Tejero-Martin D, Rezvani Rad M, McDonald A, Hussain T. Beyond traditional coatings: a review on thermal-sprayed functional and smart coatings. *Journal of Thermal Spray Technology*. 2019;28:598–644.
21. Jeyaprakash N, Yang CH, Karuppasamy SS. Laser cladding of Colmonoy 6 particles on Inconel 625 substrate: microstructure and corrosion resistance. *Surface Review and Letters*. 2022;29(08):2250102.
22. Siddiqui AA, Dubey AK. Recent trends in laser cladding and surface alloying. *Optics & Laser Technology*. 2021;134:106619.
23. Jeyaprakash N, Yang CH, Ramkumar KR. Microstructure and wear resistance of laser cladded Inconel 625 and Colmonoy 6 depositions on Inconel 625 substrate. *Applied Physics A*. 2020;126:1–1.
24. Abboud JH, Benyounis KY, Olabi AG, Hashmi MS. Laser surface treatments of iron-based substrates for automotive application. *Journal of Materials Processing Technology*. 2007;182(1–3):427–31.
25. Jeyaprakash N, Yang CH, Karuppasamy SS, Duraiselvam M. Stellite 6 cladding on AISI Type 316L stainless steel: microstructure, nanohardness and corrosion resistance. *Transactions of the Indian Institute of Metals*. 2023;76(2):491–503.
26. Arif ZU, Khalid MY, ur Rehman E, Ullah S, Atif M, Tariq A. A review on laser cladding of high-entropy alloys, their recent trends and potential applications. *Journal of Manufacturing Processes*. 2021;68:225–73.
27. Jeyaprakash N, Yang CH, Susila P, Karuppasamy SS. Laser cladding of NiCrMoFeNbTa particles on Inconel 625 alloy: microstructure and corrosion resistance. *Transactions of the Indian Institute of Metals*. 2023;76(2):599–612.
28. Liu Y, Ding Y, Yang L, Sun R, Zhang T, Yang X. Research and progress of laser cladding on engineering alloys: a review. *Journal of Manufacturing Processes*. 2021;66:341–63.
29. Jeyaprakash N, Yang CH, Karuppasamy SS, Dhineshkumar SR. Evaluation of microstructure, nanoindentation and corrosion behavior of laser cladded Stellite-6 alloy on Inconel-625 substrate. *Materials Today Communications*. 2022;31:103370.
30. Zhu L, Xue P, Lan Q, Meng G, Ren Y, Yang Z, Xu P, Liu Z. Recent research and development status of laser cladding: a review. *Optics & Laser Technology*. 2021;138:106915.
31. Karuppasamy SS, Jeyaprakash N, Yang CH. Microstructure, nanoindentation and corrosion behavior of colmonoy-5 deposition on SS410 substrate using laser cladding process. *Arabian Journal for Science and Engineering*. 2022;47:8751–67.
32. Weng F, Chen C, Yu H. Research status of laser cladding on titanium and its alloys: a review. *Materials & Design*. 2014;58:412–25.
33. Natarajan J, Yang CH, Karuppasamy SS. Investigation on microstructure, nanohardness and corrosion response of laser cladded colmonoy-6 particles on 316l steel substrate. *Materials*. 2021;14(20):6183.
34. Tamanna N, Crouch R, Naher S. Progress in numerical simulation of the laser cladding process. *Optics and Lasers in Engineering*. 2019;122:151–63.
35. Jeyaprakash N, Yang CH, Karuppasamy SS, Rajendran DK. Correlation of microstructural with corrosion behaviour of Ti-6Al-4V specimens developed through selective laser melting technique. *Proceedings of the Institution of Mechanical Engineers, Part E: Journal of Process Mechanical Engineering*. 2022;236(5):2240–51.

36. Liu J, Yu H, Chen C, Weng F, Dai J. Research and development status of laser cladding on magnesium alloys: a review. *Optics and Lasers in Engineering*. 2017;93:195–210.
37. Jeyaprakash N, Yang CH, Tseng SP. Characterization and tribological evaluation of NiCrMoNb and NiCrBSiC laser cladding on near-α titanium alloy. *The International Journal of Advanced Manufacturing Technology*. 2020;106:2347–61.
38. Sexton CL, Byrne G, Watkins KG. Alloy development by laser cladding: an overview. *Journal of Laser Applications*. 2001;13(1):2–11.
39. Jeyaprakash N, Duraiselvam M, Raju R. Modelling of Cr3C2–25% NiCr laser alloyed cast iron in high temperature sliding wear condition using response surface methodology. *Archives of Metallurgy and Materials*. 2018;63:1303–15.
40. Zhou S, Xu T, Hu C, Wu H, Liu H, Ma X. Effect of different topologies on microstructure and mechanical properties of multilayer coatings deposited by laser cladding with Inconel 625 wire. *Surface and Coatings Technology*. 2021;421:127299.
41. Bathias C, Pineau A, editors. *Fatigue of Materials and Structures* (pp. 1–511). Wiley. Print ISBN:9781848210516. Online ISBN:9781118623435. DOI:10.1002/9781118623435. DOI:10.1002/9781118623435
42. Klesnil M, Lukác P. *Fatigue of Metallic Materials*. Elsevier; 1992.
43. Klobčar D, Tušek J, Taljat B. Thermal fatigue of materials for die-casting tooling. *Materials Science and Engineering: A*. 2008;472(1–2):198–207.
44. Fissolo A, Amiable S, Ancelet O, Mermaz F, Stelmaszyk JM, Constantinescu A, Robertson C, Vincent L, Maillot V, Bouchet F. Crack initiation under thermal fatigue: an overview of CEA experience. Part I: thermal fatigue appears to be more damaging than uniaxial isothermal fatigue. *International Journal of Fatigue*. 2009;31(3):587–600.
45. Nicholas T. *High Cycle Fatigue: A Mechanics of Materials Perspective*. Elsevier; 2006.
46. Murakami Y, Miller KJ. What is fatigue damage? A view point from the observation of low cycle fatigue process. *International Journal of Fatigue*. 2005;27(8):991–1005.
47. Robertson SW, Pelton AR, Ritchie RO. Mechanical fatigue and fracture of Nitinol. *International Materials Reviews*. 2012;57(1):1–37.
48. Constantinescu A, Charkaluk E, Lederer G, Verger L. A computational approach to thermomechanical fatigue. *International Journal of Fatigue*. 2004;26(8):805–18.
49. Larrosa NO, Akid R, Ainsworth RA. Corrosion-fatigue: a review of damage tolerance models. *International Materials Reviews*. 2018;63(5):283–308.
50. Nowell D, Dini D, Hills DA. Recent developments in the understanding of fretting fatigue. *Engineering Fracture Mechanics*. 2006;73(2):207–22.
51. Madge JJ, Leen SB, Shipway PH. The critical role of fretting wear in the analysis of fretting fatigue. *Wear*. 2007;263(1–6):542–51.
52. Karuppasamy SS, Yang CH. Adapting the time-domain synthetic aperture focusing technique (T-SAFT) to laser ultrasonics for imaging the subsurface defects. *Sensors*. 2023;23(19):8036.
53. Mixson JS, Roussos LA. *Acoustic Fatigue: Overview of Activities at NASA Langley*. NASA, Langley Research Center, USA.
54. Sangid MD. The physics of fatigue crack initiation. *International Journal of Fatigue*. 2013;57:58–72.

55. Jeyaprakash N, Yang CH, Tseng SP. Wear Tribo-performances of laser cladding Colmonoy-6 and Stellite-6 Micron layers on stainless steel 304 using Yb: YAG disk laser. *Metals and Materials International*. 2021;27:1540–53.
56. Vundru C, Paul S, Singh R, Yan W. Numerical analysis of multi-layered laser cladding for die repair applications to determine residual stresses and hardness. *Procedia Manufacturing*. 2018;26:952–61.
57. Jeyaprakash N, Duraiselvam M, Aditya SV. Numerical modeling of WC-12% Co laser alloyed cast iron in high temperature sliding wear condition using response surface methodology. *Surface Review and Letters*. 2018;25(07):1950009.
58. Goodarzi DM, Pekkarinen J, Salminen A. Analysis of laser cladding process parameter influence on the clad bead geometry. *Welding in the World*. 2017;61:883–91.
59. Prabu G, Duraiselvam M, Jeyaprakash N, Yang CH. Microstructural evolution and wear behavior of AlCoCrCuFeNi high entropy alloy on Ti–6Al–4V through laser surface alloying. *Metals and Materials International*. 2021;27:2328–40.
60. Alvarez P, Montealegre MÁ, Pulido-Jiménez JF, Arrizubieta JI. Analysis of the process parameter influence in laser cladding of 316L stainless steel. *Journal of Manufacturing and Materials Processing*. 2018;2(3):55.
61. Wang K, Liu W, Hong Y, Sohan HS, Tong Y, Hu Y, Zhang M, Zhang J, Xiang D, Fu H, Ju J. An overview of technological parameter optimization in the case of laser cladding. *Coatings*. 2023;13(3):496.
62. Vetter PA, Engel T, Fontaine J. Laser cladding: the relevant parameters for process control. In PA Vetter, Thierry Engel, Joël Fontaine (Eds), *Proceedings of SPIE - The International Society for Optical Engineering*, 1994 (Vol. 2207, pp. 452–462). SPIE.
63. Jeyaprakash N, Yang CH, Ramkumar KR. Microstructural, mechanical and wear behaviour of Inconel-718 produced through laser-powder bed-fused additive manufacturing. *Materials Science and Technology*. 2021;37(3):326–37.
64. Nenadl O, Ocelík V, Palavra A, De Hosson JT. The prediction of coating geometry from main processing parameters in laser cladding. *Physics Procedia*. 2014;56:220–7.
65. Jeyaprakash N, Yang CH, Ramkumar KR. Correlation of microstructural evolution with mechanical and tribological behaviour of SS 304 specimens developed through SLM technique. *Metals and Materials International*. 2021;27:5179–90.
66. Wirth F, Wegener K. A physical modeling and predictive simulation of the laser cladding process. *Additive Manufacturing*. 2018;22:307–19.
67. Zhao Y, Guan C, Chen L, Sun J, Yu T. Effect of process parameters on the cladding track geometry fabricated by laser cladding. *Optik*. 2020;223:165447.
68. Ya W, Pathiraj B, Liu S. 2D modelling of clad geometry and resulting thermal cycles during laser cladding. *Journal of Materials Processing Technology*. 2016;230:217–32.
69. Jeyaprakash N, Yang CH. Comparative study of NiCrFeMoNb/FeCrMoVC laser cladding process on nickel-based superalloy. *Materials and Manufacturing Processes*. 2020;35(12):1383–91.
70. Reddy L, Preston SP, Shipway PH, Davis C, Hussain T. Process parameter optimisation of laser clad iron based alloy: predictive models of deposition efficiency, porosity and dilution. *Surface and Coatings Technology*. 2018;349:198–207.
71. Xu Z, Yuan J, Wu M, Arif AF, Li D. Effect of laser cladding parameters on Inconel 718 coating performance and multi-parameter optimization. *Optics & Laser Technology*. 2023;158:108850.

72. Ma M, Xiong W, Lian Y, Han D, Zhao C, Zhang J. Modeling and optimization for laser cladding via multi-objective quantum-behaved particle swarm optimization algorithm. *Surface and Coatings Technology*. 2020;381:125129.
73. Sun Y, Hao M. Statistical analysis and optimization of process parameters in Ti6Al4V laser cladding using Nd: YAG laser. *Optics and Lasers in Engineering*. 2012;50(7):985–95.
74. Yu T, Yang L, Zhao Y, Sun J, Li B. Experimental research and multi-response multi-parameter optimization of laser cladding Fe313. *Optics & Laser Technology*. 2018;108:321–32.
75. Cavaliere P, Silvello A. Laser cladding: fatigue properties. In Pasquale Cavaliere (Ed), *Laser Cladding of Metals*, 2021 (pp. 161–84). Springer, Cham.
76. Liu Y, Jin H, Xu T, Xu Z, Du F, Yu M, Gao Y, Zhang D. Effect of a substrate's preheating temperature on the microstructure and properties of Ni-based alloy coatings. *Lubricants*. 2024;12(1):21.
77. Hutasoit N, Luzin V, Blicblau A, Yan W, Brandt M, Cottam R. Fatigue life of laser clad hardfacing alloys on AISI 4130 steel under rotary bending fatigue test. *International Journal of Fatigue*. 2015;72:42–52.
78. Alam MM, Powell J, Kaplan AF, Tuominen J, Vuoristo P, Miettinen J, Poutala J, Näkki J, Junkala J, Peltola T. Surface pore initiated fatigue failure in laser clad components. *Journal of Laser Applications*. 2013;25(3):032004.
79. Barr C, Rashid RA, Da Sun S, Easton M, Palanisamy S, Orchowski N, Matthews N, Walker K, Brandt M. Role of deposition strategy and fill depth on the tensile and fatigue performance of 300 M repaired through laser directed energy deposition. *International Journal of Fatigue*. 2021;146:106135.
80. Altus E, Konstantino E. Optimum laser surface treatment of fatigue damaged Ti–6Al–4V alloy. *Materials Science and Engineering: A*. 2001;302(1):100–5.
81. Stamm H, Holzwarth U, Boerman DJ, Marques FD, Olchini A, Zausch R. Effect of laser surface treatment on high cycle fatigue of aisi 316L stainless steel. *Fatigue & Fracture of Engineering Materials & Structures*. 1996;19(8):985–95.
82. Jeyaprakash N, Yang CH. Microstructure and wear behaviour of SS420 micron layers on Ti–6Al–4V substrate using laser cladding process. *Transactions of the Indian Institute of Metals*. 2020;73:1527–33.
83. Nalla RK, Altenberger I, Noster U, Liu GY, Scholtes B, Ritchie RO. On the influence of mechanical surface treatments—deep rolling and laser shock peening—on the fatigue behavior of Ti–6Al–4V at ambient and elevated temperatures. *Materials Science and Engineering: A*. 2003;355(1–2):216–30.
84. Garimella L, Liaw PK, Klarstrom DL. Fatigue behavior in nickel-based superalloys: a literature review. *Jom*. 1997;49:67–71.
85. Pineau A, Antolovich SD. High temperature fatigue of nickel-base superalloys–a review with special emphasis on deformation modes and oxidation. *Engineering Failure Analysis*. 2009;16(8):2668–97.
86. Reuchet J, Remy L. High temperature fatigue behaviour of a cast cobalt-base superalloy. *Fatigue & Fracture of Engineering Materials & Structures*. 1979;2(1):51–62.
87. Zhao T, Jiang Y. Fatigue of 7075-T651 aluminum alloy. *International Journal of Fatigue*. 2008;30(5):834–49.
88. Fleck C, Eifler D. Corrosion, fatigue and corrosion fatigue behaviour of metal implant materials, especially titanium alloys. *International Journal of Fatigue*. 2010;32(6):929–35.

89. Da Sun S, Liu Q, Brandt M, Luzin V, Cottam R, Janardhana M, Clark G. Effect of laser clad repair on the fatigue behaviour of ultra-high strength AISI 4340 steel. *Materials Science and Engineering: A*. 2014;606:46–57.
90. Tuominen J, Näkki J, Poutala J, Miettinen J, Peltola T, Vuoristo P, Rasehorn I, Alam MM, Kaplan AF. Fatigue behavior of laser clad round steel bars. *Journal of Laser Applications*. 2015;27(1):012006.
91. Roy T, Lai Q, Abrahams R, Mutton P, Paradowska A, Soodi M, Yan W. Effect of deposition material and heat treatment on wear and rolling contact fatigue of laser cladded rails. *Wear*. 2018;412:69–81.
92. Wang G, Zhang J, Shu R, Yang S. High temperature wear resistance and thermal fatigue behavior of Stellite-6/WC coatings produced by laser cladding with Co-coated WC powder. *International Journal of Refractory Metals and Hard Materials*. 2019;81:63–70.
93. Xu GJ, Kutsuna M. Cladding with Stellite 6+ WC using a YAG laser robot system. *Surface Engineering*. 2006;22(5):345–52.

12

Laser Cladding: Industrial Application

Jayanth N, Sathies T, Mallikarjuna B, and Ananthakumar K

12.1 Introduction

Laser cladding is a multi-disciplinary process that makes use of laser, CAD/CAM, robotics, sensors and control, and powder metallurgy. It makes use of laser source to coat a thin layer of a required material on the substrate material's surface. The coating material can be supplied to the substrate using different methods such as powder or wire feeding and uniform spreading of powder on the substrate using rollers. The powder feeding using coaxial nozzle is the most efficient method and the powder particles and the substrate material are heated by the laser beam to form a melt pool. Hence, a thin coating of a material is applied on the substrate material's surface. The pictorial illustration of laser cladding is given in Figure 12.1. The laser cladding methods have the capability to produce layer thickness between 0.05 and 2mm. In addition to the coating applications, it can be used in additive manufacturing of complex parts with improved mechanical characteristics

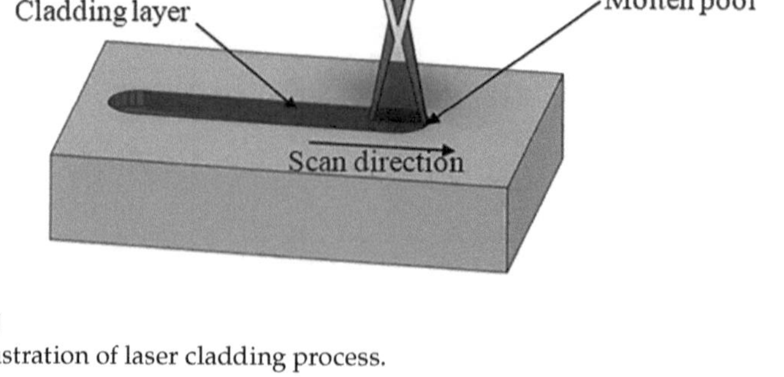

FIGURE 12.1
Pictorial illustration of laser cladding process.

DOI: 10.1201/9781032693576-12

and also used in repairing worn-out components like turbine blades, moulds and dies, boiler parts, etc., laser cladding has advantages such as better coating with less dilution and distortion, enhanced surface quality and tiny heat affected zone (HAZ) on the substrate material surface.

12.2 Aerospace Applications

The laser cladding in aerospace industry is used to repair broken or worn-out parts, improve surface properties and deposit protective coatings to enhance part strength and performance. The Inconel 625 and Rene 142 alloy powders were deposited on different substrate materials such as Inconel 792, C1023, Rene 80, Rene 125 and DS Rene 142, which are used in aerospace applications. It was found that the laser-clad samples have small dilution, porosity and better fusion bonding when compared to TIG-clad samples. The laser-deposited samples have a higher uniform coating compared to the TIG-deposited samples based on the hardness value obtained in the longitudinal and transverse direction [1]. Functionally graded Ni-WC/Al-Ni material is deposited on the AZ91D Mg alloy using laser cladding process to obtain better corrosion and wear-resistant properties. The characterization result showed that the better mechanical properties, corrosion and wear resistance were obtained for the Ni-WC/Al-Ni coating [2]. The laser cladding on Ti-6Al-4V alloy base material using 60% WC-40% NiCrBSi metal matrix composite was performed, and it was found that the microhardness value of the coating is three times higher than the substrate material. The wear-resistant property of the clad layer is 400 times of the base alloy [3]. The effect of depositing NiCrMoFeNbTa powder on Inconel 625 Alloy substrate is investigated and found that the coated layer shows better hardness and surface roughness values when compared to the interface and the substrate material. Also, better corrosion resistance is obtained for coated samples than the uncoated substrate and the passive layer formation is shown in Figure 12.2 [4].

Nickel-based super alloy CM247LC is mainly used in making aircraft turbine blades and engine parts. The effect of texture formation and grain orientation on the microstructure and mechanical behaviour of laser-deposited CM247LC alloy were studied. It was observed that the optimization of laser parameters helps in achieving the required microstructure and controlling elemental precipitations. Also, the laser-melted region of CM247LC alloy had larger surface roughness and nanohardness which is due to the formation of fine grains [5]. Laser cladding can be utilized to fix the 7xxx series Al alloy parts. AA7075-T651 substrate is coated with Al-12%Si powder using laser cladding, and it was observed that the clad layer has better corrosion resistance than HAZ and the substrate. But the fatigue life and tensile properties of the laser-cladded sample were less than those of the substrate [6]. The

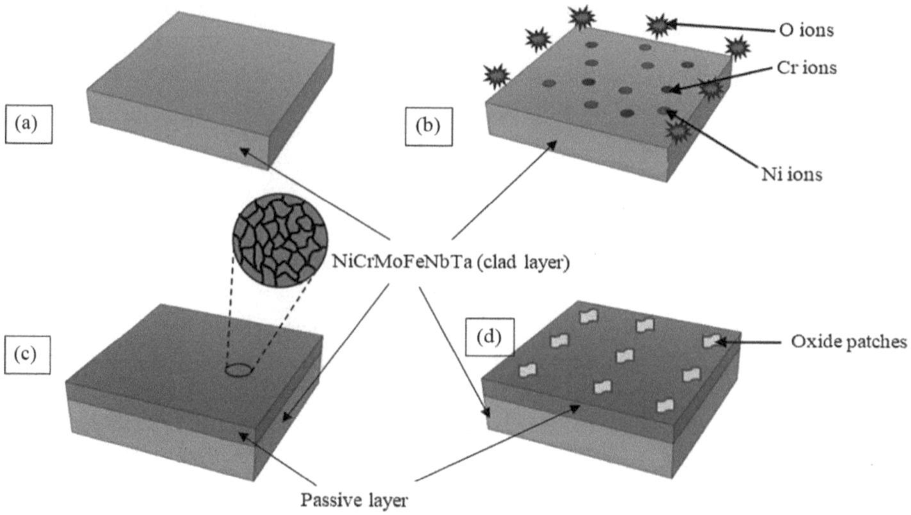

FIGURE 12.2
Passivation layer formation mechanism (a) NiCr-MoFeNbTa deposited layer, (b) reaction of Nickel and Chromium ions in the coating with O ions in the corrosion process, (c) even passive film creation on the coating due to secondary phase effect and (d) the brittle patches leave the surface with an increase in the time.

damaged area of anti-rotation bracket made up of 17-4 PH (AISI 630), a precipitation-hardened martensitic stainless steel, is repaired by laser cladding of mixed SS420 and SS316 powders and the microstructure investigations revealed that there are no cracks, lack of fusion and pores in the deposited layer. The cladding of mixed powders with 60% SS420 and 40% SS316 has the same hardness values as that of the 17-4 PH substrate. But for repairing the forward hanger assembly flange, the best powder mixture was 30% SS420 and 70% 17-4 PH to match its microhardness in the repaired region [7]. 30CrMnSiNi2A steel aircraft landing gear is remanufactured by laser deposition of Ni-based WC composite, and it was found that the hardness value improved with a rise in the WC content. Also, the wear-resistant properties of the coating are enhanced with the addition of WC [8]. A study on the effect of various heat input on the microstructure and mechanical behaviour of repaired 300M steel using laser cladding process is carried out. It was found that the raise in the heat input increased the HAZ region and repair efficiency and reduced the repair time. The post-heat treatment removed the microstructure variations caused by the different heat input values, and also, the fracture mode changed from ductile-brittle mixed fracture to plastic fracture [9]. The laser cladding of SS 420 alloy particles is performed on Ti6Al4V substrate via large power Yb: YAG laser and it was found that SS 420 coating is free from cracks and pits. The appropriate bonding is obtained between the coating and the base material. The base material's surface is protected from the loads applied externally by the coating having Fe in the martensite phase. The higher hardness values are obtained for SS 420 coated samples due to the combined outcomes like hard phases, flaw free and optimum laser cladding parameters. The main wear mechanisms obtained for the substrate were abrasive, adhesion and delamination. However, the coated samples demonstrate slight abrasive and oxide layers. The SS 420 coating decreased the average surface roughness value due to a smaller wear rate [10].

12.3 Automotive Applications

Laser cladding in automotive applications is utilized to refurbish or repair worn-out engine components such as camshafts and cylinder heads. Hence, the life and functionality of those components were improved using laser cladding. The Stellite 6 coating on AISI type 316L SS material was done via laser cladding. It was found that the microstructure of the deposited layer contains a dendritic structure enclosed by interdendritic phases. The former structure had high Co content, while the latter one had Cr-rich carbides. The Stellite 6 coating had better hardness, corrosion resistance and superior roughness value than the substrate material [11]. The Inconel 625 is deposited with Colmonoy 6 particles using laser cladding to increase its anti-corrosion behaviour. The microstructure analysis shows that the laser-deposited Colmonoy 6 layer has dendritic and interdendritic areas, which contain Cr borides and carbides formed after laser cladding. These precipitates lead to a higher hardness value of coated specimens than the base metal. The coated samples exhibited lower surface roughness values and larger polarization resistance due to the passive film formation on the cladded surface. Also, it enhanced the anti-corrosion property of the Inconel 625 substrate [12]. Laser deposition of different alloy powders such as AISI 4140, SS420 and Stellite 6 was done on grey cast iron for automotive disc brake application and their wear and friction characteristics have been studied. It was found that the AISI 4140 surface exhibited the maximum hardness and wear resistance, followed by SS420 and Stellite 6. Stable and uniform friction curves were obtained for AISI 4140, Stellite 6 and grey cast iron. But for SS420, it was continuously increasing and non-uniform. The wear behaviour of laser-cladded discs initially showed two body mechanisms, and it gradually converted into three-body mechanisms, as shown in Figure 12.3. The wear behaviour of Stellite 6, SS420 and AISI 4140 are shown in Figure 12.4a–c, respectively [13].

A study on laser deposition of Ni-based alloy which consists of more Ni content on ductile cast iron was done at varying laser power. It was observed that the clad prepared at 2.4 kW showed the maximum corrosion resistance. The rise in the laser power reduced the degree of separation in

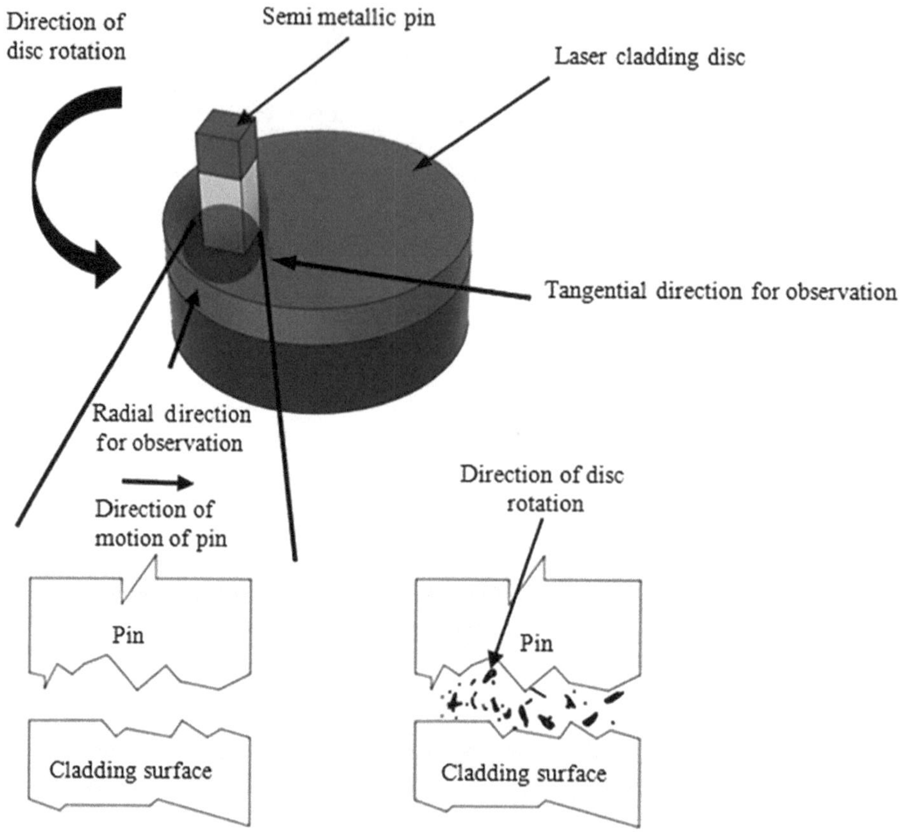

FIGURE 12.3
The change of two- to three-body wear behaviour in various LC discs.

the clad layer and uniform composition is obtained [14]. The laser deposition of copper-based powder alloys on 45# steel was performed to increase the wear resistance. The SEM analysis of the friction pairs shows that there are no cracks or flaws on the surface, and the deposited layer was metallurgically fused to the base metal. The mixing of Cu-based alloy material Ni276 self-fluxing particles enhanced the carbon atoms compatibility in the melt pool and reduced the graphite melting rate. The laser cladding produced parts with better wear resistance than conventional pressed parts, which is due to the compactness of the deposited layer [15]. An experimental study on annular laser cladding (ALC) of CuPb10Sn10 alloy powder on 42CrMo steel substrate was done. It was found that the optimized ALC parameters improved the absorptivity of laser and decreased the porosity of coating. The ALC process decreased the coefficient of friction by half, and the consistent coating shear strength was obtained. Hence, the ALC process can

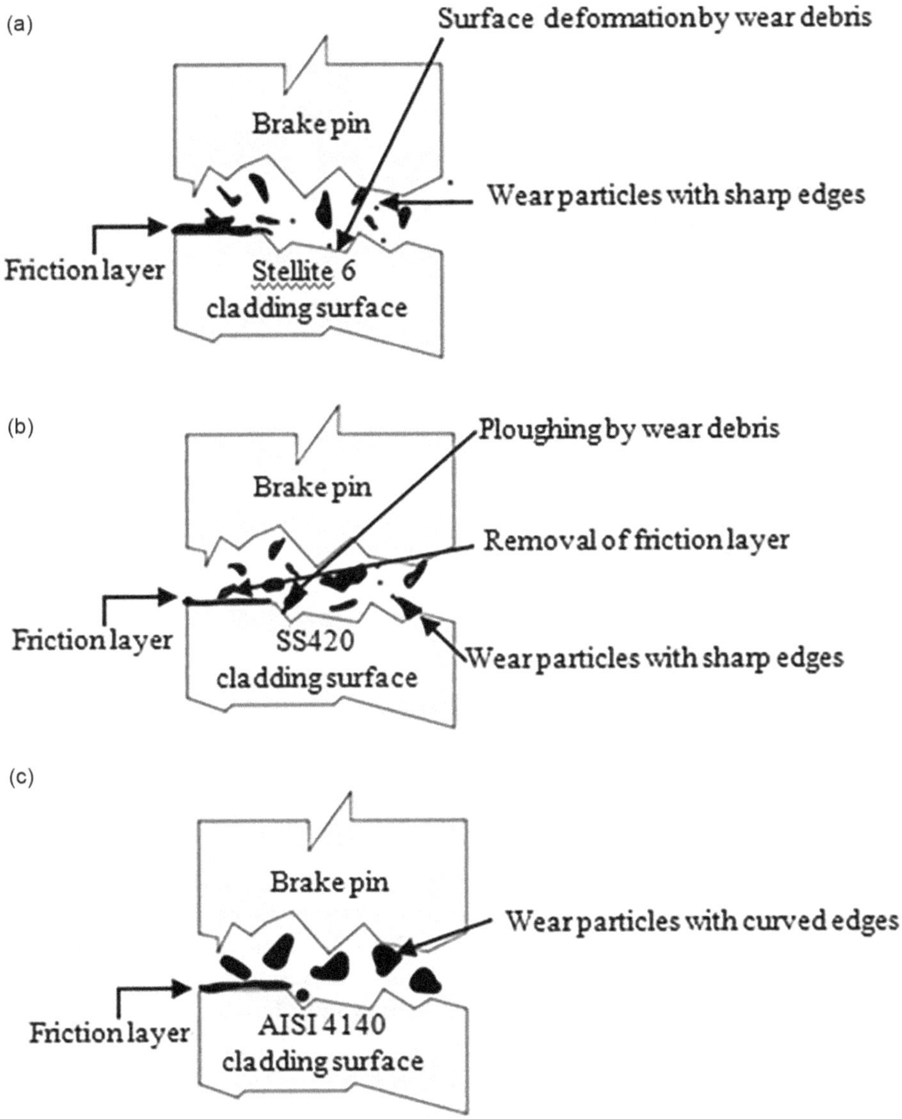

FIGURE 12.4
(a) Mechanism of sliding wear in Stellite 6 coated disc, (b) sliding wear, friction layer removal, and ploughing in SS420 coated surface, and (c) rolling three-body mechanism in AISI 4140 coated disc.

be used in manufacturing wear-resistant parts for automotive applications. The laser-powder-substrate coupling mode of ALC and laser spot-powder spot coupling mode are given in Figure 12.5a and b, respectively. The self-lubricating behaviour of lead having copper-rich phase wear-resistant coatings is given in Figure 12.6 [16]. The remanufacturing of broken gear

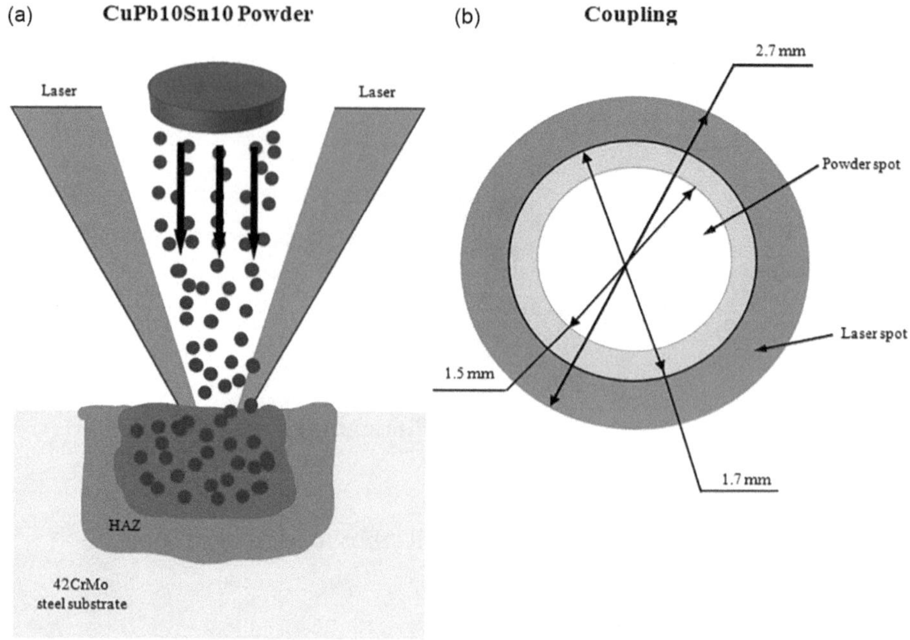

FIGURE 12.5
(a) ALC's laser-powder-substrate coupling mode and (b) coupling mode of laser spot-powder spot.

made up of 45 steel is carried out using laser cladding of H13 steel particles. The worn-out tooth was remanufactured by reverse engineering process and it was machined to get the required dimensions. It was found that the remanufactured tooth had better hardness, wear resistance and least tensile strength than the original tooth. The laser-cladded specimens have larger coefficient of friction and small mass loss when compared to the 45 steel parts. Also the mechanism of wear changed from abrasive wear (45 steel) to adhesive wear (H13 laser remanufactured part) [17]. The laser cladding of stainless steel 304 substrate using pre-placed 30%TiC/Inconel 625 powder of three different particle sizes was done, and it was observed that the maximum dilution, smaller melt pool lifetime and highest porosity are obtained for larger particles and medium-sized particles have minimum dilution and maximum melt pool lifetime which leads to the dislocation of TiC particles. This will result in the formation of dendrites, which in turn will increase the hardness value [18]. The aluminium-copper-based in-situ reinforced composite coating was applied on the AlSi10Mg base material using laser cladding. It was observed that the Al-Cu intermetallic compound gradient distribution in the composite coating causes an increase in the microhardness and anti-friction behaviour. The base material's wear behaviour changed from abrasive wear to adhesive wear when the load was

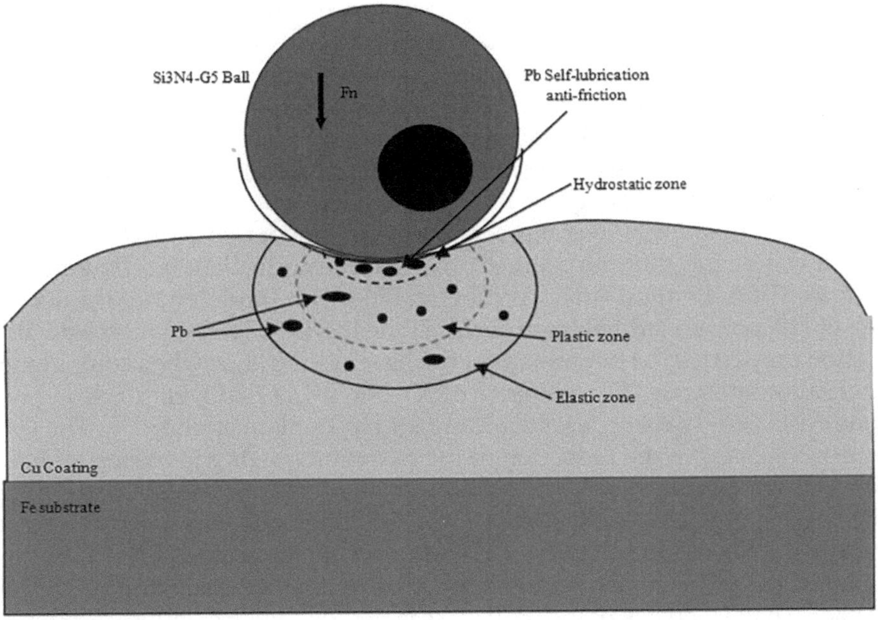

FIGURE 12.6
Self-lubricating behaviour of lead having Copper-rich phase wear-resistant coatings.

increased. While the Al-Cu coating exhibited abrasive wear mechanism for all the loading conditions [19]. An experimental investigation on the friction, wear and corrosion behaviour of laser-deposited grey cast iron using Co-based Stellite™ 6 alloy powder was performed. The microstructure of the Stellite™ 6 coating improved the wear resistance of the brake pad and led to the third body layer formation, which mainly contains the brake pad parts. Also, it forms the Cr-oxides passive layer on the surface, which improves the anti-corrosion behaviour [20]. Hence, the laser cladding can be used to deposit a protective layer and repair and remanufacture broken or worn-out automotive parts with enhanced mechanical, wear, corrosion and thermal properties.

12.4 Medical Applications

Laser cladding in the medical field is mostly used for making and repairing medical implants and prosthetics. It is suitable for the accurate coating of biocompatible materials and increases the durability and functionality of these devices. Ti6Al4V substrates were coated using bovine-bone-based hydroxyapatite (HA) by laser cladding process with pre-placed ceramic slurry, and it led to the creation of biocompatible $CaTiO_3$ phase for every deposited layer irrespective of the laser power. The HA phase is only preserved for the specimens made at lower laser beam with non-diluted slurry. The raise in the laser power leads to the loss of $CaTiO_3$ phase and the microhardness values decreased with raise in the laser power above 600 W. Hence the naturally derived ceramic coating can be controlled using laser power and HA ratio for the required biomedical application [21]. An experiment was conducted on laser cladding of multilayer bio-ceramic Ca/P powder on Ti6Al4V alloy surface. It was found that the coating provided the equivalent biocompatibility as that of Ti6Al4V alloy. Also it had better bioactivity and induced the deposition of HA on the cladded layer surface. The higher microhardness and antifriction properties were obtained for the bio-ceramic layer when compared to the Ti alloy substrate [22]. A bioactive SiO_2-Na_2O-CaO-P_2O_5 glass coating was done on Ti-6Al-4V using laser cladding for bio-implant application. The laser processing leads to the reduction in the bioactivity owing to the loss of Na, P elements and an increase in the accumulation of Al and Ti in the coating [23]. The coating of S520 bioactive glass on pure Ti substrate was done using laser cladding. It was found that the redistribution of elements such as Si, Na and P caused the enhancement in the coating adhesion on the substrate material's surface. The bioactivity of the S520 glass is proved by the creation of spherical cauliflower-like structures indicating the characteristics of the HCA. Hence, it is possible to produce non-toxic biomaterial for load-bearing biomedical implants using laser cladding [24]. An investigation on laser metal deposition of Fe-based metallic glass (FeCrMoCB) on ASTM F2229 Ni-free SS was done and it was found that the deposited layers were crack-free with large amorphous content. Better dilution between coating and substrate is obtained at lower scan speeds. The maximum hardness values are obtained for cladded samples, and it decreases by lowering the scanning speed due to the formation of a higher degree of crystallinity [25]. Bioactive Ca-P layer was deposited on quickly solidified ZTA eutectic bioceramics using laser cladding. It was found that the maximum content of Ca-P is obtained in the coating at 800W laser power and a scanning speed of 650 mm/min. The maximum hardness value is obtained for the laser-cladded Ca-P coatings when compared to plasma and thermal-sprayed Ca-P coatings [26]. The laser cladding of β-TiNb alloy powder on the Ti6Al4V surface was performed, and it was found that the β-phase was created due to the fast melting by laser cladding process that improved the microhardness of coating and anti-friction behaviour. The passive film densification and stabilization lead to a higher anti-corrosion behaviour of the coating. The coating has superior biocompatibility compared to Ti6Al4V due to the greater wettability and the bioactive sites of Ti and Nb on the alloy surface. Hence, the TixNb layer with better mechanical characteristics and biocompatibility made using laser cladding can be used as an orthopaedic biomaterial [27]. The Mo-reinforced Ti6Al4V composites were prepared using laser-based additive manufacturing, and they have higher elongation and smaller Young's modulus when compared to additive-manufactured Ti6Al4V parts. The mixing of powders reduced the cost and led to the fabrication of low-modulus/low-cost Ti6Al4V composites with enhanced mechanical properties [28]. The cluster-plus-glue-atom model has been used to develop a type of (TiZrNb)14SnMo refractory high entropy alloy (RHEA). This material is made as a coating as well as bulk using laser cladding and arc melting respectively. The exceptional thermal stability is obtained for RHEA 613°C. Both the bulk RHEA and coating made by laser cladding have better hardness, corrosion resistance and wear resistance. The biocompatibility of RHEA is better than that of Ti6Al4V alloy, which is proved by the good adhesion of MG-63 cells on the RHEA surface [29]. The formation of composite coating derived from Si_3N_4 powder in a metallic nanocrystalline matrix and amorphous silicon is obtained due to the laser cladding of Si_3N_4 on the ZrO_2 substrates. The Si_3N_4 coating improved the adhesion of the cells and the formation of bone tissue. Hence the laser cladding process is suitable for designing complex structures that can be used in dental applications [30]. Also, it is possible to develop non-toxic coatings having better mechanical strength, wear resistance and corrosion resistance with biocompatibility and bioactive properties using laser cladding process and can be used in orthopaedic, dental and biomedical applications.

12.5 Petrochemical Applications

The oil and gas industry makes use of laser cladding to refurbish and protect drilling equipment, valves and other critical components of machinery which are prone to challenging working environments. It helps in enhancing their service life and reducing downtime. The coating of nickel-based alloys and cobalt-based alloys was done on commercial API5CT grade L80 type 1 steel that is mostly used in petrochemical applications. The Ni-based alloys have hardness values similar to those of the steel substrate, which is advantageous for various applications, but the harder HAZ leads to heterogeneity, which is harmful for applications such as sulphide stress corrosion. The Co-based alloy has better hardness when compared to Ni-based alloy [31]. The laser metal deposition of CoCrW on Ti6Al4V substrate was carried out, and it was observed that the CoCrW coating had the lowest corrosion rate when compared to Ti6Al4V alloy [32]. The laser metal deposition of Ni60+WC particles on the SS ball valve was carried out. A rectangular beam spot was used in this experiment, which had the merits of big processing area, a small temperature gradient and a better cladding effect when compared to circular beam spot. The highest microhardness is obtained for the Ni60+WC powder-coated samples [33]. The Colmonoy-6 coating was applied on 316L steel base material using laser cladding, and it was observed that the 42h clad specimens have the smallest current density and highest corrosion resistance, which was caused by the creation of stable passive film. The presence of Cr-boride precipitate in the clad layer improved the hardness of the coated surface. The EIS analysis shows that the coated specimens have bigger arc radius and larger area-covered curves when compared to the base material. The highest corrosion resistance for a 42h clad sample is characterized by maximum polarization resistance (R_p) obtained [34]. Tribaloy 800 (T-800) alloy was mostly utilized by steel and petrochemical industry due to its exceptional wear and corrosion resistance at higher temperatures. Induction heating assisted laser cladding (IHLC) of T-800 coating was carried out, and it was found that the coating consists of Cr6Fe18Mo5 and γ-Co supersaturated solid solution having lattice defects. The fraction of the Cr6Fe18Mo5 phase was increased to 51.19% by increasing the preheating temperature. Also, the rise of preheating temperature leads to T-800 coating without any crack which is due to the decrease in the γ-Co fraction having lattice defects and residual tensile stress in the coating [35]. The laser cladding of 316L stainless steel substrate using CoCrFeNi HEA/WC composite coatings with varying WC mass fraction was carried out, and it was found that the WC particles get decomposed partially and minor decomposition in the laser-clad composite layer due to the contact with the laser beam and melt pool and it formed hard carbide which leads to the strong solid solution. But, the spherical-shaped WC particles stopped the growth of HEA grains, which in turn forms the disordered refined grains. The hardness value is improved by coating the HEA/WC composite, and the largest hardness value was obtained for coating having 5% WC. The electrochemical corrosion tests proved the improvement of the corrosion potential of HEA/WC composite coatings, and the coating with 5% WC had the largest corrosion potential [36]. The CoCr2.5FeNi2Tix HEA coating with different x values was made on Q235 stainless steel substrates using laser cladding process, and it was observed that for $x=0$, a single solid solution having a BCC structure was formed. But for the other values of x solid solutions with two phases are formed (BCC and FCC). The laser-deposited CoCr2.5FeNi2Tix HEA has better corrosion resistance than the Q235 steel substrate. The Nyquist plot values are initially enhanced and then reduced by adding Ti, and the largest value is obtained for the coatings with $x=1$. The abrasion resistance and hardness value of the coatings enhanced with the addition of Ti atoms. The highest microhardness and lowest wear rate are obtained for $x=1/1.5$ [37]. The super alloys such as Inconel 625, Inconel 825 and Hastelloy C 276 were deposited on T 95 grade carbon steel using hot wire laser cladding. The samples are examined in very sour wells for three successive months. The well-A makes a large concentration of H_2S (35,000 ppm) and well-B makes 14,000 ppm of H_2S. It was observed that all the cladded samples exhibited considerable resistance to FeS corrosion. Inconel 625 had the highest corrosion resistance, followed by Hastelloy C 276 and Inconel 825 when compared to substrate material. Hence, the life of the wells can be increased by high-power laser cladding at a cheaper cost and reduce the frequency of replacing pipes due to corrosion [38]. The laser cladding of two types of materials, namely, NiCrMoNb and

NiCrBSiC, were done on the Ti alloy substrate, and it was found that both the coatings have better bonding with Ti substrate and it is free from cracks and porosity. The coatings have homogeneous structure due to the distribution of elements and hardness. The higher hardness values are obtained for NiCrSiBC coating than for NiCrMoNb coating. The wear rate is higher in the substrate than in the laser-cladded samples due to the increase in the hardness after coating with NiCrSiBC and NiCrMoNb powders. The abrasive and adhesive wear mechanisms are observed for the cladded samples. The lesser average surface roughness was obtained for NiCrSiBC-coated specimens when compared to uncoated and NiCrSiBC-coated samples [39]. The laser cladding of Colmonoy-6 and Stellite-6 on SS 304 substrate was carried out. It was found that the Colmonoy-6 and Stellite-6 coatings have carbides embedded in a nickel- and cobalt-rich solid solution having dendritic structure, respectively. The hardness value increased up to 2–2.5 times for the coated samples because of the finer grain size at the surface of the coatings when compared to the substrate. Colmonoy-6-coated SS 304 samples have higher wear resistance than the Stellite-6-coated SS 304 samples. The coefficient of friction values of 0.41–0.47 and 0.46–0.51 were obtained for Colmonoy-6 and Stellite-6 samples, respectively. The coated samples exhibited abrasive and adhesive wear mechanisms. Also, the delamination wear occurred, which is proved by the presence of plate-like wear debris. The surface finish was improved 2.3 and 3 times for the coated specimens than the substrate due to the increase in hardness value and wear resistance. Hence, the laser cladding process can be used to deposit anti-friction and anti-corrosion coatings on the tools, machine parts, tubes or pipelines, which will improve their lifetime. Also, it will reduce the cost of replacing parts by repairing the worn-out or broken parts by depositing material with similar or better mechanical and tribological properties [40].

12.6 Applications in Energy Industry

The energy sector make use of laser cladding to renovate and repair turbine blades, boiler components and other high-temperature, high-stress parts. Their anti-corrosion, anti-friction properties and thermal fatigue were improved. The Stellite-6 alloy powder is uniformly coated over Inconel-625 base material using laser metal deposition process. The better hardness values and wear resistances were obtained for Stellite-6 coated samples when compared to the substrate. Also, the electrochemical corrosion tests proved that the coated samples have superior corrosion resistance because of the formation of Cr_2O_3 film, which leads to the larger passive region [41]. A study to improve corrosion-resistant property of SS410, which is used in making critical components such as turbines and nuclear reactor core was carried out by the laser metal deposition of Colmonoy-5 alloy powder. It was found that the hardness value improved by the Cr borides and carbides present in the matrix of γ-nickel. The passive film formed due to chromium-rich borides improved the corrosion behaviour of the SS410 substrate [42]. Nickel-based superalloy was coated with NiCrFeMoNb and FeCrMoVC layers using laser metal deposition, and it was observed that the NiCrFeMoNb had a γ-nickel phase and FeCrMoVC had martensitic phase. The fine microstructure of the FeCrMoVC coating leads to better hardness values than the NiCrFeMoNb coating. The VC present in FeCrMoVC coating provides better wear resistance and hardness values than the substrate and NiCrFeMoNb coating [43]. The SS 420 and Colmonoy 5 powder coating was done on SS 410 using laser cladding, and it was found that the Colmonoy 5 clad samples had higher hardness than SS 420 clad samples due to the hard laves phase present in the soft Ni matrix. The SS 420 clad samples and SS 410 substrate exhibited adhesive and abrasive wear mechanisms. But, only wear track traces were found in Colmonoy 5 clad samples. Hence, it is suitable for preventing sliding wear in pump shaft and valve applications [44]. Laser surface modification of pearlitic ductile irons (PDI) used in automotive and nuclear energy applications was done by alloying WC-12%Co powder by means of fibre laser having higher power. The hardness values improved upto $HV_{0.3}$ 1200 for the laser surface-modified samples [45]. The laser deposition of Al-102 alloy powder on Al 4047 was done, and it was observed that the width and depth of cladding joint would rise with the rise in the laser power. The dilution rate was enhanced with a rise in laser power, and it was reduced with the rise in the powder feed. The higher hardness values were attained for the coated specimens when compared to the base metal. Hence, the laser cladding process can used in repairing and restoring worn-out components, and it will reduce the cost of replacing the parts such as turbine [46]. The NiCrMoFeNbTa alloy coating on Inconel 625 Alloy substrate was carried out using laser cladding, and it was observed from the microstructural characterization results that the cellular/dendritic structure having secondary phases of interdendritic structure scattered in the matrix of γ-nickel. The presence of Nb and Cr carbides in the secondary phases changes the hardness and surface finish of the coated specimens. The highest hardness values were obtained for cladded layers when compared to the interface and substrate. Maximum R_p and R_a values are obtained for cladded samples, and the coatings have outstanding corrosion resistance when compared to the substrate. The NiCrMoFeNbTa alloy coating can be used for laser cladding of components with improved working life due to its higher hardness and corrosion resistance, which can be used in harsh environments [4]. The AISI type 316L SS material is utilized in applications such as heat exchangers, sewage water treatments, and power plants. The coating of Stellite 6 powder on AISI type 316L SS substrate was done via laser cladding process, and it was found that the coating layer microstructure contains dendritic structure enclosed by interdendritic phases. The dendritic and interdendritic structures had a higher content of cobalt and chromium-rich carbides, respectively. The separation of these carbides and the Co-rich boundaries improved the nanohardness. The oxide films consist of Cr_2O_3 and CoO layers, which leads to the excellent corrosion resistance of the laser-cladded layers. The 70h clad Stellite 6 coated samples exhibited the highest resistance owing to its smaller value of current density and larger R_p value [11].

12.7 Applications in Mould and Die Industry

The mould and die industry makes use of laser cladding processes to alter and restore broken or worn-out moulds and dies. As the laser can deposit materials accurately, it decreases the requirement for extensive rework and boosts productivity. The failure of moulds happens due to the corrosion of molten Al alloys. The laser cladding of Tribaloy T400 on H13 steel was carried out, and its corrosion behaviour was investigated using A356 Al alloys at 993K for varying time intervals and it was found that the Co-rich FCC phase and laves phase were obtained for the Tribaloy T400 coating. The coating exhibited excellent corrosion resistance to the molten Al alloy. Tribaloy T400 coating was not affected after molten Al alloy corrosion. The higher content of Mo than Co present in Al alloy after solidification leads to the corrosion of laves phase. The microhardness value increased due to the age hardening of Tribaloy T400 coating [47]. The laser metal deposition of CrNiW and CrNiFeAlZr coatings on H13 tool steel was carried out, and it was found that the cladding layers have superior bonding and it is free from flaws like cracks and pores. The element distribution provided a homogenous structure. The higher hardness values are obtained for CrNiW and CrNiFeAlZr coated samples when compared to the substrate. The considerable amount of Cr_7C_3 and $Cr_{23}C_6$ carbides present in the cladding leads to the wear and sliding-resistant structure. Also, this combination improved the thermal stability and oxidation resistance. The laser power is an important factor influencing clad depth. The powder feed rate and scanning speed were the most influencing parameters affecting clad width and height [48]. The laser cladding of Ni60-based alloy coating on GG70L cast iron, which is used in making automobile panel dies and moulds, was carried out. The optimal laser cladding parameters have been obtained using Taguchi analysis-based Principal Component Analysis Grey Relational Analysis (PCA-GRA) multi-response optimization method. The optimal parameters for obtaining superior performance of the coatings were 1.0 kW laser power, 11 mm/s scanning speed, 1.2 r/min powder feeding rate, and 3 mm defocusing distance. The major increase in the aspect ratio, dilution rate and depth of HAZ, microhardness value and coating efficiency were obtained for laser cladding samples at optimal parameters. The wear resistance also improved due to uniform refined grain structure, distributed hard particles of Cr carbide and boride and solid solutions of γ-Ni and γ-NiFe [49]. The hybrid additive-subtractive manufacturing (HASM) method was used to repair dies and moulds of automobile panels. Multi-indicator evaluation and selection of material based on the forming quality, friction property, thermophysical properties and milling machinability of the laser-cladded layers using the TOPSIS method. It was found that the Fe-based coatings have better wear resistance and forming quality than the Ni-based coatings. But the surface finish and tool wear rate of machined part is very poor. Hence, Ni60 powder was selected to repair dies, which provided adequate performance and manufacturing efficiency for HASM [50]. The laser cladding is the most efficient process to deposit wear and corrosion-resistant coatings on the dies and moulds which are corroded due to molten metal at higher temperatures. Also, it is a suitable technology for repairing and modifying damaged or worn-out dies and moulds. The coatings will increase the useful life time of the dies and moulds by providing excellent metallurgical bonding with the substrate.

12.8 Conclusions

The laser cladding or laser metal deposition process is used to deposit protective and functionally graded coating on components and tools which are used in aerospace, automotive, medical, petrochemical, energy industries, etc. The coated surfaces obtained from the laser cladding have higher microhardness values, corrosion resistance, wear resistance, surface finish and least HAZ when compared to other coating processes such as chemical vapour deposition, physical vapour deposition, plasma or thermal spray coating methods. The laser cladding process can also be used to deposit bioactive materials and non-toxic coatings on orthopaedic and dental implants. The coaxial powder-fed laser cladding head coupled with a robotic arm can be used in additive manufacturing of functionally graded materials and can produce complex parts with different compositions. This technique is also used in repairing and remanufacturing of aircraft parts, automobile parts, turbines, tools, dies and moulds. Hence, the cost of replacing the components can be reduced, which increases the useful lifetime of repaired parts.

References

1. Sexton, L., Lavin, S., Byrne, G., & Kennedy, A. (2002). Laser cladding of aerospace materials. *Journal of Materials Processing Technology*, 122(1), 63–68. https://doi.org/10.1016/S0924-0136(01)01121-9.

2. Wang, G., Wang, S., Yin, Z., Yang, X., Wen, D., & Sun, Y. (2023). Synthesis of Ni-WC/Al-Ni functionally graded coating with advanced corrosion and wear resistance on AZ91D Mg alloy by laser cladding. *Materials Letters*, 333, 133645. https://doi.org/10.1016/j.matlet.2022.133645.
3. Al-Sayed Ali, S. R., Hussein, A. H. A., Nofal, A. A. M. S., Hasseb Elnaby, S. E. I., Elgazzar, H. A., & Sabour, H. A. (2017). Laser powder cladding of Ti-6Al-4V α/β alloy. *Materials*, 10(10), 1178. https://doi.org/10.3390/ma10101178.
4. Jeyaprakash, N., Yang, C. H., Susila, P., & Karuppasamy, S. S. (2023). Laser cladding of NiCrMoFeNbTa particles on Inconel 625 alloy: Microstructure and corrosion resistance. *Transactions of the Indian Institute of Metals*, 76(2), 599–612. https://doi.org/10.1007/s12666-022-02701-7.
5. Jeyaprakash, N., Prabu, G., Yang, C. H., Banda, M. P., & Mohan, E. (2023). Structural and nanohardness of laser melted CM247LC nickel-based superalloy. *Transactions of the Indian Institute of Metals*, 76(2), 287–295. https://doi.org/10.1007/s12666-022-02734-y.
6. Liu, Q., Janardhana, M., Hinton, B., Brandt, M., & Sharp, K. (2011). Laser cladding as a potential repair technology for damaged aircraft components. *International Journal of Structural Integrity*, 2(3), 314–331. https://doi.org/10.1108/17579861111162914.
7. Liu, Q., Sharp, P. K., Brandt, M., Matthews, N., Sun, S., & Djugum, R. (2014). Repair of aircraft components by laser cladding process. In *International Congress on Applications of Lasers & Electro-Optics* (Vol. 2014, No. 1, pp. 308–313), San Diego, California, USA. Laser Institute of America. https://doi.org/10.2351/1.5063074.
8. Chen, Z., Zhou, H., Zhu, Z., Xu, C., & Zhou, Y. (2023). Laser cladding remanufacturing of aircraft landing gear based on 30CrMnSiNi2A steel. *Optik*, 283, 170902. https://doi.org/10.1016/j.ijleo.2023.170902.
9. Zheng, Y., Liu, F., Gao, J., Liu, F., Huang, C., Zheng, H., Wang, P., & Qiu, H. (2023). Effect of different heat input on the microstructure and mechanical properties of laser cladding repaired 300M steel. *Journal of Materials Research and Technology*, 22, 556–568. https://doi.org/10.1016/j.surfcoat.2019.125090.
10. Jeyaprakash, N., & Yang, C. H. (2020). Microstructure and wear behaviour of SS420 micron layers on Ti–6Al–4V substrate using laser cladding process. *Transactions of the Indian Institute of Metals*, 73, 1527–1533. https://doi.org/10.1007/s12666-020-01927-7.
11. Jeyaprakash, N., Yang, C. H., Karuppasamy, S. S., & Duraiselvam, M. (2023). Stellite 6 cladding on AISI Type 316L stainless steel: Microstructure, nanohardness and corrosion resistance. *Transactions of the Indian Institute of Metals*, 76(2), 491–503. https://doi.org/10.1007/s12666-022-02731-1.
12. Jeyaprakash, N., Yang, C. H., & Karuppasamy, S. S. (2022). Laser cladding of Colmonoy 6 particles on Inconel 625 substrate: Microstructure and corrosion resistance. *Surface Review and Letters*, 29(08), 2250102. https://doi.org/10.1142/S0218625X22501025.
13. Manoj, A., Saurabh, A., Narala, S. K. R., Saravanan, P., Natu, H. P., & Verma, P. C. (2023). Surface modification of grey cast iron by laser cladding for automotive brake disc application. *Wear*, 532, 205099. https://doi.org/10.1016/j.wear.2023.205099.
14. Wang, R., Ouyang, C., Li, Q., Bai, Q., Zhao, C., & Liu, Y. (2022). Study of the microstructure and corrosion properties of a Ni-based alloy coating deposited onto the surface of ductile cast iron using high-speed laser cladding. *Materials*, 15(5), 1643. https://doi.org/10.3390/ma15051643.

15. Li, J., Gu, J., Zhang, Y., Wang, Z., & Zhang, G. (2023). Study on laser cladding process and friction characteristics of friction pairs of copper-based powder metallurgy materials. *Tribology International*, 177, 107953. https://doi.org/10.1016/j.triboint.2022.107953.
16. Wan, L., Cheng, M., Fu, G., Wei, C., Shi, T., & Shi, S. (2023). Annular laser cladding of CuPb10Sn10 copper alloy for high-quality anti-friction coating on 42CrMo steel surface. *Optics & Laser Technology*, 158, 108878. https://doi.org/10.1016/j.optlastec.2022.108878.
17. Zhu, L., Wang, S., Pan, H., Yuan, C., & Chen, X. (2020). Research on remanufacturing strategy for 45 steel gear using H13 steel powder based on laser cladding technology. *Journal of Manufacturing Processes*, 49, 344–354. https://doi.org/10.1016/j.jmapro.2019.12.009.
18. Bhatnagar, S., & Mullick, S. (2023). A study on the influence of reinforcement particle size in laser cladding of TiC/Inconel 625 metal matrix composite. *Optics & Laser Technology*, 161, 109115. https://doi.org/10.1016/j.optlastec.2023.109115.
19. Wang, J., Jin, G., Feng, X., Cui, X., Wen, X., & Zhao, Y. (2023). Al–Cu-based in-situ reinforced gradient coating on aluminum alloy by laser cladding: Tribological behavior under gradient loading and thermal conditions. *Materials Chemistry and Physics*, 301, 127595. https://doi.org/10.1016/j.matchemphys.2023.127595.
20. Kiehl, M., Scheid, A., Graf, K., Ernst, B., & Tetzlaff, U. (2023). Coaxial laser cladding of cobalt-base alloy StelliteTM 6 on gray cast iron/investigations on friction, wear versus commercial brake pad, and corrosion characteristics. *Proceedings of the Institution of Mechanical Engineers, Part D: Journal of Automobile Engineering*, 09544070221145512. https://doi.org/10.1177/09544070221145512
21. Mocanu, A. C., Miculescu, F., Stan, G. E., Tite, T., Miculescu, M., Țierean, M. H., Pascu, A., Ciocoiu, R. C., Butte, T. M., & Ciocan, L. T. (2023). Development of ceramic coatings on titanium alloy substrate by laser cladding with pre-placed natural derived-slurry: Influence of hydroxyapatite ratio and beam power. *Ceramics International*, 49(7), 10445–10454. https://doi.org/10.1016/j.ceramint.2022.11.227.
22. Liu, B., Deng, Z., & Liu, D. (2021). Preparation and properties of multilayer Ca/P bio-ceramic coating by laser cladding. *Coatings*, 11(8), 891. https://doi.org/10.3390/coatings11080891.
23. Kuo, P. H., Joshi, S. S., Lu, X., Ho, Y. H., Xiang, Y., Dahotre, N. B., & Du, J. (2019). Laser coating of bioactive glasses on bioimplant titanium alloys. *International Journal of Applied Glass Science*, 10(3), 307–320. https://doi.org/10.1111/ijag.12642.
24. Bajda, S., Liu, Y., Tosi, R., Cholewa-Kowalska, K., Krzyzanowski, M., Dziadek, M., Kopyscianski, M., Dymek, S., Polyakov, A. V., Semenova, I. P., & Tokarski, T. (2021). Laser cladding of bioactive glass coating on pure titanium substrate with highly refined grain structure. *Journal of the Mechanical Behavior of Biomedical Materials*, 119, 104519. https://doi.org/10.1016/j.jmbbm.2021.104519.
25. Ibrahim, M. Z., Sarhan, A. A., Kuo, T. Y., Yusof, F., & Hamdi, M. (2019). Characterization and hardness enhancement of amorphous Fe-based metallic glass laser cladded on nickel-free stainless steel for biomedical implant application. *Materials Chemistry and Physics*, 235, 121745. https://doi.org/10.1016/j.matchemphys.2019.121745.
26. Su, H., Fan, G., Dong, D., Liu, Y., Shen, Z., Zhao, D., ... & Guo, M. (2023). In-situ synthesis and characterization of calcium phosphate coatings on rapidly solidified zirconia toughened alumina eutectic bioceramics by laser cladding. *Materials & Design*, 232, 112173. https://doi.org/10.1016/j.matdes.2023.112173.

27. Zheng, Y., Xu, P., Li, L., Liu, Q., & Guo, S. (2023). Enhancing biocompatibility, mechanical properties, and corrosion resistance of laser cladding β-TiNb coatings. *Journal of Vacuum Science & Technology A*, 41(6), 063103. https://doi.org/10.1116/6.0002885.
28. Zhou, M. M., Wang, P., Xu, L., Deng, L., Qi, J. F., & Zhang, L. C. (2023). Biomedical Mo particles reinforced titanium alloy fabricated by laser additive manufacturing. *Journal of Alloys and Compounds*, 960, 170512. https://doi.org/10.1016/j.jallcom.2023.170512.
29. Guo, Y., Li, X., & Liu, Q. (2020). A novel biomedical high-entropy alloy and its laser-clad coating designed by a cluster-plus-glue-atom model. *Materials & Design*, 196, 109085. https://doi.org/10.1016/j.matdes.2020.109085.
30. Marin, E., Zanocco, M., Boschetto, F., Santini, M., Zhu, W., Adachi, T., Ohgitani E., McEntire, B. J., Bal, B. S., & Pezzotti, G. (2020). Silicon nitride laser cladding: A feasible technique to improve the biological response of zirconia. *Materials & Design*, 191, 108649. https://doi.org/10.1016/j.matdes.2020.108649.
31. Almeida, N. C., Cândido, L. C., Faria, G. L. D., Lima, M. S. F. D., & Trindade, V. B. (2016). Laser cladding and thermal spray coatings on steel pipe serving the oil and gas industry. *Journal of Material Science & Engineering*, 5, 2169–0022. https://doi.org/10.4172/2169-0022.1000279.
32. Sun, Y. B., Niu, H. J., Wang, J. Y., Dong, G. F., & Lin, C. X. (2023). Microstructure and corrosion property of prepared CoCrW coatings on the TC4 surface by laser cladding. *Coatings*, 13(10), 1687. https://doi.org/10.1016/j.surfcoat.2023.129430.
33. Zhu, H., Lin, X., Xue, X., Zhang, Y., Wang, L., Ning, Y., Dong, Y., & Fang, X. (2024). Laser cladding of stainless-steel ball valves by a high-power diode laser source with a rectangular beam spot. *Optics & Laser Technology*, 169, 110123. https://doi.org/10.1016/j.optlastec.2023.110123.
34. Natarajan, J., Yang, C. H., & Karuppasamy, S. S. (2021). Investigation on microstructure, nanohardness and corrosion response of laser cladded colmonoy-6 particles on 316L steel substrate. *Materials*, 14(20), 6183. https://doi.org/10.3390/ma14206183.
35. Yong, Z., Xie, D., Liu, Y., Shen, L., Jiang, S., Xing, F., Chang, L., & Tian, Z. (2023). Validation of Cr6Fe18Mo in tribaloy T-800 coating fabricated by induction heating assisted laser cladding. *Journal of Alloys and Compounds*, 944, 169195. https://doi.org/10.1016/j.jallcom.2023.169195.
36. Huang, J., Zhu, Z., Wang, H., Li, K., Shi, W., & Jiao, T. (2023). Effect of WC content on microstructure and properties of CoCrFeNi HEA composite coating on 316L surface via laser cladding. *Materials*, 16(7), 2706. https://doi.org/10.3390/ma16072706.
37. Gu, Z., Xi, S., & Sun, C. (2020). Microstructure and properties of laser cladding and CoCr2. 5FeNi2Tix high-entropy alloy composite coatings. *Journal of Alloys and Compounds*, 819, 152986. https://doi.org/10.1016/j.jallcom.2019.152986.
38. Assiri, W., Asiri, A., San Roman-Alerigi, D., Batarseh, S., & Salman, M. (2020). A new horizon in corrosion prevention using state of the art high power laser cladding technology. In *SPE Annual Technical Conference and Exhibition* (p. D021S004R002), Virtual. SPE. https://doi.org/10.2118/201617-MS.
39. Jeyaprakash, N., Yang, C. H., & Tseng, S. P. (2020). Characterization and tribological evaluation of NiCrMoNb and NiCrBSiC laser cladding on near-α titanium alloy. *The International Journal of Advanced Manufacturing Technology*, 106, 2347–2361. https://doi.org/10.1007/s00170-019-04755-2.

40. Jeyaprakash, N., Yang, C. H., & Tseng, S. P. (2021). Wear tribo-performances of laser cladding Colmonoy-6 and Stellite-6 micron layers on stainless steel 304 using Yb: YAG disk laser. *Metals and Materials International*, 27, 1540–1553. https://doi.org/10.1007/s12540-019-00526-6.
41. Jeyaprakash, N., Yang, C. H., Karuppasamy, S. S., & Dhineshkumar, S. R. (2022). Evaluation of microstructure, nanoindentation and corrosion behavior of laser cladded Stellite-6 alloy on Inconel-625 substrate. *Materials Today Communications*, 31, 103370. https://doi.org/10.1016/j.mtcomm.2022.103370.
42. Karuppasamy, S. S., Jeyaprakash, N., & Yang, C. H. (2022). Microstructure, nanoindentation and corrosion behavior of colmonoy-5 deposition on SS410 substrate using laser cladding process. *Arabian Journal for Science and Engineering*, 47, 8751–8767. https://doi.org/10.1007/s13369-021-06398-6
43. Jeyaprakash, N., & Yang, C. H. (2020). Comparative study of NiCrFeMoNb/FeCrMoVC laser cladding process on nickel-based superalloy. *Materials and Manufacturing Processes*, 35(12), 1383–1391. https://doi.org/10.1080/10426914.2020.1779933.
44. Jeyaprakash, N., Yang, C. H., Ramkumar, K. R., & Sui, G. Z. (2020). Comparison of microstructure, mechanical and wear behaviour of laser cladded stainless steel 410 substrate using stainless steel 420 and Colmonoy 5 particles. *Journal of Iron and Steel Research International*, 27, 1446–1455. https://doi.org/10.1007/s42243-020-00447-4.
45. Jeyaprakash, N., Yang, C. H., & Tseng, S. P. (2020). Optimization of tribological parameters over WC-12% Co laser alloyed pearlitic ductile iron using Taguchi based Grey relational analysis. *Journal of Central South University*, 27(3), 736–751. https://doi.org/10.1007/s11771-020-4327-9.
46. Kim, I., Kim, M., Kim, H., Lee, M. G., & Jeon, Y. (2023). Laser cladding of Al 102 powder on Al 4047 with direct energy deposition. *Metals*, 13(5), 856. https://doi.org/10.3390/met13050856.
47. Wang, Y., Zhu, H., & Huang, J. (2023). Corrosion behaviour of Tribaloy T400 coating prepared by laser cladding in molten aluminium alloys. *Advances in Engineering Technology Research*, 5(1), 38–38. https://doi.org/10.56028/aetr.5.1.38.2023.
48. Martin Vinoth, S., & Dinesh Babu, P. (2023). Comparing and analysing the characterisation of laser cladding of CrNiW and CrNiFeAlZr powders on H13 tool steel. *Proceedings of the Institution of Mechanical Engineers, Part C: Journal of Mechanical Engineering Science*, 237(13), 2969–2989. https://doi.org/10.1177/09544062221141267.
49. Liu, M., Duan, C., Li, G., Cai, Y., Wang, F., & Li, L. (2023). Multi-response optimization of Ni-based laser cladding via principal component analysis and grey relational analysis. *Optik*, 287, 171122. https://doi.org/10.1016/j.ijleo.2023.171122.
50. Liu, M., Duan, C., Li, G., Cai, Y., Li, L., & Wang, F. (2023). Multi-indicator evaluation and material selection of hybrid additive-subtractive manufacturing to repair automobile panel dies and molds. *The International Journal of Advanced Manufacturing Technology*, 127, 1675–1690. https://doi.org/10.1007/s00170-023-11621-9.

Index

FSC
www.fsc.org
MIX
Papier | Fördert gute Waldnutzung
FSC® C083411

Kolibri 360 GmbH
Ferdinand-Jühlke-Straße 7
99095 Erfurt, Deutschland
produktsicherheit@kolibri360.de